T0387964

Fluid Waves

Fluid Waves

Richard Manasseh

CRC Press is an imprint of the
Taylor & Francis Group, an **informa** business

First edition published 2022
by CRC Press
6000 Broken Sound Parkway NW, Suite 300, Boca Raton, FL 33487-2742

and by CRC Press
2 Park Square, Milton Park, Abingdon, Oxon, OX14 4RN

CRC Press is an imprint of Taylor & Francis Group, LLC

ISBN: 978-0-367-27164-0 (hbk)
ISBN: 978-1-032-11319-7 (pbk)
ISBN: 978-0-429-29526-3 (ebk)

DOI: 10.1201/9780429295263

Typeset in CMR10
by KnowledgeWorks Global Ltd.

To Irena and Dylan

Contents

Nomenclature

Number type $\mathbb{R}$: real; $\mathbb{C}$: complex. MKS: metre-kilogram-second units; SI: standard Système International units. To save space with multiple units, slashes are used instead of the power notation used elsewhere in this book, so kg/m/s^2 is kg m^{-1} s^{-2}. Dashes (-) indicate dimensionless quantities. A few symbols, e.g. m and T, have two meanings; if so, the meanings are quite different and should be apparent from the context. Symbols only used briefly during a derivation or explanation are not below, since their definition is on the same page or very close.

Roman letters

		Type	Units	
			MKS	SI
$\bullet$	Used to denote any variable	any	any	
A	Amplitude of a wave or oscillator	$\mathbb{C}$	any	
$A_\times$	Area, cross-sectional, in general	$\mathbb{R}$	m^2	
A_h	Area, cross-sectional, of pipe or cylinder	$\mathbb{R}$	m^2	
a	Amplitude of an oscillator	$\mathbb{R}$	m	
a_i	Constant used in various derivations, e.g. a_1, a_2	$\mathbb{R}$	any	
a_a	Acoustic absorption co-efficient	$\mathbb{R}$	-	
a_u	Amplitude of solution to KdV equation	$\mathbb{R}$	-	
$\boldsymbol{a}$	Acceleration vector	$\mathbb{C}$	m/s^2	
B	Parameter in solution to KdV equation	$\mathbb{R}$	-	
b	Constant used in various derivations, e.g. b_1, b_2	$\mathbb{R}$	any	
c	Speed of wave propagation	$\mathbb{R}$	m/s	
c_0	Speed of wave propagation, linear approx.	$\mathbb{R}$	m/s	
D	Length constant, typically a diameter	$\mathbb{R}$	m	
D	Prefix indicating material ('total') derivative			
d	Length constant, typically engineering depth	$\mathbb{R}$	m	
E_v	Bulk modulus	$\mathbb{R}$	kg/m/s^2	Pa
F	Force	$\mathbb{C}$	kg m/s^2	N
$\boldsymbol{F}$	Force vector	$\mathbb{C}$	kg m/s^2	N
$\mathcal{F}$	Force amplitude per unit mass	$\mathbb{R}$	m/s^2	
$\hat{\mathcal{F}}$	Displacement amplitude of forcing	$\mathbb{R}$	m	
$\mathfrak{F}$	Function of azimuthal angle φ only	$\mathbb{C}$	-	
Fr	Froude number	$\mathbb{R}$	-	

f	Frequency (in cycles per second)	$\mathbb{R}$	s^{-1}	Hz
f_0	Frequency, natural (in cycles per second)	$\mathbb{R}$	s^{-1}	Hz
f_d	Frequency shift due to Doppler effect	$\mathbb{R}$	s^{-1}	Hz
$f(\bullet)$	Function of $\bullet$	any	any	
$f_\oplus$	Approx. to Coriolis parameter at given latitude	$\mathbb{R}$	rad/s	
g	Acceleration due to gravity, surface of Earth	$\mathbb{R}$	m/s^2	
$\mathbf{g}$	Acceleration due to gravity vector	$\mathbb{R}$	m/s^2	
g'	Reduced gravity in a two-layer system	$\mathbb{R}$	m/s^2	
g'_c	Reduced gravity in a gravity current	$\mathbb{R}$	m/s^2	
$g(\bullet)$	Function of $\bullet$	any	any	
H	Height or depth constant, typically some engineered object, such as a ship draught	$\mathbb{R}$	m	
$\hat{H}$	Wave height when the waves may be nonlinear	$\mathbb{R}$	m	
H_{m0}	Significant wave height	$\mathbb{R}$	m	
h	Height or depth constant, typically some natural feature such as depth of the sea	$\mathbb{R}$	m	
I	Intensity of wave energy	$\mathbb{R}$	kg/s^3	W/m^2
$\Im(\bullet)$	Imaginary part of $\bullet$			
Ir	Iribarren number	$\mathbb{R}$	-	
i	Imaginary number ($\sqrt{-1}$)	$\mathbb{C}$	-	
i	Integer index	$\mathbb{R}$	-	
J_m	Bessel function of the first kind, order m	$\mathbb{R}$	-	
j	Integer index	$\mathbb{R}$	-	
K	Stiffness of a spring	$\mathbb{R}$	kg/s^2	
k	Wavenumber, general, or vertical (z-direction)	$\mathbb{R}$	m^{-1}	
$\boldsymbol{k}$	Wavenumber vector	$\mathbb{R}$	m^{-1}	
L	Length constant, or length scale	$\mathbb{R}$	m	
$L_\oplus$	Radius of deformation (or Rossby radius)	$\mathbb{R}$	m	
L_p	Sound pressure level	$\mathbb{R}$	-	dB
L_v	Length scale in the vertical direction	$\mathbb{R}$	m	
$\mathcal{L}$	Dimensionless wavelength	$\mathbb{R}$	-	
ℓ	Wavenumber in the y-direction	$\mathbb{R}$	m^{-1}	
m	Mass	$\mathbb{R}$	kg	
m	Wavenumber in azimuthal direction	$\mathbb{R}$	rad^{-1}	

n	Integer number	$\mathbb{R}$	-	
N	Buoyancy frequency	$\mathbb{R}$	rad/s	
N	Integer number, usually maximum in series	$\mathbb{R}$	-	
P	Pressure, total (or 'absolute')	$\mathbb{R}$	kg/m/s^2	Pa
P_0	Pressure, total, in some initial steady state	$\mathbb{R}$	kg/m/s^2	Pa
$P_\emptyset$	Pressure, ambient	$\mathbb{R}$	kg/m/s^2	Pa
P_{atm}	Pressure, atmospheric	$\mathbb{R}$	kg/m/s^2	Pa
P_h	Pressure, hydrostatic	$\mathbb{R}$	kg/m/s^2	Pa
P_s	Pressure, engineering static	$\mathbb{R}$	kg/m/s^2	Pa
$\mathcal{P}$	Function of spatial variables (e.g. x, y, z) only	$\mathbb{R}$	kg/m/s^2	Pa
$\bar{\mathbb{P}}_{\hat{\eta}}$	Power, cycle-averaged, per unit crest length of of single-frequency wave	$\mathbb{R}$	kg m/s^3	W/m
$\bar{\mathbb{P}}_H$	Power, cycle-averaged, per unit crest length of of ocean-wave spectrum	$\mathbb{R}$	kg m/s^3	W/m
$\bar{\mathbb{P}}_P$	Power, cycle-averaged, engineering system	$\mathbb{R}$	kg m^2/s^3	W
p	Pressure, dynamic	$\mathbb{C}$	kg/m/s^2	Pa
$\hat{p}$	Pressure amplitude	$\mathbb{R}$	kg/m/s^2	Pa
p_R	Real part of p	$\mathbb{R}$	kg/m/s^2	Pa
$p_\emptyset$	Pressure, dynamic, reference (acoustics)	$\mathbb{R}$	kg/m/s^2	Pa
p_∞	Pressure from external driver (acoustics)	$\mathbb{C}$	kg/m/s^2	Pa
p_σ	Pressure, surface tension	$\mathbb{R}$	kg/m/s^2	Pa
Q	Pressure constant, usually a pressure scale	$\mathbb{R}$	kg/m/s^2	Pa
R	Radius, time-varying, of an object	$\mathbb{R}$	m	
R_0	Radius constant	$\mathbb{R}$	m	
$\mathcal{R}$	Function of r only	$\mathbb{R}$	-	
$\Re(\bullet)$	Real part of $\bullet$			
Re	Reynolds number	$\mathbb{R}$	-	
r	Distance in the radial direction	$\mathbb{R}$	m	
St	Strouhal number	$\mathbb{R}$	-	
s	Distance in an arbitrary direction	$\mathbb{R}$	m	
T	Time constant, typically one period (cycle)	$\mathbb{R}$	s	
T	Temperature, absolute	$\mathbb{R}$	K	K
$\mathcal{T}$	Function of t only	$\mathbb{C}$	-	
t	Time	$\mathbb{R}$	s	
t	Dimensionless time	$\mathbb{R}$	-	

U	Speed constant, usually a velocity scale	$\mathbb{R}$	m/s
$\boldsymbol{U}$	Spatial part of velocity vector	$\mathbb{C}$	m/s
u	Horizontal (x) or radial (r) velocity component	$\mathbb{C}$	m/s
$\boldsymbol{u}$	Velocity vector (u, v, w)	$\mathbb{C}$	m/s
u	Dimensionless speed in x-direction	$\mathbb{R}$	-
$\hat{u}$	Amplitude of x-direction velocity component	$\mathbb{R}$	m/s
u_0	Linear solution for fluid velocity (x-direction), or zeroth eigenmode of the linear solution	$\mathbb{C}$	m/s
u_i	ith eigenmode of of the linear solution	$\mathbb{C}$	m/s
$\tilde{u}$	Particle-location velocity (x-direction)	$\mathbb{C}$	m/s
u_R	Real part of u	$\mathbb{R}$	m/s
u_S	Drift speed (Stokes drift)	$\mathbb{R}$	m/s
V	Volume	$\mathbb{R}$	m^3
$\mathcal{V}$	Volume per unit mass	$\mathbb{R}$	m^3/kg
$\mathcal{V}_0$	Volume per unit mass when fluid at rest	$\mathbb{R}$	m^3/kg
v	Horiz. (y) or azimuthal (φ) velocity component	$\mathbb{C}$	m/s
w	Vertical (z) or axial velocity component	$\mathbb{C}$	m/s
x	Distance in x-direction, usually horizontal	$\mathbb{R}$	m
$\boldsymbol{x}$	Distance vector (x, y, z)	$\mathbb{R}$	m
$\tilde{x}$	Location of a particle in the x-direction	$\mathbb{R}$	m
$\mathcal{X}$	Function of $\boldsymbol{x}$ only (in 3D) or x only (in 1D)	$\mathbb{C}$	-
y	Distance in y-direction, usually horizontal	$\mathbb{R}$	m
$\mathcal{Y}$	Function of y only	$\mathbb{C}$	-
z	Distance in z-direction, usually vertically up; for rotating-fluids, z is along the axis of rotation	$\mathbb{R}$	m
$\tilde{z}$	Location of a particle in the z-direction	$\mathbb{C}$	m
$\hat{Z}$	Acoustic impedance, specific characteristic	$\mathbb{R}$	$kg/m^2/s$
$\mathcal{Z}$	Function of z only	$\mathbb{R}$	m

Greek letters

α	Angle	$\mathbb{R}$	rad	
α_T	Thermal diffusivity	$\mathbb{R}$	m^2/s	
β	Angle	$\mathbb{R}$	rad	
γ	Adiabatic index (ratio of specific heats)	$\mathbb{R}$	-	
Δ	Prefix indicating a difference in a variable			
δ	Perturbation in bubble radius	$\mathbb{C}$	m	
δ_ω	Stokes-layer thickness	$\mathbb{R}$	m	
ϵ	Dimensionless small parameter in derivations	$\mathbb{R}$	-	
ζ	Damping ratio	$\mathbb{R}$	-	
ζ_P	Damping ratio due to useful-power extraction	$\mathbb{R}$	-	
ζ_r	Damping ratio due to wave radiation	$\mathbb{R}$	-	
ζ_T	Damping ratio due to thermal losses (bubbles)	$\mathbb{R}$	-	
ζ_μ	Damping ratio, linearised fluid-dynamical (wave-energy conversion) or viscosity (bubbles)	$\mathbb{R}$	-	
η	Surface elevation of surface gravity waves	$\mathbb{C}$	m	
η_R	Real part of η	$\mathbb{R}$	m	
$\hat{\eta}$	Amplitude of surface gravity waves; if waves are nonlinear, defined as $\hat{H}/2$	$\mathbb{R}$	m	
θ	Angle, or latitude of the Earth	$\mathbb{R}$	rad	
κ	Wavenumber in x-direction (Cartesian); in r-direction (spherical polars)	$\mathbb{R}$	m^{-1}	
κ_p	Polytropic index	$\mathbb{R}$	-	
Λ	Wavenumber in r-direction (cylindrical polars)	$\mathbb{R}$	m^{-1}	
λ	Wavelength	$\mathbb{R}$	m	
μ	Dynamic viscosity	$\mathbb{R}$	kg/m/s	Pa s
ν	Kinematic viscosity	$\mathbb{R}$	m^2/s	
ξ	Displacement of a fluid particle (real) or of mechanical oscillator in a fluid (complex)	$\mathbb{C}$	m	
ρ	Density	$\mathbb{R}$	kg/m^3	
ρ_0	Time-averaged, constant density (may vary slowly in z)	$\mathbb{R}$	kg/m^3	
ϱ	Small variation in density from ρ_0	$\mathbb{C}$	kg/m^3	
σ	Surface tension coefficient	$\mathbb{R}$	kg/s^2	N/m
τ	Stress component	$\mathbb{C}$	$kg/m/s^2$	Pa
$\boldsymbol{\tau}$	Stress tensor	$\mathbb{C}$	$kg/m/s^2$	Pa
Φ	Phase angle	$\mathbb{R}$	rad	
ϕ	Velocity potential	$\mathbb{C}$	m^2/s	
φ	Azimuthal angle	$\mathbb{R}$	rad	
Ω	Angular rotation rate	$\mathbb{R}$	rad/s	
$\Omega_\oplus$	Angular rotation rate of the Earth	$\mathbb{R}$	rad/s	
ω	Frequency	$\mathbb{R}$	rad/s	
ω_0	Frequency, natural (undamped)	$\mathbb{R}$	rad/s	
$\omega_{0\zeta}$	Damped natural frequency	$\mathbb{R}$	rad/s	

Operators

$\boldsymbol{\nabla}$	Vector differential operator, 'del' or 'nabla'	$\mathbb{R}$	m^{-1}
∇^2	$\nabla \cdot \nabla$, 'Laplacian'	$\mathbb{R}$	m^{-2}
$\nabla_{\perp}^2$	$\nabla \cdot \nabla$ in horizontal only	$\mathbb{R}$	m^{-2}
$\overline{\bullet}$	Time-mean of $\bullet$	$\mathbb{R}$	any

Preface

Aims and motivation

The aims of this book are:

1. to list useful key results for selected fluid-wave problems;
2. to show full derivations of selected key results;
3. to illustrate diverse fluid-wave applications in engineering and nature.

An immense variety of phenomena involving mechanical waves occur in fluids. Fluid waves animate the ocean, provide us with music and affect the climate, while many vital new technologies for medicine, industry and infrastructure all involve fluid waves. Calculations are needed to pursue these applications, and it is hoped this book may be a convenient entry-point both for the calculations and for an appreciation of the applications.

This book is written for students undertaking the later years of a university degree, or post-graduate and professional readers. The main prerequisite to understanding this text is differential and integral calculus. Students should ideally also have had some exposure to vector calculus and the principle of perturbation methods. However, readers whose mathematical knowledge is limited to algebra should still be able to use this book

How to use this book

The reader may consult first the *Summary of key points* that begins each Chapter, where it is possible that a formula or result of use to the reader may be found immediately. The page on which each key formula was derived is indicated. It is highly recommended that at least the section leading up to the key formula is read, to ensure the formula is not applied unawares of the assumptions behind it, which may lead to erroneous results. Important reference texts are also indicated in each Summary.

References are collected at the end of the book rather than after each chapter. This is because some references are relevant to multiple chapters.

A few problems follow each chapter in the first part of the book. These may be useful in the teaching of a subject based on this book. They include very elementary questions that experience has shown can trip up students undertaking calculations under time pressure, some multiple-choice questions to test students' comprehension and a few harder questions. The majority of questions should not take longer than a few minutes. The focus is on problems placed in a practical-applications context.

Structure

Part I addresses Aims 1 and 2. It comprises selected introductory fundamental topics and results on fluid waves. These are topics thought to be of perpetual relevance, such as surface gravity waves and sound waves. It is not comprehensive, omitting some significant topics such as waves due to instabilities in fluid flows, many classes of geophysical waves affected by stratification and rotation, and shock waves in compressible flows. Within each chapter, the introductory nature of this book has also meant that many sub-topics are omitted. The references provided cover the missing material. Meanwhile, a few topics are subjected to full, if rather lengthy, mathematical derivations so that there are few gaps left as 'exercises for the student'. Thus, the reader is made aware of the many steps required to address a typical problem in this field and is fully equipped to tackle similar problems. Inevitably, however, many detailed derivations are not undertaken.

Part II addresses Aim 3. It is intended to showcase the breadth of relevance of wave phenomena in fluids. Thus, Part II is rather like a collection of specialised review papers, which in most cases refer to sources right up to the date of publication of this book. Some chapters involve significant derivations, but most are explanatory. In some cases, the applications, which are as diverse as renewable energy, cancer treatment, climate change, and the search for habitable exoplanets, are pursued to a depth such that the relevance to fluid waves may appear tangential; nevertheless, without the ability to calculate fluid-wave problems, rigorous pursuit of the applications would not be possible.

Approach and acknowledgements

Where references are made to significant equations in other chapters, equations are generally repeated, rather than forcing the reader to turn back through many pages and lose one's train of thought. A clear exception is the two-layer internal wave derivation, for which many pages of derivation would be identical to that of surface gravity waves.

The disadvantage of this 'stand-alone' approach, apart from verbosity, is that some principles common to many types of wave are introduced in the chapter where they may be of most benefit, possibly obscuring their relevance to other types of wave. For example, refraction and reflection are introduced in the context of sound waves, but these are relevant to all the waves in this book; and conversely, beats, admittedly important in music, are instead derived for water waves. Wherever possible, cross-references are made to counter this issue.

I am very grateful to Filippo Nelli, Danica Tothova, Elissa Goodrich, Justin Leontini and Shaung Zhu for helpful and insightful comments on the text, and finally, to Gagandeep Singh and his team at Taylor & Francis for being patient and helpful publishers.

Part I

Theory and classical applications

1

Fundamentals

1.1 Summary of key points

- The ambient **pressure** in a fluid, $P_{\emptyset}$, is given by (1.2) on page 8,

$$\boxed{P_{\emptyset} = P_{\mathrm{atm}} + P_h + P_s},$$

where P_{atm} is the atmospheric pressure, conventionally taken to be 101 325 Pa, P_s is some static pressure that may be applied by an engineering system, and P_h is the hydrostatic pressure, given by (1.1) on page 8,

$$\boxed{P_h = \rho_0 \mathrm{g} h},$$

where ρ_0 is the assumed-constant density of the fluid of depth h above the point of interest and g is the acceleration due to gravity, which is defined as 9.80665 m s^{-2} (Bureau international des poids et measures, Paris, 2006), with more precise values of g calculable as a function of latitude (Moritz, 2000) and height above sea level (Li and Götze, 2001). For water, $\rho_0 = 998$ kg m^{-3} and for air, $\rho_0 = 1.2$ kg m^{-3} (Streeter and Wylie, 1979). The total pressure P in the fluid, including any dynamic pressure p due to fluid motion, is given by (1.3) on page 9,

$$\boxed{P = P_{\emptyset} + p}.$$

- The **Ideal Gas Law**, relating the absolute pressure P and the volume per unit mass, $\mathcal{V}$, to their initial values, P_0 and $\mathcal{V}_0$, is given by (1.5) on page 10,

$$\boxed{P\mathcal{V}^{\kappa_p} = P_0 {\mathcal{V}_0}^{\kappa_p}},$$

where κ_p is the polytropic index ($\kappa_p = \gamma$ for an adiabatic or isentropic process, where the adiabatic index γ is very close to 1.4 for air).

- The **Viscosity Equation** (Newton's Law of Viscosity), (1.8) on page 13, relates the shear stress τ to the gradient in velocity by

$$\boxed{\tau = \mu \frac{\mathrm{d}u}{\mathrm{d}y}},$$

DOI: 10.1201/9780429295263-1

where μ is the dynamic viscosity, and u is the speed of flow in the x-direction with the y-direction at right angles. For water, $\mu = 1.0016 \times 10^{-3}$ Pa s at 20°C and for air, $\mu = 1.81 \times 10^{-5}$ Pa s at 20°C (Streeter and Wylie, 1979).

- The increased pressure on the concave side of a surface due to **surface tension** is given by (1.11) on page 15,

$$\boxed{p_\sigma = \sigma \left(\frac{1}{r_1} + \frac{1}{r_2} \right)},$$

where σ is the surface tension coefficient which takes a value of 0.07275 ± 0.00036 N m^{-1} for a water-air surface at 20°C (Vargaftik et al., 1983) and r_1 and r_2 are the radii of curvature of the surface in the two planes at right angles to the surface.

- The Law of Conservation of Mass, or **Continuity Equation**, is given by (1.15) on page 17,

$$\boxed{\frac{\partial \rho}{\partial t} = -\boldsymbol{\nabla} \cdot (\rho \boldsymbol{u})},$$

where ρ is the density that could vary in space and time, t is time, $\boldsymbol{\nabla}$ represents gradients in space and $\boldsymbol{u}$ is the velocity vector.

- The Law of Conservation of Momentum, or **Momentum Equation**, for a fluid (the Navier-Stokes momentum equation), is given by (1.19) on page 20,

$$\boxed{\frac{\mathrm{D}(\rho \boldsymbol{u})}{\mathrm{D}t} = \boldsymbol{\nabla} \cdot \boldsymbol{\tau} + \rho \mathbf{g}},$$

where $\mathrm{D}/\mathrm{D}t$ denotes the material (or 'total') derivative, $\boldsymbol{\tau}$ is the stress tensor and $\mathbf{g}$ is the gravitational acceleration vector; for an incompressible fluid, (1.19) reduces to (1.20) on page 20,

$$\boxed{\frac{\mathrm{D}\boldsymbol{u}}{\mathrm{D}t} = -\frac{1}{\rho_0} \boldsymbol{\nabla} P + \nu \nabla^2 \boldsymbol{u} + \mathbf{g},},$$

where the kinematic viscosity is given by $\nu = \mu / \rho_0$.

- For all waves, **radian frequency**, **frequency**, **period**, **wavenumber** and **wavelength** are given by three relations, firstly (1.68) on page 36,

$$\boxed{f = \frac{\omega}{2\pi}},$$

where f is the frequency in Hertz (or cycles per second) and ω is the frequency in radians per second (often just called the 'frequency'); secondly, the *period* of the waves, T, where $T = 1/f$, is given by (1.69) on page 36

$$\boxed{T = \frac{2\pi}{\omega}};$$

and, according to (1.70) on page 36,

$$\boxed{\lambda = \frac{2\pi}{k}},$$

where k is the wavenumber in m^{-1} and λ is the wavelength in metres.

- The relations between **wave speed**, frequency, wavenumber and wavelength are given by (1.71) and (1.72) on page 36,

$$\boxed{c = \frac{\omega}{k}},$$

where c is the wave speed in metres per second, or equivalently by

$$\boxed{c = f\lambda}.$$

- **Dimensionless numbers** relevant to fluid waves include those derived in §1.2.4 and listed below, for a length scale of L, velocity scale U and time scale T (for oscillations, $T = 2\pi/\omega$), and Stokes boundary-layer thickness $\delta_\omega = \sqrt{2\nu/\omega}$,

Traditional name of number	Abbreviation	Function	Ratio of forces
Reynolds	Re	$\frac{UL}{\nu}$	Nonlinear-inertia to viscous
Froude	Fr	$\frac{U}{\sqrt{gL}}$	Nonlinear-inertia to gravity
Strouhal	St	$\frac{L}{UT}$	Oscillatory- to nonlinear-inertia
Keulegan–Carpenter	K_C	$\frac{UT}{L}$	Nonlinear-inertia to oscillatory
Womersley	Wo	$L\sqrt{\omega/\nu}$ or $\sqrt{2}\,L/\delta_\omega$	Oscillatory-inertia to viscous
Mach	Ma	$\frac{U}{c}$	Nonlinear-inertia to bulk-stiffness (flow speed to sound speed)

- The **forced response** of any linear oscillator is given by (1.83) on page 38,

$$\boxed{\begin{aligned}\frac{a}{\hat{\mathcal{F}}} &= \frac{1}{\sqrt{\left(1-\omega'^2\right)^2+\left(2\zeta\omega'\right)^2}},\\ \Phi &= -\tan^{-1}\left(\frac{2\zeta\omega'}{1-\omega'^2}\right)\end{aligned}}.$$

where a is the real amplitude of the response, Φ is the phase, $\hat{\mathcal{F}}$ is the real forcing amplitude in the same units as a, the damping ratio is ζ, and $\omega' = \omega/\omega_0$ where ω_0 is the natural frequency and ω is the forcing frequency.

- The additional **pressure due to surface tension**, p_σ, is given by (1.11) on page 15,

$$\boxed{p_\sigma = \sigma\left(\frac{1}{r_1}+\frac{1}{r_2}\right)},$$

where, for a water-air surface, $\sigma = 0.07275 \pm 0.00036$ N m^{-1} at 20°C (Vargaftik et al., 1983) and r_1 and r_2 are the radii of curvature of the surface.

- Useful **textbooks** include Batchelor (1973) for an applied-mathematical approach to fluid dynamics and Lamb (1932) for many classical fluid dynamics derivations. Batchelor (1973) includes an appendix in which the equations of motion for a fluid are given in cylindrical and spherical co-ordinates. Detailed derivations of many fluid-wave problems are in Lighthill (1978). Streeter and Wylie (1979) provide an engineering approach to fluid mechanics, which includes detailed calculations on flow in pipes and ducts, and turbomachinery and aerospace design. There are very many other engineering textbooks on fluid mechanics, most of which follow the same pattern. Useful relations for the density of the atmosphere and of seawater are given in Gill (1982).

1.2 Basic fluid dynamics

1.2.1 Fluid mechanics and fluid dynamics

A fluid is a substance that flows. This statement, while easily grasped, is imprecise. The formal definition of a fluid requires an understanding of stresses, which will be outlined shortly. A fluid is defined as a substance that *deforms indefinitely in response to shear stress.* Both liquids and gases are fluids. The difference between a liquid and a gas is that a given mass of liquid forms a distinct surface bounding a finite volume, whereas a given mass of gas expands to fill whatever container it is housed in. Despite this difference, the same fluid-dynamical laws - and almost all of the same fluid-wave phenomena - apply equally to liquids and to gases.

Fluid mechanics is the study of forces and motions in fluids, and of the forces and motions caused by fluids on solid objects and vice versa. Together with solid mechanics, fluid mechanics forms the branch of classical physics called *continuum mechanics*. The term *fluid dynamics* is often used interchangeably with 'fluid mechanics', but is usually reserved for those fluid mechanical situations where there is some motion. For example, the calculation of the force on a dam wall due to the pressure of motionless water on the wall is a fluid mechanics problem, but not a fluid dynamics problem. Since fluid waves inherently involve motion, they are fluid-dynamical phenomena.

Continuum mechanics, and thus fluid mechanics and fluid dynamics, is only valid in a *continuum*. This is a substance that is definitely composed of indivisible 'particles', such as molecules or atoms, but whose behaviour only matters to us for volumes very much larger than the individual particles. Thus, it is possible to average over the very many particles inside the smallest volume of interest to us and ignore the reality that the particles exist. Thus, in fluid dynamics, individual molecules or atoms, which move in a statistical fashion, confer their averaged quantities such as pressure, velocity, density or temperature to continuum equations, allowing us to use the rules of calculus to determine the behaviour of the continuum. It is worth noting that if the smallest volume of interest is large enough, a continuum can also be considered to be composed of 'particles' that are much larger than atoms or molecules. For example, sand or grains of wheat can be observed to flow rather like a fluid, and subject to suitable approximations, the flow of such *granular materials* could be treated as a fluid dynamics problem. Meanwhile, the particles of dust that compose the accretion disk of a forming solar system (discussed in §11.2.1), even though there are great distances between the particles, can be treated as a continuum when the smallest volume of interest is the size of a planet, so there are still very many particles in the smallest volume.

1.2.2 Constitutive relations for fluid continua

1.2.2.1 Stress

Before introducing what is sometimes called the 'Three Laws of Fluid Dynamics', it is important to consider what we might call the 'Zeroth Law of Fluid Dynamics'. This is a set of relations, the constants in which depend on the particular fluid being considered, so that the constants are different, for example, for water, molten steel and blood, and different for cold air and hot exhaust gas. The relations mostly connect stress to strain or rate of strain, and to appreciate these relations, it is first necessary to be clear about stress and about strain.

Stress is force per unit area and has units of kg m^{-1} s^{-2}, with the SI name Pascal, abbreviated Pa. It is divided into two types, *normal stress* and *shear stress.* The normal stresses create the *pressure* in the fluid, though the normal stresses are not uniquely related to the pressure unless the fluid is motionless or incompressible. Stress, in general, is usually denoted with the symbol τ. Since there are three dimensions of space (in Cartesian coordinates, the x, y and z directions), forces could be acting in each of the three directions, but the force in each direction could be acting on areas normal to all of the x, y and z directions, giving nine possible combinations. Hence, τ is a 3×3 tensor. The three normal stresses are τ_{xx}, τ_{yy} and τ_{zz}, whereas the six shear stresses are τ_{xy}, τ_{xz}, τ_{yx}, τ_{yz}, τ_{zx} and τ_{zy}. We will return to the shear stresses in §1.2.2.4 when considering viscosity, but first, the normal stresses creating pressure must be understood.

A bewildering variety of quantities are called 'pressure' in fluid dynamics, engineering, physics and medicine. It is common for students to confuse these and get the wrong result, so it is helpful to begin with a precise set of definitions. The *hydrostatic pressure* is the pressure due to the weight of the fluid above. If the fluid above is a liquid and thus almost incompressible, the hydrostatic pressure P_h is given by

$$\boxed{P_h = \rho_0 \mathrm{g} h}, \tag{1.1}$$

where ρ_0 is the liquid density, g is the acceleration due to gravity, which has the standard value of 9.80665 ms^{-2} (Bureau international des poids et measures, Paris, 2006), and h is the depth of liquid above. For water at 20° C, $\rho_0 = 998$ kg m^{-3} and for air at 20° C, $\rho_0 = 1.2$ kg m^{-3} (Streeter and Wylie, 1979).

However, if the fluid above is a gas, such as the gases of the Earth's atmosphere, a more detailed relation than (1.1) is needed. Fortunately, many practical applications do not involve movements between the Earth's surface and high altitude, so that *atmospheric pressure*, P_{atm}, does not vary as much as other sorts of pressure to be discussed shortly. In addition, it is possible that the fluid could be mechanically pressurised by some static pressure P_s above the hydrostatic and atmospheric pressure and that flows are being considered relative to the sum of these unchanging pressures. These three pressures added together give what we will call the *ambient pressure*, $P_\emptyset$, (often called the *static pressure*), which is thus given by

$$\boxed{P_\emptyset = P_{\mathrm{atm}} + P_h + P_s}. \tag{1.2}$$

Thus, $P_\emptyset$ is the pressure experienced when there is no fluid motion. Finally, the symbol P will be used for *total pressure* (sometimes called the *absolute pressure*), which includes the ambient pressure $P_\emptyset$ plus any *dynamic pressure* due to fluid motion, p (total minus atmospheric pressure is called *gauge pressure* in some engineering applications). The dynamic pressure can be negative, but the *total pressure is always positive.*

Because the dynamic pressure is due to fluid motion, it will be of most relevance in fluid-wave calculations, but the ambient pressure will be needed too. In many circumstances, the dynamic pressure, p, is small relative to the ambient pressure, $P_\emptyset$, and because $P_\emptyset$ does not vary significantly, the term 'pressure' is often used to refer only to the pressure due to fluid motion. When we consider fluid waves, it is variations in p that will be created by waves. Thus

$$\boxed{P = P_\emptyset + p}. \tag{1.3}$$

In a motionless fluid, $p = 0$, so the total and ambient pressures are equal and are equal to the average of the three normal stresses (and the normal stresses are negative since the total pressure is positive), so $P = P_\emptyset = -\frac{1}{3}(\tau_{xx} + \tau_{yy} + \tau_{zz})$. If the fluid is moving but is incompressible, only the total pressure P is given by this average, so $P = -\frac{1}{3}(\tau_{xx} + \tau_{yy} + \tau_{zz})$. If the fluid is moving and is compressible, some part of the normal stresses is due to viscous resistance to the rate of change of volume with time, which will be outlined in §1.2.2.4 below.

1.2.2.2 Strain

Linear strain is the ratio of one of the lengths of the substance when normal stress is applied along that length, to its length when not under stress. Engineers typically study linear strain when learning how metal stretches under load, for example, but for fluid flows, variations in the volume are paramount.

Volumetric strain is the ratio of the volume, V, of the substance when under some additional pressure to its volume when not under such pressure, $V_\emptyset$. If the contributions to normal stresses from fluid motion are all negative (so that the dynamic pressure is positive), the substance is said to be *compressed* by the motion. If these contributions are all positive (so that the dynamic pressure is negative), the motion is said to put the substance under *rarefaction* or sometimes 'expansion'. Note that the 'pressure' we have just referred to is the dynamic pressure p that is due to fluid motion, and thus the compression or rarefaction is *relative* to the inherent compression due to the ambient pressure. If a fluid is compressed, its density, ρ, will be higher than its ambient density ρ_0, and if a fluid is rarefied, ρ will be lower than ρ_0. For example, in the propagation of sound waves, ρ varies cyclically about ρ_0, being higher than ρ_0 during the compression half of the oscillation and less than ρ_0 during the rarefaction half of the oscillation.

Shear strain is the angle by which a substance is distorted owing to shear stress.

1.2.2.3 Relation between pressure and volumetric strain

The general relation between pressure and volumetric strain is given by the *bulk modulus* of the fluid, E_v, usually expressed as

$$E_v = \rho \frac{\partial P}{\partial \rho}, \tag{1.4}$$

in which the volumetric strain can be identified once we realise that $\rho/\Delta\rho = \Delta V/V$. It can be thought of as the 'stiffness' of the fluid since the higher the bulk modulus, the more pressure is needed to cause a given volumetric strain. In general, the bulk modulus is a function of temperature and fluid composition. For liquids, the relation is often extracted from empirical data. The bulk modulus will be used in §3.3.2 in the derivation of the speed of sound, and in practice, the bulk modulus is actually determined from speed-of-sound measurements (Fine and Millero, 1973). For an incompressible liquid, which of course exists only as a theoretical approximation (albeit a very useful approximation that we will employ many times), the bulk modulus is infinite. For water $E_v \simeq 2.2 \times 10^9$ Pa.

For gases, however, a much more convenient relation is available, and, moreover, it can be derived from the fundamental theory of the mechanics of gas molecules, called the *kinetic theory of gases*. The *Ideal Gas Law* relates any value of total pressure, P, and the *specific volume* or volume per unit mass, $\mathcal{V}$, to any other 'initial' values of the total pressure and volume per unit mass of the same gas, P_0 and $\mathcal{V}_0$. Here, care has been taken to introduce a symbol for the initial state of the pressure, P_0, that is different to that of the ambient pressure, $P_\emptyset$, since some unchanging or 'steady' fluid flow might be occurring (steady flows are defined in §1.2.4 below), causing some background dynamic pressure; but if there is no such background flow, $P_0 = P_\emptyset$. Note that $\mathcal{V} = 1/\rho$ and $\mathcal{V}_0 = 1/\rho_0$. The Ideal Gas Law is given by

$$\boxed{P\mathcal{V}^{\kappa_p} = P_0 {\mathcal{V}_0}^{\kappa_p}}, \tag{1.5}$$

where κ_p is the *polytropic index* whose value depends on the nature of the heat transfer occurring during compression or expansion of the gas. If the compression or expansion is *isothermal*, meaning that the temperature of the gas does not alter during the compression or expansion, $\kappa_p = 1$. An isothermal compression or expansion requires that heat flows out of or into the gas across the boundaries of whatever is containing the gas without any restriction, so that the temperature can stay constant. The bulk modulus for a gas is given by $E_v = \kappa_p P_0$ and, as just-noted, it will be used in §3.3.2 in the derivation of the speed of sound. For isothermal air at atmospheric pressure and 20° C, $E_v \simeq 10^5$ Pa.

If the compression or expansion is *adiabatic* or *isentropic*, meaning that there is zero heat flow across the boundaries, $\kappa_p = \gamma$, where γ is the adiabatic index. Adiabatic conditions need not imply the gas is in some container, like

FIGURE 1.1
A set of oceanographic probes and sample bottles is winched into the ocean from a research ship to a depth of over 1000 m. At such depths, water cannot be considered incompressible, so the bulk modulus (1.4) is needed. For seawater, the bulk-modulus relation between stress and volumetric strain requires an empirical relation, determined in practice by speed-of-sound measurements; see Fine and Millero (1973) or Gill (1982). Photograph by Richard Manasseh.

the cylinder of a bicycle pump. If the expansions and contractions are very rapid and are very small in magnitude, which is typical of most sound waves, there is insufficient time for significant heat to flow one way on the crest of the wave before it has to flow the opposite way on the trough, so in practice, the heat flow is extremely small, if not exactly zero. The two cases of isothermal and adiabatic behaviour represent the two idealised limiting cases for heat transfer, so that $1 < \kappa_p < \gamma$.

Now, the adiabatic index is related to the molecular structure of the gas and is given by the formula

$$\gamma = 1 + \frac{2}{n_f}, \tag{1.6}$$

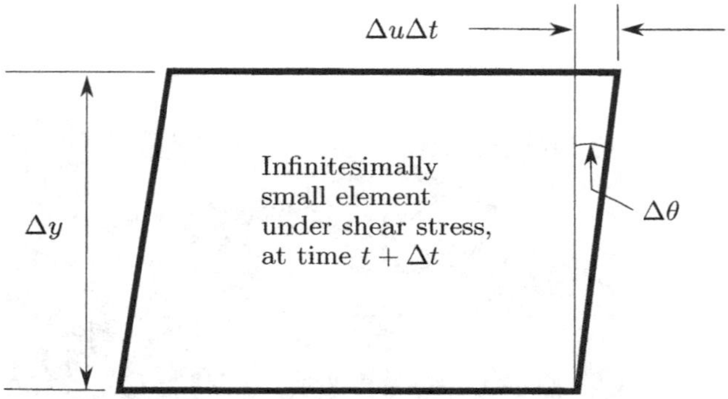

FIGURE 1.2
The definition of the rate of shear strain

where n_f is the number of *degrees of freedom* of the gas molecule. This is the number of ways the atoms in the molecule can move and also move relative to each other. In moving, the molecule possesses kinetic energy and thus heat. In principle, the number of degrees of freedom of the molecule is purely a geometric property of the molecule, but in practice, some degrees of freedom only occur at high temperatures. For a gas such as argon (Ar) consisting of just one atom, $n_f = 3$ since the atom can move in the three dimensions of space. For a diatomic molecule, for example, nitrogen (N_2) or oxygen (O_2), two added degrees of freedom are active at room temperature, so $n_f = 3+2 = 5$. Hence, for both N_2 and O_2, $\gamma = 1 + 2/5 = 1.4$, and since the Earth's atmosphere is about 78% N_2 and 21% O_2, the air has a value of γ very close to 1.4. Meanwhile, carbon dioxide (CO_2) is a molecule with three atoms arranged in a line. It has a value of γ of about 1.3 at room temperature. The adiabatic index is also is equal to the ratio of *specific heats* of the gas, C_P/C_V, sometimes called the 'heat capacity ratio'.

1.2.2.4 Relation between shear stress and the rate of shear strain

If a fluid is flowing in the x-direction with speed u, and this speed u varies in the y-direction (at right angles to the x-direction), the fluid is in a state of shear strain. Imagine that u is increasing in the y-direction. The shear strain is then constantly increasing with time as the faster fluid at higher y constantly overtakes the slower fluid. The ability of a fluid to continually suffer an ever-increasing shear strain is the attribute that defines a fluid; as noted in §1.2.1, it deforms indefinitely in response to shear stress.

The definition of the rate of change of shear strain can be understood with the aid of figure 1.2. The rate of change of shear strain is the rate at which the 'tilt angle', $\Delta\theta$, changes with time, i.e. $\Delta\theta/\Delta t$, where for small $\Delta\theta$,

$$\Delta\theta \simeq \tan(\Delta\theta) = \frac{\Delta u \Delta t}{\Delta y}$$

$$\Rightarrow \frac{\Delta\theta}{\Delta t} \simeq \frac{\Delta u}{\Delta y} \tag{1.7}$$

Therefore, when the usual limits are taken as Δy and Δu tend to zero, the rate of change of shear strain becomes $\mathrm{d}u/\mathrm{d}y$.

However, the fluid does not experience stress without resistance. There are chemical and physio-chemical forces between the particles of the fluid; for example, there are attractive forces between the molecules of H_20 in flowing water, which transfer momentum from the faster-flowing portions of fluid to the slower portions, slowing down the faster flows. Similarly, as faster-flowing portions of gas overtake the slower portions, the molecules may collide, likewise transferring momentum. Owing to the random orientation of molecules, these microscopic interactions inevitably transfer energy from the kinetic energy of the bulk flow to tiny and random molecular motions, and since the amount of vibration of atoms and molecules is related to the temperature, the fluid is thus heated. If the fluid is to continue deforming indefinitely, there must be a shear stress continually applied to it that balances the shear stress due to these inter-particle interactions.

The bulk fluid property that controls this transfer of momentum across fluid layers is called *dynamic viscosity* (or sometimes 'molecular viscosity'), and is denoted with the symbol μ. The relation between the rate of shear strain, $\mathrm{d}u/\mathrm{d}y$, and the shear stress due to viscosity, τ, was first understood by Isaac Newton; it is

$$\boxed{\tau = \mu\frac{\mathrm{d}u}{\mathrm{d}y}}, \tag{1.8}$$

which is called the *Viscosity Equation* and traditionally called *Newton's Law of Viscosity*. For water, $\mu = 1.0016 \times 10^{-3}$ Pa s at 20°C, and for air, $\mu = 1.81 \times 10^{-5}$ Pa s at 20°C (Streeter and Wylie, 1979).

Now, a fluid flowing only in the x-direction with speed u could have gradients in u in the z-direction as well as the y-direction, and indeed this is occurring as fluid flows in any conduit (pipe, duct or channel); whatever the value of the speed u in the middle of the conduit, u must be zero for those fluid molecules physio-chemically bound to the walls. Thus the τ in (1.8) is more properly written τ_{yx}, the shear stress in the x-direction acting on a plane normal to the y-direction. Including the other directions gives

$$\tau_{yx} = \mu\frac{\partial u}{\partial y},$$

$$\tau_{zx} = \mu\frac{\partial u}{\partial z}. \tag{1.9}$$

FIGURE 1.3
A thin film of water flows down a glass feature window, forming waves affected by surface tension. Photograph by Richard Manasseh.

Furthermore, the speed u could also vary in the x-direction, and as the flow tries to compress (or expand) the fluid in the x-direction, energy could also be transferred to heat, giving a form of normal stress in addition to that due to pressure, the volume-viscous stress, with the corresponding parameter being called the *dilational viscosity* or sometimes the 'second coefficient of viscosity'.

The fluid could also be flowing in any and all of the x-, y- and z-directions with velocity given by the vector $\boldsymbol{u}$, where $\boldsymbol{u} = (u, v, w)$. Once flows in the other directions are possible, for small strains, considering the geometry of the deformed fluid element, (1.9) becomes

$$\begin{aligned} \tau_{yx} &= \mu \frac{1}{2}\left(\frac{\partial u}{\partial y} + \frac{\partial v}{\partial x}\right), \\ \tau_{zx} &= \mu \frac{1}{2}\left(\frac{\partial u}{\partial z} + \frac{\partial w}{\partial x}\right), \end{aligned} \tag{1.10}$$

and similarly for τ_{xy}, τ_{zy}, τ_{xz} and τ_{yz}. This gives a total of six shear stresses. Plus, in addition to the variation of u in the x-direction, possible variations of v in the y-direction and w in the z-direction add to the volume viscous stress. Thus, as noted in §1.2.2.1, there are nine stresses in the stress tensor, $\boldsymbol{\tau}$, comprising the parts of the three normal stresses due to pressure plus the three normal stresses due to volume viscosity, and the six shear stresses due to Newtonian viscosity.

1.2.2.5 Surface tension

Surface tension is important for many fluid dynamical and fluid-wave phenomena, such as ripples (§2.3.5.4) and bubble-acoustic vibrations (§9.3.1). A liquid was defined in §1.2.1 to be a fluid where a distinct surface bounds its volume. That is because attractive forces hold the molecules or atoms of the liquid together, whereas, for gas, the molecules or atoms move comparatively freely relative to each other and constantly bounce off each other. At the liquid surface, the attractive force is defined to have a value of σ Newton per metre, so that an imaginary line of length L 'drawn' on the surface would have a force of σL Newtons acting at right angles to it. For a water-air surface, $\sigma = 0.07275 \pm 0.00036$ N m^{-1} at 20°C (Vargaftik et al., 1983), or, slightly more precisely, $\sigma = 0.07268 \pm 0.00018$ N m^{-1} at 22°C (Berry et al., 2015), and decreases with temperature (see Vargaftik et al., 1983, for values at various temperatures). If the surface is curved, it can be shown in a few lines (Batchelor, 1973) that these forces develop a component normal to the surface, causing an increase in pressure, p_σ, *on the concave side* of the curved surface given by

$$\boxed{p_\sigma = \sigma\left(\frac{1}{r_1} + \frac{1}{r_2}\right)}, \tag{1.11}$$

where r_1 and r_2 are the radii of curvature of the surface in the two planes at right angles to the surface. Clearly, if length scales are small, so r_1, or r_2, or both are small, surface tension becomes important; conversely, in flows with large length scales, surface tension is negligible. Examples of such cases are in §2.3.5.4 and §9.3.1.

1.2.3 Conservation laws

Conservation laws are at the heart of fluid dynamics and thus fluid wave problems. The laws were determined independently by different scientists over history, but they can all be expressed by a single relation, the *Reynolds Transport Theorem.* There are three conservation laws of fluid mechanics and fluid dynamics:

- Conservation of mass
- Conservation of momentum
- Conservation of energy

Of these, the conservation of mass and momentum are more frequently used in fluid wave problems, and so will be detailed below. The Law of Conservation of Energy is often used when fluid dynamics is applied in an engineering context, and an example of this will occur in §8.5.2.

1.2.3.1 Conservation of mass

The statement that mass is conserved is one of the most fundamental statements in classical physics, which can be expressed as 'mass can neither be created nor destroyed'. Those superficially familiar with nuclear reactions might disagree since both nuclear fusion and fission reactions result in the conversion of a small fraction of mass into energy, hence in principle forever 'destroying'

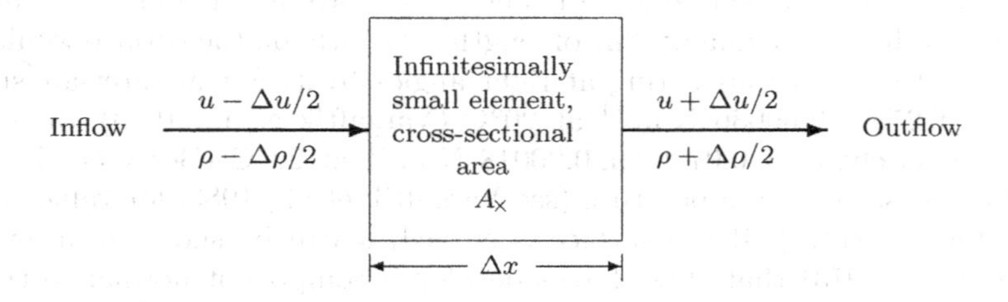

FIGURE 1.4
Transport of mass in one dimension for an infinitesimally small element of fluid.

that mass. However, the principle of mass and energy equivalence, elucidated by Albert Einstein in 1905, and encapsulated by the relation $E = mc^2$ where E is energy, m is mass and c is the speed of light, effectively unites the Law of Conservation of Mass with the Law of Conservation of Energy. Excluding nuclear reactions, however, it is convenient to express the conservation of mass and energy with separate laws.

The conservation of mass in a fluid-dynamics context is easily expressed by first imagining a one-dimensional flow in which fluid with density ρ is transported by a flow with speed u along a channel in the x-direction, as shown in figure 1.4. The channel has a cross-sectional area $A_\times$, say. After travelling an infinitesimally small distance along the channel, Δx say, imagine that the speed has changed from $u - \Delta u/2$ at the entry to the infinitesimally small element, to $u + \Delta u/2$ at the exit, and the density has likewise changed from $\rho - \Delta\rho/2$ to $\rho + \Delta\rho/2$. The volume of this short segment is $A_\times \Delta x$. Therefore, mass is exiting this short segment of the channel of length Δx at a rate $(\rho + \Delta\rho/2)(u + \Delta u/2)A_\times$ while it is entering at a rate $(\rho - \Delta\rho/2)(u - \Delta u/2)A_\times$, and the net change of mass in transiting the short segment would be $[(\rho - \Delta\rho/2)(u - \Delta u/2) - (\rho + \Delta\rho/2)(u + \Delta u/2)]A_\times$.

Since $\Delta\rho$ and Δu are by definition small, their product is negligible, so the net change of mass becomes $-(\Delta\rho u + \rho\Delta u)A_\times$. In general, $\Delta\rho$ and Δu could be positive or negative. Meanwhile, it is possible that the density inside the short segment is changing with time, so the rate of net increase in mass inside the short segment is $(\Delta\rho/\Delta t)A_\times \Delta x$. Since mass can neither be created nor destroyed, it can be neither gained nor lost in its journey along the short segment of the channel, so that

$$(\Delta\rho/\Delta t)A_\times \Delta x = -(\Delta\rho u + \rho\Delta u)A_\times . \tag{1.12}$$

Dividing (1.12) by the volume $A_\times \Delta x$ of the small segment and taking the limits as the small quantities tend to zero gives

$$
\begin{aligned}
\frac{\partial \rho}{\partial t} &= -\frac{\partial \rho}{\partial x}u - \rho\frac{\partial u}{\partial x}, \\
\Rightarrow \frac{\partial \rho}{\partial t} &= -\frac{\partial}{\partial x}(\rho u).
\end{aligned} \tag{1.13}
$$

This reasoning is an application of the Reynolds Transport Theorem to mass.

Now, imagine that rather than being a segment of a channel, the small volume is in the middle of a flowing mass of fluid, and, adopting Cartesian coordinates, imagine that mass is able to flow into and out of the volume in the y and z directions as well, with velocity components u, v and w in the x, y and z directions respectively. The analysis is the same: dividing by the volume $\Delta x \Delta y \Delta z$ at the equivalent of the step that led to (1.13) gives

$$
\frac{\partial \rho}{\partial t} = -\frac{\partial}{\partial x}(\rho u) - \frac{\partial}{\partial y}(\rho v) - \frac{\partial}{\partial z}(\rho w), \tag{1.14}
$$

or, using standard vector-calculus notation, use the operator $\boldsymbol{\nabla}$ for the derivatives in space, which has the advantage that the coordinate system does not need to be specified. Then (1.14) has the compact form

$$
\boxed{\frac{\partial \rho}{\partial t} = -\boldsymbol{\nabla} \cdot (\rho \boldsymbol{u}).} \tag{1.15}
$$

This is often called the *Continuity Equation*.

1.2.3.2 Conservation of momentum

The Law of Conservation of Momentum is traditionally known as Newton's Second Law, and it was first expressed by Isaac Newton in 1687. It is the statement that force is equal to mass multiplied by acceleration, i.e.

$$
\boldsymbol{F} = m\boldsymbol{a}, \tag{1.16}
$$

where $\boldsymbol{F}$ is the force vector, m is mass, and $\boldsymbol{a}$ is the acceleration vector. The right-hand side of (1.16) is the rate of change of momentum with time. To prepare for fluid-dynamical calculations, first, make a trivial re-arrangement of (1.16),

$$
\boldsymbol{a} = \frac{\boldsymbol{F}}{m}. \tag{1.17}
$$

The purpose of this re-arrangement is to highlight that in fluid dynamics, the greatest challenge is often posed by the determination of the motion of the fluid for a given applied force. The calculation of the velocity resulting from the acceleration can be extremely difficult, as we will see shortly.

In contrast, the determination of motion under Newton's Second Law is easy for a solid. For example, if a solid mass of 1 kg is acted upon by a horizontal force of 10 N (recall that is 10 kg m s^{-2}), Newton's Second Law (1.17) predicts that it will accelerate in the horizontal direction at 10 m s^{-2}. After 0.5 s, it will be travelling horizontally at 5 m s^{-1}, and if the force ceases to be applied to it after 0.5 s, it will continue to move horizontally at 5 m s^{-1}. The Law of Inertia (Newton's First Law) states that it will continue to move horizontally at 5 ms $^{-1}$ forever unless another force is applied to it. Furthermore, far more complicated systems of accelerating solid masses, such as the parts of an engine or the motions of planets and spacecraft, are also amenable to direct integration of (1.17). The latter example results in predictions accurate centuries or even millennia in advance and our ability to send machines and people to other worlds.

Now imagine the same force is exerted in the same way, but to 1 kg of water in an open container. The outcome will be completely different: the mass of water will immediately become greatly distorted, break into many drops and be splattered widely.

This stark contrast illustrates a vital difference between solids and fluids. A portion of fluid can accelerate, not just by changing its velocity with time at a fixed point but also by moving into a region within the fluid mass where the velocity is different. This ability caused the fluid mass to distort and thus break up. A less-dramatic example is water flowing steadily in a wide pipe that feeds all its flow into a narrow pipe; as the water moves into the narrow section, its speed increases, even through at every point in the pipe, the speed is not changing with time. It is mathematically expressed by recognising that the velocity is a function of space as well as time and that the differentiation of velocity with respect to time must necessarily include the use of the chain rule of differentiation. For example, for motion of an incompressible fluid in the x-direction only, the rate of change of the x component of velocity, u, is given by

$$\frac{\mathrm{D}u}{\mathrm{D}t} = \frac{\partial u}{\partial t} + u\frac{\partial u}{\partial x}. \tag{1.18}$$

The term $\partial u/\partial t$ is called the *local acceleration* and represents the change in velocity with time only: exactly the sort of acceleration a solid mass would experience. The operator $\mathrm{D}/\mathrm{D}t$ is called the *material derivative* or *total derivative* though many other terms are used for it.

The term $u\partial u/\partial x$ has appeared because of the ability of a fluid to change its velocity by moving into a region where the velocity is different. It is of immense importance. It is a nonlinear term, usually called the *advective acceleration* (or sometimes 'convective' although it is nothing to do with heat-driven convection). This term, and its full three-dimensional equivalent, accounts for most of the complexity and beauty of fluid flows. This nonlinearity causes phenomena such as the *vortex shedding* that causes the flap of a flag and generates the fundamental notes of many musical instruments, and it creates the chaos of turbulence. The unpredictability of the weather is a consequence of nonlinearity in the conservation of momentum. Concomitantly, the existence of this nonlinear term also makes the equations of fluid dynamics unsolvable in general.

While waves on the surface of the water (chapter 2), sound waves (chapter 3), internal waves (chapter 4), and rotating-fluid waves (chapter 5) are all derived by neglecting this nonlinear term, in chapters 6 and 7, effects of nonlinearity will be considered; and the effects of nonlinearity feature in many practical applications in Part II of this book.

A more physical derivation of the acceleration, and also more general since we will not assume incompressibility, can be achieved by considering the transport of momentum, just as for the transport of mass, as shown in figure 1.5. Indeed, this is simply the application of the Reynolds Transport Theorem to momentum, and exactly the same reasoning is applied to momentum as to mass. Momentum is exiting the short segment of length Δx in

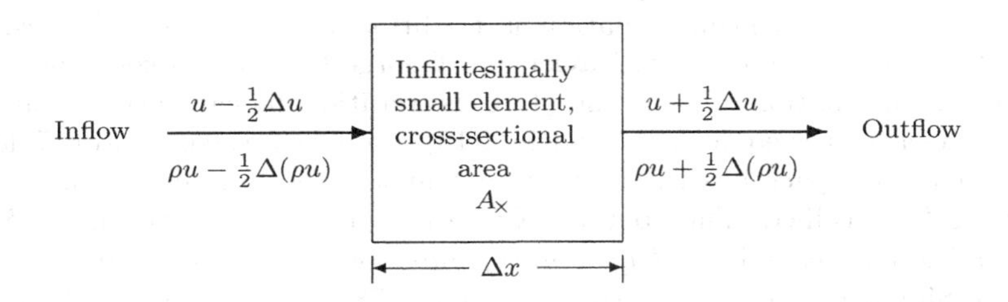

FIGURE 1.5
Transport of momentum in one dimension for an infinitesimally small element of fluid.

figure 1.5 at a rate $(\rho u + \frac{1}{2}\Delta(\rho u))(u + \frac{1}{2}\Delta u)A_\times$ while it is entering at a rate $(\rho u - \frac{1}{2}\Delta(\rho u))(u - \frac{1}{2}\Delta u)A_\times$. Neglecting the product of the two small quantities, the net change of momentum becomes $-[\Delta(\rho u)u + \rho u \Delta u]A_\times$. Just as for the transport of mass that led to (1.13), take limits as the small quantities tend to zero, giving a change of momentum equal to

$$\frac{\mathrm{D}(\rho \boldsymbol{u})}{\mathrm{D}t}.$$

Considering the force in (1.16), the force is composed of two parts; the forces acting on the *body* of the infinitesimal element of fluid, and the forces acting on the *surface* of an infinitesimal element of fluid. Of the body forces, the most significant is usually that due to gravity, with a value per unit volume of $\rho\mathbf{g}$; this is the key to waves on the surface of the sea, for example. There may be other body forces, such as magnetic forces relevant to the magnetic fields of planets (discussed in §11.2.3) or nuclear-fusion reactors, but here only gravity will be considered. The surface force per unit volume is the derivative of the stress with respect to distance, conveniently expressed as the divergence of the stress tensor, or $\boldsymbol{\nabla}\cdot\boldsymbol{\tau}$. Dividing by the volume of the infinitesimal element means that the conservation of momentum expression becomes

$$\boxed{\frac{\mathrm{D}(\rho\boldsymbol{u})}{\mathrm{D}t} = \boldsymbol{\nabla}\cdot\boldsymbol{\tau} + \rho\mathbf{g}}, \tag{1.19}$$

and if the fluid is incompressible, so ρ is a constant, say $\rho = \rho_0$, (1.19) reduces to

$$\boxed{\frac{\mathrm{D}\boldsymbol{u}}{\mathrm{D}t} = -\frac{1}{\rho_0}\boldsymbol{\nabla}P + \nu\nabla^2\boldsymbol{u} + \mathbf{g}}, \tag{1.20}$$

where ν is the *kinematic viscosity* defined as $\nu = \mu/\rho_0$. Although all of the precursors to (1.19) equally express the conservation of momentum, in fluid dynamics (1.19) is generally called the *Momentum Equation.* This version (1.19) of the momentum equation is traditionally called the *Navier-Stokes equation*, after Claude-Louis Navier and George Gabriel Stokes, whose work led to its formulation in 1822, and (1.20) is traditionally called the incompressible Navier-Stokes equation. As with (1.15), the use of the operator $\boldsymbol{\nabla}$ for the derivatives in space has the advantage that the coordinate system does not need to be specified. The operator ∇^2 results from the operation $\boldsymbol{\nabla}\cdot\boldsymbol{\nabla}$ and is traditionally called the *Laplacian operator* or simply the 'Laplacian' after Pierre-Simon Laplace who introduced it in 1783. (Sometimes ∇^2 is denoted by Δ but in this book that notation will be avoided to preclude confusion with the use of Δ for 'infinitesimal difference' in calculus.)

1.2.4 Scaling of equations and dimensionless groups

In many analyses in fluid dynamics, and in fluid-wave analyses, terms are dropped from equations because they are negligible. Often, this seems to be done on the basis of intuition, or possibly because it is 'usually done that way', and the student may be left wondering about the rationale. However, there is a rigorous way of determining which terms are negligible. As an illustration, consider the one-dimensional, incompressible momentum equation (1.20). Recalling (1.3), the typical assumption has been made that the only part of the ambient pressure, $P_\emptyset$, that varies is the hydrostatic pressure, which is perfectly balanced by the term due to gravity, so the only pressure remaining is the dynamic pressure, p, giving

$$\frac{\partial u^*}{\partial t^*} + u^*\frac{\partial u^*}{\partial x^*} = -\frac{1}{\rho_0}\frac{\partial p^*}{\partial x^*} + \nu\frac{\partial^2 u^*}{\partial x^{*2}}. \tag{1.21}$$

In (1.21), a superscript * has been applied to all variables, both the independent variables x^* and t^* and the dependent variables u^* and p^*. This is to emphasise that they are *dimensional variables*; for example, t^* has the units of seconds and u^* has the units of m s^{-1}. Quantities such as ρ_0 are also dimensional (ρ_0 has units of kg m^{-3}), but they are *constants* in the equation being dealt with, so do not have an * superscript.

The next step is to *scale* the dimensional variables. This means that the dimensional variable is expressed as some *scaling constant*, which has the same units as the dimensional variable, multiplied by a *dimensionless* version of the variable (sometimes called a 'scaled' variable). For the example above, a scaling may be

$$\begin{aligned} x^* &= Lx, \\ t^* &= Tt, \\ u^* &= Uu, \\ p^* &= Qp, \end{aligned} \tag{1.22}$$

where the scaling constant L is a fixed length, for example, 100 m, that is characteristic of the flow investigated, T is a fixed timescale, say 10 s, U is a characteristic fixed velocity, say 1 m s^{-1}, and Q is a characteristic fixed pressure, say 10 000 Pa. Usually, the scaling constants are some power of 10 rather than a very precise number. The 'art' in scaling is to pick the values of the scaling constants, so that they represent the nature of the flow being investigated. Clearly, we cannot say exactly what values are taken by the velocity, for example, until the equation is actually solved. The scaling constants are just there to express our guess of the order of magnitude of typical values, and we can often guess what factor of 10 the true value lies within. In some cases, some of the scaling constants could be defined as combinations of dimensional constants and the other scaling constants; for example, it might be convenient to say $Q \equiv \rho_0 U^2$; in another scenario, it may be that $U \equiv L/T$.

Provided one is careful, it is possible to temporarily apply the * superscripts to unambiguously identify dimensional quantities and undertake the scaling analysis below to identify which term or terms can safely be dropped from the equation. Two courses of action are then possible.

(i) The 'Return to dimensional equations then analyse' approach. One could then return to the dimensional form of the equation by absorbing the scaling constants L, T, etc, in the surviving terms back into the original dimensional quantities identified with * superscripts, then finally drop the * superscripts. The result is a dimensional equation that has been simplified by a rigorous process. Further calculations are then done with the simplified dimensional equation.

(ii) The 'Dimensionless analysis and dimensionless results' approach Alternatively, one could undertake calculations on the simplified equation in its dimensionless form. This is a fairly common approach, with the advantage that the results are universal: they are valid for any system for which the scaling assumptions were reasonable. However, if calculations are done in dimensionless form, it is important to explain how to translate the final results back into the dimensional variables. There are few situations more frustrating for a practitioner than a complicated formula resulting from a lengthy calculation without clarity on how to translate it to physical reality.

To avoid confusion, formulae in the summaries of this book are in dimensional units unless otherwise stated.

For the momentum-equation example of (1.21), the substitution of (1.22) gives

$$\frac{U}{T}\frac{\partial u}{\partial t} + \frac{U^2}{L}u\frac{\partial u}{\partial x} = -\frac{Q}{\rho_0 L}\frac{\partial p}{\partial x} + \frac{\nu U}{L^2}\frac{\partial^2 u}{\partial x^2}. \tag{1.23}$$

It is now possible to 'weigh' up the different terms in the equation and see which ones can be neglected under the circumstances of the practical situation to which the equation is to be applied.

In the first example of scaling analysis, to compare the nonlinear advective term with the viscous stress term, divide (1.23) by U^2/L, giving

$$\frac{L}{UT}\frac{\partial u}{\partial t} + u\frac{\partial u}{\partial x} = -\frac{Q}{\rho_0 U^2}\frac{\partial p}{\partial x} + \frac{\nu}{UL}\frac{\partial^2 u}{\partial x^2}, \tag{1.24}$$

or

$$\frac{L}{UT}\frac{\partial u}{\partial t} + u\frac{\partial u}{\partial x} = -\frac{Q}{\rho_0 U^2}\frac{\partial p}{\partial x} + \frac{1}{\mathrm{Re}}\frac{\partial^2 u}{\partial x^2}, \tag{1.25}$$

where

$$\mathrm{Re} = \frac{UL}{\nu} \tag{1.26}$$

is the *Reynolds number*, a dimensionless ratio written with the abbreviation 'Re' that represents the relative importance of forces due to nonlinear fluid inertia (the advective acceleration term) to viscous forces. It is named after Osborne Reynolds, who discovered in 1883 that this ratio predicts the transition from laminar to turbulent flow in a pipe. If the Reynolds number is large, the flow is *turbulent*. If it is small, the flow is *laminar*. For a given type of fluid, flows that are high-speed and large in scale will have a large Reynolds number. If the Reynolds number is very large, the viscous term can be neglected and the nonlinear term retained, which might be an appropriate approximation for a high-speed flow such as the turbulent flow of air around an aircraft or submarine. If the viscosity were zero, the Reynolds number would be infinite, and the flow is said to be *inviscid*.

The inviscid approximation is used in the derivations in Part I of this book: of surface-gravity waves (chapter 2), sound waves (chapter 3), internal waves (chapter 4), and rotating-fluid waves (chapter 5).

Since viscosity represents the dissipation of energy, one might wonder if a very large Re means that a high-speed flow does not lose energy. It does lose energy, because the nonlinear term causes the creation of turbulence, which removes momentum from the flow by transforming it into the motion of ever-smaller vortices, from which energy can not in practice be recovered. Conversely, if the Reynolds number is very small, the nonlinear term can be neglected and the viscous term retained, which might be an appropriate approximation for a low-speed flow such as water moving through thin tubes, or indeed the flow of blood in our smaller arteries and veins.

If the starting point for our scaling analysis, (1.21), were written in the vertical (z) direction instead, it would look the same, with u replaced by w and x replaced by z, but an additional term would appear on the right-hand side, $-\mathrm{g}$, for the acceleration due to gravity. Then, the same steps that converted (1.24) to (1.25) would yield an additional dimensionless ratio on the right-hand side, $1/\mathrm{Fr}^2$, where

$$\mathrm{Fr} = \frac{U}{\sqrt{\mathrm{g}L}} \tag{1.27}$$

is called the *Froude number*, owing to William Froude's studies in 1861. Recalling that L is now a length in the vertical direction, this represents the ratio of forces due to fluid inertia to the force of gravity. It is important in the design of ships and of channels in which unchanging ('steady') water flow and water-surface waves both occur, and the scale $\sqrt{\mathrm{g}L}$ turns out to be a wave speed. This will be discussed in §2.3.12.

Another, commonly-used ratio can be seen multiplying the first term in (1.25); it is the *Strouhal number*,

$$\mathrm{St} = \frac{L}{UT}, \tag{1.28}$$

which is relevant in circumstances where a significant oscillatory aspect to the fluid flow exists in addition to an unchanging flow of speed U; in this case, the timescale T would be the period of the oscillation (defined mathematically in §1.5.4). It is named after Vincenc Strouhal, who discovered its significance in 1878. A flow in which the timescale, T, over which the velocity changes is very large has a very small St. If the timescale is infinite, St is zero, and that is another way of saying the local acceleration is zero; the flow is then said to be a *steady flow*. Conversely, if St is large relative to unity, the local acceleration dominates over the mathematically difficult nonlinear advective acceleration. This $\mathrm{St} \gg 1$ approximation, together with the just-noted inviscid approximation $\mathrm{Re} \gg 1$, is used in the derivations in Part I of this book. The Strouhal number is best known in classical aerodynamics, where the period T is that of the vortex shedding mentioned in §1.2.3.2 above.

Yet another dimensionless number can be found by returning to the scaled momentum equation (1.24) and dividing by $\nu/(UL)$, giving a factor of $L^2/(\nu T)$ multiplying the local acceleration. Imagine T is the period of some oscillation with radian frequency ω, so that $\omega = 2\pi/T$ (we will return to this definition in §1.5.4). After multiplying the momentum equation by 2π, the square root of this factor can be written

$$\mathrm{Wo} = L\sqrt{\frac{\omega}{\nu}}, \tag{1.29}$$

and is called the *Womersley number* after John Womersley, who identified its importance in the study of blood flow in the early 20th century. It represents the ratio of forces due to oscillations to viscous forces. It can also be thought of as $\sqrt{2}$ times the ratio of the fundamental length scale of the flow, L, to a length δ_ω given by

$$\delta_\omega = \sqrt{\frac{2\nu}{\omega}}, \tag{1.30}$$

which is traditionally called the *Stokes boundary layer* thickness, which like the Navier-Stokes equations is due to the work of George Gabriel Stokes on viscosity. This layer is found adjacent to a solid boundary with fluid oscillating relative to it; within this layer, viscosity strongly affects the motion. This will be detailed in §13.3.1 in the context of medical ultrasound.

Very many other scalings are possible. For example, the flow could be divided into time-averaged and oscillatory parts (this will be done in §7.3 to illustrate phenomena like ocean rip currents and acoustic streaming) and each part given different length and velocity scales. If the assumption of incompressibility is not made, one dimensionless number that emerges is the ratio $U/\sqrt{Q/\rho_0}$, a ratio of the nonlinear fluid inertia to the 'stiffness' of the fluid (represented by the bulk modulus), which can be written

$$\mathrm{Ma} = \frac{U}{c}, \tag{1.31}$$

where c, the origin of which will be explained in §1.5.3, is, in this case, the speed of sound, and Ma is the *Mach number* named after Ernst Mach's work on supersonic motion in 1887. Including surface tension generates still more dimensionless numbers.

In the second example of scaling analysis, consider a situation that might occur with waves on the surface of the sea. Some thought leads us to conclude that there are actually two length scales in this situation. Introduce another length scale, $\hat{\eta}$, as well as the horizontal length scale L, where $\hat{\eta}$ is the *amplitude* (half the height) of the waves. In a wave-dominated incompressible flow, where the relevant length scale is the wave amplitude, and the relevant timescale is the period of the waves (to be defined in §1.5.4), the velocity of the fluid is governed by this length scale and timescale. Thus, $U \simeq \hat{\eta}/T$. (To be precise and insist that U is the velocity amplitude of the waves, means that $U = 2\pi\hat{\eta}/T$.) Meanwhile, the variations with x are still governed by the horizontal length scale L, which will be related to the wavelength. Furthermore, the forces giving the pressure variations are governed by the acceleration due to gravity, g, so the scaling constant for pressure should be defined as $Q \equiv \rho_0 g\hat{\eta}$. In this case, (1.23) becomes

$$\frac{\hat{\eta}}{T^2}\frac{\partial u}{\partial t} + \frac{\hat{\eta}^2}{LT^2}u\frac{\partial u}{\partial x} = -\frac{g\hat{\eta}}{L}\frac{\partial p}{\partial x} + \frac{\nu\hat{\eta}}{L^2T}\frac{\partial^2 u}{\partial x^2}, \tag{1.32}$$

which, dividing by $\hat{\eta}$ and multiplying by T^2, gives

$$\frac{\partial u}{\partial t} + \frac{\hat{\eta}}{L}u\frac{\partial u}{\partial x} = -\frac{gT^2}{L}\frac{\partial p}{\partial x} + \frac{\nu T}{L^2}\frac{\partial^2 u}{\partial x^2}, \tag{1.33}$$

Now suppose that $\hat{\eta}/L$ is very small because the wavelength L is very large, which corresponds to the slope of the sea surface being very small. The troublesome nonlinear term disappears; in fact, replacing U in the Strouhal number (1.28) with the present scaling $U = \hat{\eta}/T$ leads to the same point noted earlier: if $\hat{\eta}/L$, is very small, St $\gg 1$ and the nonlinear term is negligible. Furthermore, if L is very large, the ratio $\nu T/L^2$ would be very small (therefore, the Womersley number would be very large). We can check this by using a few typical numbers. If $L = 100$ m and $T = 10$ s, noting that for water under atmospheric conditions, the constant ν is very close to 10^{-6} m^2 s^{-1}, the ratio $\nu T/L^2$ is certainly extremely small compared to unity. Therefore, (1.32) becomes

$$\frac{\partial u}{\partial t} = -\frac{gT^2}{L}\frac{\partial p}{\partial x}, \tag{1.34}$$

a considerable - and solvable - simplification of the original momentum equation.

In maritime-engineering situations where waves are involved, the reciprocal of the Strouhal number is more common and is often denoted $\mathrm{K_C}$, i.e.

$$\mathrm{K_C} = \frac{UT}{L}; \tag{1.35}$$

it is called the *Keulegan-Carpenter number* after Garbis Keulegan and Lloyd Carpenter, who emphasised its significance in 1958. However, the length scale L in $\mathrm{K_C}$ is not the ocean wavelength but is usually the measure of some engineered system, such as the diameter of a pier footing or wind-turbine pylon in the ocean. Since, as just-noted, if U is the wave amplitude, $U = 2\pi\hat{\eta}/T$ so $\mathrm{K_C} = \hat{\eta}/L$.

1.3 Flow descriptions

Standard terminology for the description of flow fields is used in the fluid-dynamics literature as well as in most textbooks in more specialised fields based on fluid dynamics, so it is important to remember what the following terms mean.

- Steady flow was defined in §1.2.4 as a state where the local acceleration, $\partial \boldsymbol{u}/\partial t$, is zero; therefore, *unsteady flow* is the state where the local acceleration is nonzero. All fluid waves are unsteady flows.

- Viscous versus inviscid flows were also defined in §1.2.4; an inviscid flow is an unphysical idealisation in which viscosity is zero. Unphysical it may be, but it is a useful approximation that will be used in all the detailed derivations in this book.

- Laminar versus turbulent flows were also defined in §1.2.4 with reference to the size of the Reynolds number: the ratio of the nonlinear advective term to the viscous term in the momentum equation. Hence, an inviscid flow is the extreme limit of infinite Re.

- One, two and three-dimensional (1D, 2D and 3D) flows are also commonly described. Of course, the real world is three dimensional, but key aspects of flow in a long pipeline, river or channel are usefully approximated as 1D, as are many gravity-wave and sound-wave propagation cases, as well as the radial expansion and contraction of a bubble (chapter 9). Flows of thin films of fluid are conveniently approximated as 2D. The concept of scaling covered in §1.2.4 applies in these approximations too; if the dimensions neglected can be rigorously shown to be small *relative* to the dimensions considered, the approximation is valid. For example, some waves made possible by the Earth's rotation over large distances (chapter 5) are considered 2D.

- A line tangent to the velocity vector is a *streamline*, a concept that is only defined for a steady flow and thus is not used much in fluid wave problems. Meanwhile, a *pathline* is literally the path taken by a particle carried along with the fluid and hence any molecule of the fluid itself. In this book, the term *particle trajectory* will be used instead of pathline since it accommodates the more general case where a solid particle (or bubble) might not follow the flow owing to its weight or buoyancy.

- The nonlinearity of most fluid flows demands a further detail, in which both location and velocity can be described in a *fixed frame* or in a *particle frame* of reference. The traditional terms for these are respectively *Eulerian* and *Lagrangian* frames of reference, the former after the 18th-century polymath Leonhard Euler and the latter after Euler's contemporary Joseph-Louis Lagrange. This will be particularly important in chapter 6, where drifts due to waves are described.

1.4 Euler and Bernoulli equations

Returning to (1.25), the assumption that Re is very large reduces the momentum equation (1.25) to

$$\frac{L}{UT}\frac{\partial u}{\partial t}+u\frac{\partial u}{\partial x}=-\frac{Q}{\rho_0 U^2}\frac{\partial p}{\partial x},$$

and adopting the return-to-dimensional-equations approach (i) in §1.2.4, the dimensional equation becomes

$$\frac{\partial u^*}{\partial t^*}+u^*\frac{\partial u^*}{\partial x^*}=-\frac{1}{\rho_0}\frac{\partial p^*}{\partial x^*},$$

which, on dropping the *s, is

$$\boxed{\frac{\partial u}{\partial t}+u\frac{\partial u}{\partial x}=-\frac{1}{\rho_0}\frac{\partial p}{\partial x}}, \tag{1.36}$$

which is the horizontal, one-dimensional, incompressible version of the momentum equation: one of the most useful equations in fluid dynamics. This relation was determined by Leonhard Euler in 1757. Armed with Newton's expression for the conservation of momentum, but without the understanding of viscous stresses later contributed by Navier and Stokes, Euler was the first to understand how to represent the acceleration of a fluid, as outlined in §1.2.3.2. Thus (1.36) is traditionally called the horizontal, one-dimensional, incompressible *Euler's equation*. Since, as mentioned in §1.2.4, viscosity is neglected in most of the derivations in Part I of this book, these derivations are effectively Euler's-equation analyses.

Exactly the same reasoning applies in all three dimensions, and assuming the gravitational term is perfectly balanced by hydrostatic pressure gives a coordinate-free form of Euler's equation,

$$\frac{\mathrm{D}\boldsymbol{u}}{\mathrm{D}t}=-\frac{1}{\rho_0}\boldsymbol{\nabla}p. \tag{1.37}$$

Taking the curl of (1.37), i.e. applying the vector-calculus operation $\boldsymbol{\nabla}\times$ and noting that the curl of a gradient is zero, gives

$$\frac{\mathrm{D}\tilde{\omega}}{\mathrm{D}t}=\mathbf{0}, \tag{1.38}$$

where $\tilde{\omega}$ defined by

$$\tilde{\omega}=\boldsymbol{\nabla}\times\boldsymbol{u} \tag{1.39}$$

is the *vorticity* of the flow. The result (1.38) is a fundamental statement in fluid dynamics that *vorticity is conserved* in a frictionless fluid; equivalent statements can be made for compressible flows as well. Vorticity is particularly useful when waves in rotating fluids are considered in chapter 5.

More generally, the full momentum equation (1.19) is typically called Euler's equation if the stress tensor is simplified such that the viscosity is zero, i.e. $\mu = 0$. Since (1.19) is a three-dimensional vector equation that is true in any coordinate system, imagine that the motion is still one-dimensional, but in some 's'-direction that is at some angle to the horizontal. Retaining the assumption of incompressibility but including gravity and corresponding hydrostatic pressure differences, Euler's equation would be

$$\frac{\partial u}{\partial t} + u\frac{\partial u}{\partial s} = -\frac{1}{\rho_0}\frac{\partial P}{\partial s} - \mathrm{g}\frac{\mathrm{d}z}{\mathrm{d}s}, \tag{1.40}$$

where $\mathrm{d}z/\mathrm{d}s$ is the angle of the s-direction to the vertical direction, z, in which gravity acts.

In many practical circumstances, the velocity and pressure at every point in a flowing fluid is not of interest; rather, what is relevant is the velocity and pressure at the entrance to and exit from some system. The simplest such system is a pipe, and the engineering of pipes and channels, practised by all human civilizations since antiquity, led to a great need to calculate the relationship between pressure and velocity at pipe entrances and exits. Furthermore, one is mostly interested in the situation where the flow is steady, so that the flow rate u is not changing with time, and nor are any other quantities. Assuming the flow is steady means the local acceleration $\partial u/\partial t$ is zero. The relationship between velocity and pressure can immediately be understood by integrating (1.40), with $\partial u/\partial t = 0$, with respect to distance s, giving

$$\boxed{\frac{1}{2}u^2 + \frac{1}{\rho_0}P + \mathrm{g}z = \text{const.}}, \tag{1.41}$$

where the right-hand side is some constant of integration. This is called *Bernoulli's equation* or sometimes Bernoulli's principle, after Daniel Bernoulli who devised it in 1738, *before* Euler derived (1.37), a point we will return to at the end of this section. Labelling the entrance to the pipe with subscript 1 and the exit with subscript 2, Bernoulli's equation takes a particularly useful form,

$$\boxed{\frac{1}{2}{u_1}^2 + \frac{1}{\rho_0}P_1 + \mathrm{g}z_1 = \frac{1}{2}{u_2}^2 + \frac{1}{\rho_0}P_2 + \mathrm{g}z_2}, \tag{1.42}$$

which immediately permits engineers to calculate the pressure differences given differences in flow rate and elevation z, or indeed differences in flow rate or elevation given the other two variables. This relation (1.42) is not limited to pipes; it is valid along any streamline in steady, incompressible and inviscid flow.

Even though fluid-wave systems are inherently unsteady, so that (1.42) cannot be used to calculate the wave motion itself, there may be situations where (1.42) could be used to estimate the *consequences* of the wave. For example, consider the calculation of the pressure momentarily exerted on the wall of a building by a strong quasi-steady wind of speed u_1. This might be caused by the passage of an atmospheric thunderstorm-generated wave over several minutes (§12.4.3). Set point 1 to be where the wind is blowing far from but directly towards the wall of a building, point 2 to be on the wall at the same elevation z, so that P reduces to p, the dynamic pressure corresponding to changes in fluid flow. This means that far from the wall, $P_1 = p_1 = 0$; meanwhile, since the wall is not moving (at least while it holds up), $u_2 = 0$ and $z_2 = z_1$, so that (1.42) reduces to

$$\boxed{\frac{1}{2}{u_1}^2 = \frac{1}{\rho_0}p_2}, \tag{1.43}$$

which determines the stress on the wall due to the wind pressure, p_2; multiplying p_2 by the area of the wall gives the horizontal force in Newton the wall would have to withstand.

It may be noticed that (1.42) has the units of energy per unit mass, which arose because acceleration (force per unit mass) was integrated over distance and a definition of energy is force multiplied by distance. As noted earlier, Bernoulli came up with (1.42) before Euler came up with (1.37), and this is because Bernoulli was thinking independently about the conservation of energy. However, Bernoulli's equation (1.42) does not properly represent the conservation of energy, the third of the conservation laws, since losses of energy to viscosity and inputs and outputs of heat are not included, and nor are inputs and outputs of mechanical work. Thus, modern practice is to teach Bernoulli's equation as an integral form of momentum conservation, not energy conservation, a distinction that continues to confuse some students. The energy equation, which will not be explicitly used in this book, includes all inputs and outputs of energy to a flowing system and may be derived by application of the Reynolds Transport Theorem to energy, just as it was applied to mass and momentum.

1.5 Wave tools

1.5.1 Complex exponentials

The complex exponential is a very convenient tool for wave problems but may make students feel that a disturbing layer of mathematical abstraction separates them from the real-world problem at hand. It permits the manipulation of the sines and cosines that are the typical solutions to wave problems with a single, convenient expression, given by Euler's formula,

$$\mathrm{e}^{\mathrm{i}x} = \cos x + \mathrm{i} \sin x. \tag{1.44}$$

It is worth noting in passing that if $x = \pi$, (1.44) becomes $\mathrm{e}^{\mathrm{i}\pi} + 1 = 0$, which is known as *Euler's Identity*, considered by mathematicians to be particularly beautiful since it contains the numbers 0, 1, π, e and i, as well as applying addition, multiplication and exponentiation. It is important to remember that the use of imaginary numbers is a mathematical convenience. It is certainly not a trivial convenience, because in some circumstances it can save a vast amount of tedious algebra and the attendant risk of errors. However, the real world is just that: real, and before any comparison with reality is made, the real part of any expression must be taken. This is usually denoted $\Re$, while the imaginary part is denoted $\Im$, so that

$$\begin{aligned} \Re(\mathrm{e}^{\mathrm{i}x}) &= \cos x, \\ \Im(\mathrm{e}^{\mathrm{i}x}) &= \sin x. \end{aligned} \tag{1.45}$$

For any complex number $A = a_R + \mathrm{i}a_I$, its modulus (or 'magnitude' or 'absolute value') is denoted $|A|$, so that $|A| = \sqrt{a_R^2 + a_I^2}$. Since $\cos^2 x + \sin^2 x = 1$, $|\mathrm{e}^{\mathrm{i}x}| = 1$

A pitfall to avoid is raising complex exponentials to powers higher than unity; this is *not* the same as raising a cosine or sine to that power. This is evident on evaluating the real part of $(\mathrm{e}^{\mathrm{i}x})^2$, which gives

$$\Re\left[(\mathrm{e}^{\mathrm{i}x})^2\right] = \Re(\mathrm{e}^{\mathrm{i}2x}) = \cos 2x \neq \cos^2 x. \tag{1.46}$$

Therefore, when calculating the energy or power of waves, which involves multiplying wave-like quantities, or when doing calculations with the nonlinear term in the momentum equation, which can also involve multiplying wave-like quantities, it is necessary to start with the real part. For example, calculating the power of waves, which is necessary for sound-wave measurements in §3.3.8.1 or electricity generation from ocean waves in §8.2.1 involves multiplying the sinusoidal quantities that are the real parts of the expressions for force and velocity, and the result is also time-varying, so it is often most relevant to consider the average over one cycle. The time average over some time T (the wave period, say) of any quantity, say $\bullet(x,t)$, can be defined as

$$\overline{\bullet} = \frac{1}{T}\int_0^T \bullet(x,t)\mathrm{d}t. \tag{1.47}$$

Recall the standard trigonometric identities,

$$\sin^2\alpha = \tfrac{1}{2}\left[1 - \cos(2\alpha)\right], \tag{1.48a}$$

$$\cos^2\alpha = \tfrac{1}{2}\left[1 + \cos(2\alpha)\right], \tag{1.48b}$$

(the double-angle formulae). The time average of the $\cos(2\alpha)$ terms in (1.48) is zero, but owing to the 1 inside the square brackets of (1.48), the overall time average of squared sinusoidal terms is $1/2$.

1.5.2 Wave equation notations

The one-dimensional wave equation was derived by Jean le Rond d'Alembert in 1746. It takes the same form for all travelling waves, whether the waves are electromagnetic radiation such as light, or 'mechanical' waves such as fluid waves. It can be written in many ways or *notations*; the notation chosen is simply the one most convenient for the task at hand. The least-simplified notation is

$$\frac{\partial^2 p}{\partial t^2} - c^2 \left(\frac{\partial^2 p}{\partial x^2} + \frac{\partial^2 p}{\partial y^2} + \frac{\partial^2 p}{\partial z^2} \right) = 0, \tag{1.49}$$

where c is a constant that will be explained at the end of §1.5.3 below. However, the appearance of the partial-derivative symbols make the equation take up a lot of space on the page, and the use of x, y and z automatically imply that a Cartesian coordinate system is being used. As in §1.2.3, use the standard vector-calculus operator $\boldsymbol{\nabla}$ for the derivatives in space, giving a coordinate-free notation; in this case, the wave equation looks like

$$\frac{\partial^2 p}{\partial t^2} - c^2 \nabla^2 p = 0, \tag{1.50}$$

and furthermore, using the common notation (originally introduced by Isaac Newton) in which an overdot represents differentiation with respect to time, it could also be written

$$\ddot{p} - c^2 \nabla^2 p = 0. \tag{1.51}$$

This notation (1.51) takes up the least space on the page. In some cases, however, the top of the symbol is needed for other modifications indicating averaging, a vector or a related quantity, and the overdots could get in the way. In this case the wave equation might be written using subscripts to denote differentiation, i.e., as

$$p_{tt} - c^2 \left(p_{xx} + p_{yy} + p_{zz} \right) p = 0, \tag{1.52}$$

in Cartesian coordinates. In coordinate-free subscript notation, the wave-equation would be

$$p_{tt} - c^2 \nabla^2 p = 0. \tag{1.53}$$

If subscripts are going to be used for other purposes, this notation could get confusing. In this book, we will mostly use notations like (1.50) when an equation is displayed (i.e. the equation has a number), but when equations are written in-line as part of the flow of the text, for example, $\ddot{p} - c^2 \nabla^2 p = 0$, it is usually better to use a compact notation like (1.51).

1.5.3 Separation of variables and d'Alembert solutions

Solutions of the wave equation are most commonly achieved by two alternative methods. The *separation of variables method* splits the desired solution into functions only of a single variable (x, y, z or t). Alternatively, the *method of d'Alembert* makes the solution a function of a new variable that anticipates wave-like behaviour.

It is very common to present an equation like (1.49) and in the next lines, say " ... and the solution is ..." However, a differential equation like (1.49) by itself is not a problem requiring a solution. It is the statement of a relationship. In the case of the wave equation, it is a relationship that expresses the laws of conservation of mass and momentum subject to the linearising assumption (and in most wave-physics systems, several other assumptions have also been made to get a wave equation). To have a problem to be solved, the equation must also be considered together with *boundary conditions*: conditions on the dependent variable (i.e., p in (1.49)) when its dependent variables (in this case, time and space) take certain known values. In many wave-physics cases, the known values are those at the start: at time $t = 0$ and at location $\boldsymbol{x} = \mathbf{0}$. In this case, the boundary conditions are called *initial conditions.* In some cases, however, the values of the dependent variable are known at both the start and end.

This emphasis on the type of boundary conditions might seem pedantic, but it has a real manifestation. Initial-value problems are typical when waves are created at one point and travel away to infinity. For example, a rock dropped in the ocean creates waves travelling away from the initial point without limit. Likewise, a sound made at a point high in the atmosphere travels away from that point without limit. Usually, this means that the possible solutions range from infinitesimally short, infinitely-high-frequency waves to infinitely long, infinitesimally-low-frequency waves, and importantly, there is an infinite number of wavelengths and corresponding frequencies permissible in between these extremes. Conversely, boundary conditions at both ends usually mean that only certain solutions are possible. Drop the rock in a bathtub, and the boundary presented by the edge of the bathtub will demand that the solution consists of only certain wavelengths. Likewise, a sound made at the end of a tube with some boundary at the other end demands that the solution consists of only certain wavelengths and corresponding frequencies; in this case, the frequencies may be musical notes. In many cases where an equation like (1.49) is presented together with an immediate solution, some of the boundary conditions have usually been assumed, and sometimes, that assumption is not spelt out. One of the most common issues in trying to model real-world wave problems is getting the boundary conditions wrong.

To use the separation of variables method, assume that the pressure p can be factorised into two functions: $\mathcal{X}$, a function *only* of space and $\mathcal{T}$, a function *only* of time i.e.,

$$p = \mathcal{X}(\boldsymbol{x})\mathcal{T}(t). \tag{1.54}$$

Consider for simplicity the one-dimensional wave equation in Cartesian coordinates in which there is only one dimension in space, x. Exactly the same reasoning below applies to the 3D wave equation. Substitute (1.54) into the 1D Cartesian version of (1.49), i.e., into $p_{tt} - c^2 p_{xx} = 0$, giving

$$\frac{\partial^2 \mathcal{T}}{\partial t^2} \mathcal{X} = c^2 \mathcal{T} \frac{\partial^2 \mathcal{X}}{\partial x^2}. \tag{1.55}$$

Next, divide both sides of (1.55) by $\mathcal{X}\mathcal{T}$, giving

$$\frac{1}{\mathcal{T}} \frac{\partial^2 \mathcal{T}}{\partial t^2} = c^2 \frac{1}{\mathcal{X}} \frac{\partial^2 \mathcal{X}}{\partial x^2}. \tag{1.56}$$

The left-hand side of (1.56) is a function of t only. Meanwhile, the right-hand side is a function of x only. For this to be true, the dependence of the left-hand side on t must cancel out and the dependence of the right-hand on x must cancel out, so that both sides must be equal to a constant. To rapidly arrive at a familiar-looking result, we can somewhat anticipate the solution, in other words, implicitly assume some boundary conditions; if so, the constant could be called $-\omega^2$.

The left-hand side of (1.56) then gives

$$\frac{\partial^2 \mathcal{T}}{\partial t^2} = -\omega^2 \mathcal{T}. \tag{1.57}$$

This is the equation of simple harmonic motion; the same equation would be derived to model the small oscillations of a pendulum, a mass bouncing on a spring, or the oscillations of an electronic circuit, and an enormous range of other systems in physics. Since (1.57) is a second-order ordinary differential equation, it has two possible linearly-independent solutions, sine and cosine. These two possibilities must both be included. Somewhat painfully, let us forego the convenience of the complex exponential and undertake the algebra - shortly, we will see how much less work the complex-exponential path is. The solution is written as

$$\mathcal{T} = a_1 \cos(\omega t) + a_2 \sin(\omega t). \tag{1.58}$$

where the sin and cos functions are multiplied by arbitrary real constants a_1 and a_2, the values of which have to be determined by the boundary conditions. Thus, until the boundary conditions are applied, a_1 and a_2 remain arbitrary and the problem is not truly solved. The right-hand side of (1.56) must equal the same constant, $-\omega^2$, so it becomes

$$\frac{\partial^2 \mathcal{X}}{\partial x^2} = -\frac{\omega^2}{c^2} \mathcal{X}, \tag{1.59}$$

which also has the two solutions of sin and cos,

$$\mathcal{X} = a_3 \cos(kx) + a_4 \sin(kx), \tag{1.60}$$

where $k = \omega/c$. The physical significance of ω, k and c will be discussed at the end of this section. Multiplying the separated factors $\mathcal{X}$ and $\mathcal{T}$ gives

$$p = [a_1 \cos(\omega t) + a_2 \sin(\omega t)] [a_3 \cos(kx) + a_4 \sin(kx)] . \tag{1.61}$$

The specification of boundary conditions would give the four real constants a_1, a_2, a_3 and a_4. This is the one-dimensional solution for waves in Cartesian coordinates.

Before considering the d'Alembert method, the application of the trigonometric identities

$$\begin{aligned} \cos(\alpha + \beta) &= \cos\alpha \cos\beta - \sin\alpha \sin\beta, \\ \sin(\alpha + \beta) &= \sin\alpha \cos\beta + \cos\alpha \sin\beta, \end{aligned} \tag{1.62}$$

to (1.61) transforms it (after some tedious algebra) to

$$\begin{aligned} p = {} & b_1 \cos[k(x - ct)) + b_2 \sin(k(x - ct)] \\ & + b_3 \cos[k(x + ct)) + b_4 \sin(k(x + ct)] , \end{aligned} \tag{1.63}$$

where b_1, b_2, b_3 and b_4 are just more real constants composed of the original constants a_1, a_2, a_3 and a_4. It is clear that the cos and sin functions have arguments $x - ct$ and $x + ct$. This result demonstrates the interchangeability of time and space. There is always a time t at a given point x_1 where the dependent variable has exactly the same value as at some other given point x_2. Equivalently, there is always a point x for a time t_1 where the dependent variable is exactly the same as at another time t_2.

Considering the argument $x - ct$, it is clear that increasing t has the same effect as decreasing x; as time passes, the level of p that was further back in x arrives. This is the essence of a travelling wave. The two new arguments of the sinusoidal functions, $x - ct$ and $x + ct$, simply mean that a disturbance has the ability to travel in both the positive and negative x directions. This is a feature of all travelling waves, be they electromagnetic waves or the fluid waves discussed in later chapters.

Again, the boundary conditions are important. If the disturbance were generated by a solid wall (a wall temporarily set into motion, say), waves could only propagate away from the wall. Depending on which direction is defined as the positive x-direction, either b_1 and b_2 would be zero, or b_3 and b_4 would be zero. However, if there were a symmetrical disturbance in the middle of the fluid, waves would propagate in both directions, represented by $b_1 = b_3$ and $b_2 = b_4 = 0$; and an asymmetrical disturbance in the middle of the fluid would be represented by $b_2 = b_4$ and $b_1 = b_3 = 0$.

It is vital that the temporal and spatial derivatives in the wave equation have opposite signs when on the same side of the equation; the relation $\ddot{p} - c^2\nabla^2 p = 0$ is a wave equation, but $\ddot{p} + c^2\nabla^2 p = 0$ is not. In mathematical language, this requirement is that the governing equation be a *hyperbolic partial differential equation* and not an *elliptic partial differential equation.*

Solutions to hyperbolic equations lie along *characteristic surfaces*, or simply 'characteristics'; in the one-dimensional example above, the characteristics are simply the lines $x \pm ct$. The characteristics are lines along which solution is constant and along which the wave propagates; an example with the familiar water-surface waves will be shown in figure 2.5, and a more exotic example in a rotating fluid will be discussed in §13.4.1.

Considering the derivation above, we can immediately see the consequence an elliptic equation would have at the stage of (1.59). The second derivative of $\mathcal{X}$ would be proportional to a positive constant times $\mathcal{X}$. This means that $\mathcal{X}$ has the solution $\mathcal{X} = a_3 e^{kx} + a_4 e^{-kx}$: depending on the boundary conditions, this would give rise to values of p that are exponentially increasing or decreasing in x, but not changing sinusoidally in x. For waves to exist, there must be sinusoidal variations *both* in time and in *at least one* spatial dimension. Of course, (1.59) was an example in which there was only one spatial dimension, x. If there were two or three dimensions, the solution in one of those dimensions could be sinusoidal, allowing waves to exist, while in the other dimensions, the solution could be exponentially increasing or decreasing. Indeed, in §2.3, we will study exactly this situation when we consider waves on the surface of water bodies: the waves are sinusoidal in the horizontal but decrease exponentially in the vertical.

To appreciate the benefit of the complex exponential, it is worth undertaking the algebra to get from (1.61) to (1.63). Then, as an alternative, allow p, $\mathcal{T}$ and $\mathcal{X}$ to be complex numbers, and write (1.58) and (1.60) as

$$\begin{aligned} \mathcal{T} &= A_t e^{i\omega t}, \\ \mathcal{X} &= A_x e^{\pm ikx}, \end{aligned} \tag{1.64}$$

where A_t and A_x are complex numbers, and substituting (1.64) into (1.54), *immediately* gives

$$p = A e^{i(\pm kx + \omega t)}, \tag{1.65}$$

where $A = A_x A_t$ is a complex constant: the amplitude. (In the equation-scaling example in §1.2.4 specific to wave-surface waves, the symbol $\hat{\eta}$ was used for the magnitude of the amplitude; thus $\hat{\eta} = |A|$.) Taking the real part of (1.65) allows A to be relatable if necessary to the b_1, b_2, b_3 and b_4 in (1.63). This is much less work than the algebra to get from (1.61) to (1.63). Moreover, if further calculations are required, (1.65) is much easier to work with than (1.63).

The method of d'Alembert more directly anticipates a travelling-wave solution by defining new independent variables $x - ct$ and $x + ct$, and thus expecting the solution to be of the form

$$p = f(x - ct) + g(x + ct), \tag{1.66}$$

where f and g are arbitrary functions. Using (1.66), the wave equation (1.49) becomes (1.63), where it is now clear that

$$f = b_1 \cos\left[k(x - ct)\right) + b_2 \sin\left(k(x - ct)\right] \quad \text{and}$$
$$g = b_3 \cos\left[k(x + ct)\right) + b_4 \sin\left(k(x + ct)\right]. \tag{1.67}$$

1.5.4 Measuring a wave

The constants ω and k should now be given some physical significance. In radians per second, the units arising naturally from the mathematics, the frequency of the waves is ω, sometimes called the *radian frequency*. The frequency usually measured in experiments, observed and experienced by humans is in cycles per second, s^{-1} or *Hertz*; this is given by

$$\boxed{f = \frac{\omega}{2\pi}}, \tag{1.68}$$

and since the frequency in cycles per second is related to the *period* (or cycle time) of the waves, T, by $f = 1/T$, (1.68) can also be written

$$\boxed{T = \frac{2\pi}{\omega}}, \tag{1.69}$$

a definition already used in the definition of the Womersley number in (1.29). After a student has successfully undertaken quite complex calculations, one of the most common mistakes is forgetting to divide or multiply by 2π when relating the result to physical reality! The constant k that appeared in (1.60) and equivalently in (1.63) is called the *wavenumber*, a sort of 'frequency in space'. Similarly to the relation between the radian frequency and the frequency in cycles per second, k is related to the *wavelength*, λ by

$$\boxed{\lambda = \frac{2\pi}{k}}. \tag{1.70}$$

The wavelength, the distance in metres from one wave crest to the next, is usually what is measured or observed, and the relation (1.70) is the exact equivalent of (1.69), so that it is clear that the wavelength is a sort of 'period in space'. The relation between ω and k was defined at the derivation of (1.59); as for all waves, it is

$$\boxed{c = \frac{\omega}{k}}, \tag{1.71}$$

or equivalently, f and λ are related by

$$\boxed{c = f\lambda}, \tag{1.72}$$

and it is evident from (1.67) that c is the *wave speed*: the speed with which the crests and troughs propagate through the fluid.

For two- or three-dimensional motion, the derivation above is simply repeated for the y and z directions. There would be three wavenumbers, which we can denote κ, ℓ and k for the x, y and z directions respectively, and it is sometimes convenient to represent these as a *wavenumber vector*, $\boldsymbol{k}$, given by $\boldsymbol{k} = (\kappa, \ell, k)$. Likewise, the wave speed becomes the *wave velocity*, a vector $\mathbf{c}$. If the wave crests and troughs are not curved, the waves are usually called *plane waves*.

In many fluid wave problems, and indeed in wave problems of any kind, it is very useful to have an idea of the wavelength because it can immediately be compared with physical objects. A wavelength much larger than a physical object can pass the object as if the object did not exist. An ocean wave with a wavelength of 100 metres will pass a narrow post, say 10 cm wide, without suffering any noticeable change. Unfortunately, the bass notes of the music from a neighbour's noisy party, with a wavelength of three metres, will similarly pass through a pane of glass, say 3 mm thick, with little attenuation.

1.5.5 Oscillators and resonance

In many fluid-wave systems, such as the wave-energy converters in chapter 8 and the medical ultrasound contrast agents in §9.6.2, the result of linearising the equations of motion is the equation for an oscillator,

$$\ddot{\xi} + 2\zeta\omega_0\dot{\xi} + \omega_0^2\xi = \mathcal{F}\mathrm{e}^{\mathrm{i}\omega t}, \tag{1.73}$$

where the dependent variable, ξ, is most typically some displacement in metres, ζ is the dimensionless *damping ratio* representing dissipative (frictional) losses, ω_0 is the *natural frequency* (in radians per second) of the oscillator (sometimes called the undamped natural frequency), $\mathcal{F}$ is the *forcing* amplitude, typically some force per unit mass, and ω is the forcing frequency. For example, ξ could be the vertical displacement of a wave-energy converter machine, discussed in chapter 8, or the radial displacement of the wall of a bubble, discussed in chapter 9. As in all such equations, the general solution is the sum of the *complementary function* and the *particular integral*. The complementary function represents the 'free' oscillations of the system, which would occur if the oscillator were temporarily displaced and then released. It is given by the solution to the *homogeneous* version of (1.73),

$$\ddot{\xi} + 2\zeta\omega_0\dot{\xi} + \omega_0^2\xi = 0, \tag{1.74}$$

obtained by setting

$$\xi = A\mathrm{e}^{\mathrm{i}\omega_{0\zeta}t - b_\zeta t}, \tag{1.75}$$

where A is a complex-number amplitude as before, so that $|A|$ is the maximum value of ξ, and $\omega_{0\zeta}$ and b_ζ are real numbers, $\omega_{0\zeta}$ being the frequency of the oscillation and b_ζ the rate of decay of the oscillation. Substitution of (1.75) into (1.74) gives

$$-\omega_{0\zeta}{}^2 + \mathrm{i}2\omega_{0\zeta}b_\zeta + b_\zeta{}^2 + \mathrm{i}2\zeta\omega_0\omega_{0\zeta} - 2\zeta\omega_0 b_\zeta + \omega_0^2 = 0. \tag{1.76}$$

Equating the imaginary parts of (1.76) yields

$$b_\zeta = \zeta\omega_0, \tag{1.77}$$

and equating real parts yields

$$\omega_{0\zeta}{}^2 = b_\zeta{}^2 - 2\zeta\omega_0 b_\zeta + \omega_0^2 = 0, \tag{1.78}$$

which, using (1.77), gives

$$\omega_{0\zeta} = (\sqrt{1-\zeta^2})\omega_0, \tag{1.79}$$

where the positive root was taken since, in this context, a negative frequency has no physical meaning. Introducing the initial condition $\xi = A$ when $t = 0$ gives the final form of the complementary function (1.75) as

$$\boxed{\xi = A\mathrm{e}^{\mathrm{i}\left(\sqrt{1-\zeta^2}\right)\omega_0 t - \zeta\omega_0 t}.} \tag{1.80}$$

The oscillation occurs with a frequency $\omega_{0\zeta} = (\sqrt{1-\zeta^2})\ \omega_0$ which is called the *damped natural frequency.* If the damping ratio ζ is small, the damped natural frequency is very close to the natural frequency ω_0.

The particular integral to (1.73) is obtained by assuming

$$\xi = a\mathrm{e}^{\mathrm{i}\omega t + \Phi}, \tag{1.81}$$

where a and Φ are real numbers and are the amplitude and *phase* of the oscillation, respectively. Alternatively, we could write (1.81) as

$$\xi = A\mathrm{e}^{\mathrm{i}\omega t}, \tag{1.82}$$

where $A = a\mathrm{e}^{\mathrm{i}\Phi}$ is a complex number; however, if no further calculations are to be undertaken beyond determining the response of the oscillator, it is the amplitude and phase that are of practical interest, so (1.81) will be used.

The particular integral is the response to forcing; it is obtained by substituting (1.81) into (1.73) and is

$$\boxed{\begin{aligned} a &= \frac{\mathcal{F}}{\sqrt{\left(\omega_0^2 - \omega^2\right)^2 + \left(2\zeta\omega\omega_0\right)^2}}, \\ \Phi &= -\tan^{-1}\left(\frac{2\zeta\omega\omega_0}{\omega_0^2 - \omega^2}\right) \end{aligned}}. \tag{1.83}$$

When $\omega \simeq \omega_0$, the value of a reaches a maximum. This is called *resonance.* Owing to the presence of damping, the actual resonant peak will occur for a value of ω slightly less than ω_0; this is called the *resonant frequency.* In chapter 8, the resonance of wave-energy converters driven by ocean waves is sought to maximise renewable energy extracted from the ocean, and in chapter 9 the resonance of microbubbles driven by ultrasound is used to explain their diagnostic and therapeutic abilities. Note that $\mathcal{F}$ is a force per unit mass (and thus acceleration, in m s^{-2}), whereas a is an amplitude in metres; to develop a general formula that will hold for any resonator, irrespective of the units, note that the amplitude of the forcing acceleration, $\mathcal{F}$, must be related to an equivalent forcing-displacement amplitude, $\hat{\mathcal{F}}$ by some constant with the units of frequency squared. Some practical examples will be given in §8.1, but in general, we can write $\mathcal{F} = \omega_0^2 \hat{\mathcal{F}}$. Therefore, defining $\omega' = \omega/\omega_0$, (1.83) becomes

$$\boxed{\begin{aligned} \frac{a}{\hat{\mathcal{F}}} &= \frac{1}{\sqrt{\left(1-\omega'^2\right)^2 + \left(2\zeta\omega'\right)^2}}, \\ \Phi &= -\tan^{-1}\left(\frac{2\zeta\omega'}{1-\omega'^2}\right) \end{aligned}}. \tag{1.84}$$

A set of curves of (1.84) for different values of ζ is shown in figure 1.6. A curve such as one of those in figure 1.6 effectively relates the output of some system (here, a simple oscillator) to input as a function of frequency and is sometimes called a *transfer function*. The representation of wave-like signals (and their transformation by devices such as the simple oscillator) as a function of frequency is intimately tied to the concept of a spectrum, which will be introduced in §1.5.6 below.

1.5.6 Introduction to spectra and Fourier transforms

A *spectrum* is a representation of a signal in what is usually called the *frequency domain* instead of the physical domains of time and space. The advantage of a spectral representation is that a spectrum immediately shows how much of the signal is composed of each frequency. The disadvantage is that the information about the relative times at which each frequency occurred is lost. For example, in designing an ocean pier, we may want to know which frequencies of ocean waves at that location have the greatest power to see if those frequencies risk resonating parts of the structure, causing fatigue damage over a long time. The timing of the different-frequency waves relative to each other is unimportant. Likewise, if designing an acoustic barrier to block noise in a factory, it is important to know at what frequencies the noise is loudest. It would be very difficult to estimate which frequencies are loudest by looking at a graph of the pressure fluctuations that are sound waves as a function of time. A spectrum will answer the question immediately. We do not care if the noise consists of a sound from one machine, followed by a sound from another, and so on, since all the noise should be eliminated, so the loss of information about the timing of the sounds is unimportant.

Most of us have become familiar with spectra well before learning the mathematical approach needed to calculate them. One way to illustrate a

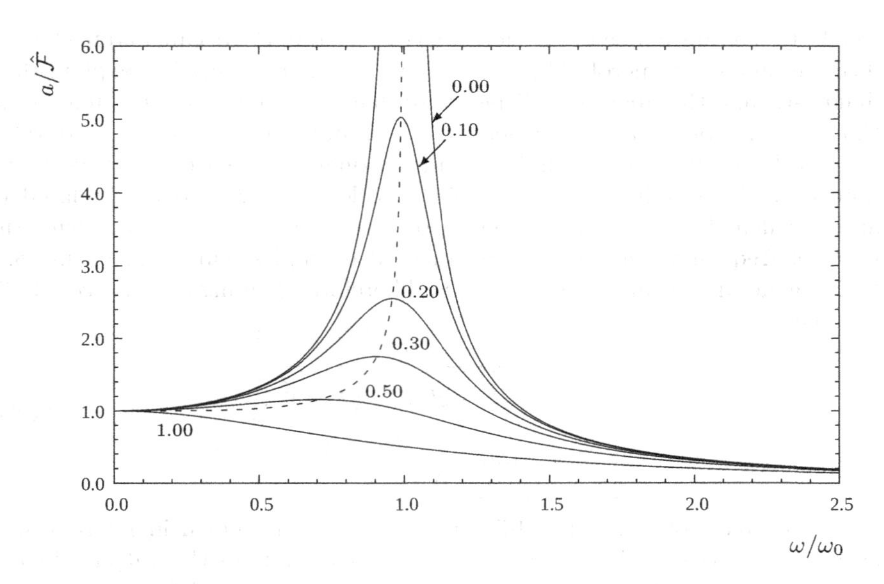

FIGURE 1.6
Response function of any linear, damped oscillator for values of ζ of 0.0, 0.1, 0.2, 0.3, 0.5 and 1.0, which are shown close to the corresponding curve. The maximum values of each curve, which are the resonant frequencies, are joined by the dashed line.

spectrum is with music. Consider the simple childrens' song shown in figure 1.7 which has been modified to have only three notes, C, D and E. For example, if some notes C, D and E were played in what is called *scientific pitch*, they would have frequencies of 512.00 Hz, 574.70 Hz and 645.08 Hz. If this were played on some electronic device that generated only pure frequencies, the resulting sound would be a time series consisting of only three frequencies. (Real musical instruments, even if tuned perfectly, always generate additional frequencies for each note played that give the instrument its characteristic sound.) A crude spectrum of the song in figure 1.7 can be determined by simply counting the number of notes of each frequency; there are six Cs (including the four-beat-long C at the end of the song), 11 Ds (including the one double-beat-length D) and 15 Es (including the two double-beat-length Es), as shown in figure 1.8(a). Since music is an important reason to manage sound (the topic of chapter 3), an understanding of the relationship between frequencies and musical notes is important. Frequencies for the octave including the International Standards Organisation (ISO) standard tuning frequency of A_4 = 400 Hz are given in Table 1.1.

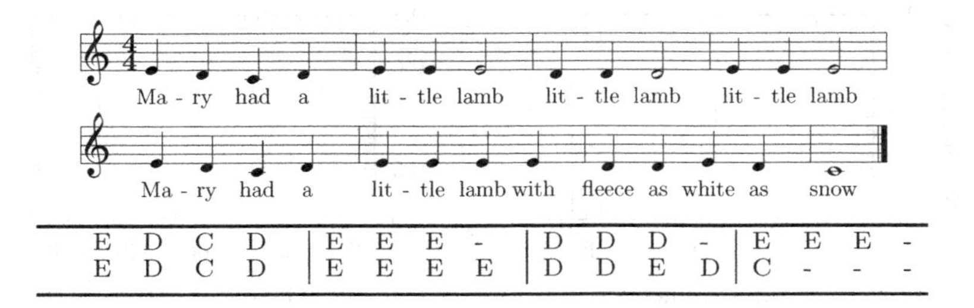

FIGURE 1.7
A version of the song *Mary had a little lamb* arranged to use only three notes: C, D and E. Time goes from left to right on a musical staff, and the second staff is simply a continuation in time. The sequence of notes in time in the two staves is printed below the staves; a dash (-) indicates that the preceding note should continue to be played for an extra beat, effectively doubling the amount of energy at that frequency.

Note	C_4	$C\sharp_4$	D_4	$D\sharp_4$	E_4	F_4	$F\sharp_4$	G_4	$G\sharp_4$	A_4	$A\sharp_4$	B_4
f (Hz)	262	277	294	311	330	349	370	392	415	440	466	494

TABLE 1.1
International Standards Organisation (ISO) notation for the frequencies, f, of the octave around $A_4 \equiv f = 440$ Hz, (ISO 16) rounded to the nearest Hz. There is no semitone between E and F nor between B and C.

Most spectra are *power spectra* since the power transmitted is most often of practical interest, and since power is proportional to amplitude squared, the differences between the heights of the three peaks (6, 11 and 15) in figure 1.8(a) would be amplified in a power spectrum (to 36, 121 and 225). Spectra are usually also *normalised*: usually, the power at each frequency is divided by the integral over the entire spectrum (here, that would be $36 + 121 + 225 = 382$, so the peaks would become 0.094, 0.318 and 0.589), as shown in figure 1.8(b). Since, unlike figure 1.8(a), figure 1.8(b) is a line graph and not a histogram, the pure frequencies appear as 'spikes' with zero width in frequency, located precisely at the frequencies of each note. Finally, so that the largest power peaks do not dominate the spectrum, it is usually plotted with the vertical (power) axis on a log scale, and often with the horizontal (frequency) axis on a log scale as well.

In general, of course, we do not begin with a musical score, in which we already, by definition, know exactly what notes are present and therefore know exactly what frequencies are present. It is most common to begin with some signal in which the frequency content is unknown. Continuing the musical example of figure 1.8, if this three-note song were played on an instrument (or

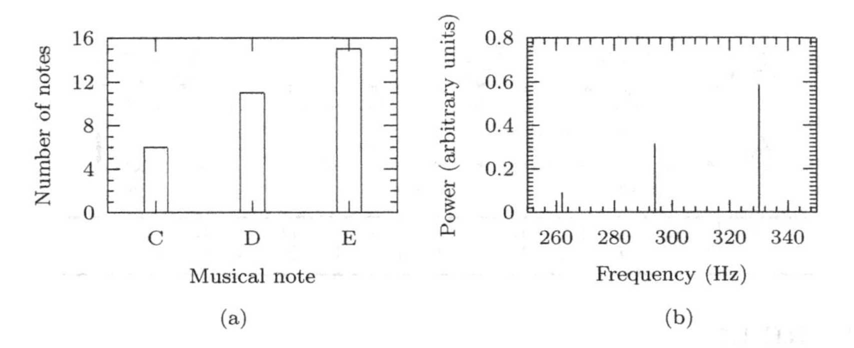

FIGURE 1.8
Conversion of the musical score of the song in figure 1.7 into a simple spectrum. (a) Statistics of the notes. (b) Normalised power spectrum, assuming the notes are C_4, D_4 and E_4 as defined in table 1.1.

sung) and a listener ignorant of the original score wanted to determine which notes were played, a recording of the signal would be converted to a spectrum. There would never be a mathematically pure frequency at each note, so that instead of the 'spikes' in figure 1.8(b), there would be a continuous curve with peaks at each of the three note-frequencies.

Thus, it is necessary to transform the signal from the time domain, where it is a function of time, t, into the frequency domain, where it is a function of frequency, ω. The mathematical procedure to do this is the Fourier transform, invented by Joseph Fourier in 1822. For a function of time, say $g(t)$, its Fourier transform, often denoted $\widehat{g}(\omega)$, is given by

$$\widehat{g}(\omega) = \int_{-\infty}^{\infty} g(t) \mathrm{e}^{\mathrm{i}\omega t} \mathrm{d}t. \tag{1.85}$$

In practice, the signal is digitised at a particular *sampling rate*; for example, for music, the standing sampling rate is 44.1 kHz. The result is simply a long list of numbers stored in a digital file; for sound, the numbers are typically the voltage output by the microphone, with one number for every (1/44 100)th of a second. For good-quality microphones, as detailed in §3.3.8.2, a *calibration* of the microphone has been done by the manufacturer, so that it is possible to relate the voltages to a physical variable such as the pressure. Both microphones and hydrophones will be discussed in §3.3.8.2. An algorithm known as the *Fast Fourier Transform* re-arranges the samples in time to create the spectrum in frequency.

Finally, for travelling waves, recall that there is a direct relationship between time and distance travelled, and that is the wave speed, c; and recall from (1.71) that c is also the relation between ω and k. Thus, equivalently to the Fourier transform from time to frequency in (1.85), one could have a Fourier transform from space to wavenumber. It depends on what is most relevant to the practical application. For sound waves, frequencies are of most interest, but for waves on the surface of the sea, the wavelength (hence the wavenumber) may be of greater interest. One could decompose a shape into wavenumbers just like one could decompose a sound into frequencies.

1.6 Problems

1. Argon gas ($\gamma = 1.667$) is adiabatically compressed by a factor of three for an underwater welding operation. It has a density of 1.78 kg m^{-3} before compression. What is its density after compression?
2. An industrial reciprocating saw is lubricated by cutting fluid with a dynamic viscosity of 0.003 kg m^{-1}s^{-1}. The gap between the saw and workpiece is 0.5 mm and the flow in the gap has an approximately constant velocity profile. What is the shear stress on each side of the saw at the peak saw speed of 0.6 m s^{-1}?
3. Water is pumped at 6 ML/min into the initially-empty upper tank of a renewable-energy storage system. The tank is cylindrical with a diameter of 50 m and a height of 10 m. How long will it take to fill? Note 1 ML = 1,000,000 litre.
4. A pressure sensor at the foundation of an ocean bulk-loading pier records a pressure peak of 20 kPa as a long surface wave passes. Satellite images show the wavelength to be 250 m. Assuming the water is inviscid as well as incompressible, what is the peak acceleration magnitude of the water surface? The density of seawater is 1026 kg m^{-3}.
5. A sound-absorbing barrier is being designed to block factory noise from reaching an office. The barrier indentations should be 1/4 of the sound wavelength. A spectrum of noise measured in the factory has a main peak frequency of 250 Hz and two other smaller peaks at 125 Hz and 500 Hz. The speed of sound in air is $c = 340$ m s^{-1}. What length should the barrier indentations be?
6. A tsunami travels from the deep ocean with a depth 4 km onto a continental shelf with a depth 150 m. What will be the percentage change in its speed?
7. A hydroelectricity dam discharges into a river of depth 1.5 m and width 15 m. What volumetric flow rate will prevent long waves from propagating back upstream, potentially causing a flood?
8. Air ($\gamma = 1.4$) is being adiabatically compressed by a factor of 10 for a factory pneumatic system. By what factor does its density change?
9. A flow has zero divergence if it

   A Has equal and opposing inflows in the y-direction

   B Has a rate of change of density with time-matched by its inflow and outflow rates

   C Has equal and opposing inflows in the x-direction

   D Has zero rate of change of density with time and inflow and outflow rates that match

10. Water drains from a pumped-storage upper tank at 2 m s^{-1} through a pipe 1 m in diameter. The cylindrical tank has a surface area of 8 000 m^2. At what rate will the water level fall?
11. The difficulty in expressing conservation of momentum for a fluid arises from

    A The fluid's ability to accelerate by changing its position

B The solid's ability to withstand a shear stress

C The fluid's bulk modulus

D The solid's Young's modulus

12. Xenon gas ($\gamma = 1.667$) is adiabatically compressed by a factor of 25 in the manufacture of high-pressure arc lamps for cinema projectors. It has a density of 5.89 kg m^{-3} before compression. What is its density after compression?

13. A sliding joint in a wave-energy converter is lubricated by a thin film of seawater with a dynamic viscosity of 0.001 kg m^{-1}s^{-1}. The film is 3 mm thick and has an approximately constant velocity profile. What is the shear stress in the joint at the peak sliding speed of 0.75 m s^{-1}?

14. Biodiesel oil is pumped into a tank at a rate of 5 L/s. The tank is empty initially and has a diameter of 2 m and a height of 2 m. How long will it take to fill?

15. An array of air-pressure sensors at an airport records a pressure peak of 800 Pa as an atmospheric wave from a thunderstorm passes. The pressure falls to zero 2 km from the location of the peak. Assuming the air at ground level is inviscid and incompressible, what is the peak acceleration magnitude of the resulting wind gust? The density of air is 1.23 kg m^{-3}.

16. A wave-energy converter is being designed to attach to a port breakwater. Ideally, this design should have a length 1/4 of the ocean wavelength. A spectrum of ocean waves measured at the breakwater has a main peak frequency of 0.4 Hz and two other smaller peaks at 0.2 Hz and 0.1 Hz. Waves at that location are known to travel with phase speed $c = 10$ m s^{-1}. What length should the wave-energy converter have?

17. A tsunami travels 500 km over deep over with depth 4 km and then 50 km over continental shelf with average depth 50 m. How long will it take to arrive?

18. Flow down the spillway of a hydroelectric scheme enters a channel of depth 1.0 m and width 30 m. What volumetric flow rate will prevent long waves from propagating back upstream, causing the channel to flood?

19. Pure carbon dioxide gas ($\gamma = 1.3$) is pumped into the ocean from the hull of a submarine at a depth of 150 m, where the seawater has a density of 1027 kg m^{-3}. Assuming it is generated at atmospheric pressure (101,325 Pa) inside the vessel with a density of 1.98 kg m^{-3} and that it is compressed adiabatically. What is its density at the point of release into the ocean? Assume g=9.78 ms^{-2}.

20. A novel glass-recycling process is proposed in which jets of air cool molten glass sheets. Each jet has a density of 1.2 kg m^{-3} and exits a nozzle at a speed of 10 m s^{-1} where the pressure is atmospheric, impacting the liquid-glass surface at right angles. The air-molten-glass surface tension is about 0.3 N m^{-1}. Estimate the radius of curvature of any ripples a jet could make in the glass.

21. Discuss whether the following fluid-wave situations would be divergence-free, i.e., satisfy $\boldsymbol{\nabla} \cdot u = 0$.

(a) Water flow in a column of the ocean extending from the sea surface to the sea bed, far from land, as a tsunami passes

(b) Molten-metal flow at the bottom of a smelter vessel with waves at the surface

(c) Air flow in a trumpet as a musician plays a note

(d) Air flow around an aircraft travelling close to the speed of sound

22. Liquid hydrazine (density 1021 kg m^{-3}) is pumped into a spacecraft manoeuvering-thruster tank during fueling on the ground. The tank has a diameter of 0.5 m and has a central cylindrical wall of length 0.5 m with spherical end-walls. It is initially filled with nitrogen gas at atmospheric pressure, and at the end of fueling, the nitrogen has been compressed isothermally to 24 times atmospheric pressure. If the hydrazine can be pumped at a constant rate no faster than 1 L/s for safety reasons, what is the minimum safe time to fill the tank?

23. Given the one-dimensional momentum equation with terms numbered (i)-(iv) below,

$$\underset{\text{(i)}}{\frac{\partial \rho u}{\partial t}} + \underset{\text{(ii)}}{u\frac{\partial \rho u}{\partial x}} = \underset{\text{(iii)}}{-\frac{\partial p}{\partial x}} + \underset{\text{(iv)}}{\mu\frac{\partial^2 u}{\partial x^2}},$$

discuss which term or terms could be neglected under the following circumstances.

(a) Ultrasonic waves passing through seawater

(b) Flow around a spacecraft entering the Earth's atmosphere

(c) Steady blood flow in a microgravity surgical device

(d) Waves due to density differences in the Earth's atmosphere

2

Water-surface waves

2.1 Summary of key points

- The **dispersion relation** for waves with wavelengths longer than about 0.1 m, giving the radian frequency ω (where $\omega = 2\pi f$ and f is the frequency in Hertz) as a function of wavenumber, is given by (2.35) on page 61,

$$\boxed{\omega^2 = \mathrm{g}k \tanh kh},$$

where g is the acceleration due to gravity (about 9.81 m s^{-2}; source of precise data given on page 3), k is the wavenumber given by $k = 2\pi/\lambda$ where λ is the wavelength, and h is the depth of the water. If waves are shorter than $\lambda \simeq 0.1$ m, surface-tension effects begin to be significant, and are included by replacing g with g_σ in all formulae, according to (2.41) on page 63,

$$g_\sigma = \mathrm{g} + \frac{\sigma k^2}{\rho_0},$$

where σ is the surface tension coefficient (about 0.072 N/m at 25°C; source of precise data given on page 4) and ρ_0 is the density of water (about 998 kg m^{-3} at 20°C; source of precise data given on page 3).

- The **phase speed**, giving the speed with which the pattern of crests and troughs moves, is given (2.44) on page 65,

$$\boxed{c = \sqrt{\left(\frac{\mathrm{g}}{k}\tanh kh\right)}}.$$

- The **surface elevation** - the vertical displacement of the water surface in metres that observers see - is given by (2.54) on page 66,

$$\boxed{\eta = -\mathrm{i}A\mathrm{e}^{\mathrm{i}(\pm\kappa x \pm \ell y + \omega t)}},$$

where A is a complex amplitude, x and y are distances in the horizontal and κ and ℓ are the corresponding wavenumbers, which represents simple sine waves in horizontal space and in time; for propagation in the

DOI: 10.1201/9780429295263-2

x-direction only and for a phase of zero, the **real surface elevation** η_R is given by (2.61) on page 68,

$$\boxed{\eta_R = \hat{\eta}\cos(kx - \omega t)},$$

where $\hat{\eta}$ is the real surface-elevation amplitude (half the wave height).

- The **velocity potential**, giving the velocity and pressure with which the water is set into motion at any point in three dimensions, (2.53) on page 66,

$$\boxed{\phi = A\frac{\mathrm{g}}{\omega}\frac{\cosh(kh + kz)}{\cosh(kh)}\mathrm{e}^{\mathrm{i}(\pm\kappa x \pm \ell y + \omega t)}},$$

where z is the distance in the vertical and is positive upwards from the undisturbed water surface. For a phase of zero, and waves propagating in the x-direction only, a simplification of (2.53) giving the **real water-velocity components** in the horizontal and vertical, u_R and w_R respectively, and well as the **real pressure**, p_R, is given by (2.64) on page 69,

$$\boxed{\begin{aligned} u_R &= \hat{\eta}k\frac{\mathrm{g}}{\omega}\frac{\cosh(kh + kz)}{\cosh(kh)}\cos(kx - \omega t), \\ w_R &= \hat{\eta}k\frac{\mathrm{g}}{\omega}\frac{\sinh(kh + kz)}{\cosh(kh)}\sin(kx - \omega t), \\ p_R &= \hat{\eta}\rho_0\mathrm{g}\frac{\cosh(kh + kz)}{\cosh(kh)}\cos(kx - \omega t). \end{aligned}}$$

- Although (2.35), (2.44) and (2.53) above are valid for all depths, the following **approximations** are often used since they make calculations easier.

Approximation	Deep water		Shallow water	
Depth condition	$h > 0.5\lambda$	Eq.	$h < 0.05\lambda$	Eq.
Frequency	$\omega^2 = \mathrm{g}k$	(2.74)	$\omega^2 = \mathrm{g}hk^2$	(2.80)
Phase speed	$c = \sqrt{\mathrm{g}/k}$	(2.75)	$c = \sqrt{\mathrm{g}h}$	(2.81)
Group speed	$c_g = \frac{1}{2}c$	(2.78)	$c_g = c$	(2.82)
Potential	$\phi = A(\mathrm{g}/\omega)\mathrm{e}^{kz}\mathcal{X}\mathcal{Y}\mathcal{T}$,		$\phi = A(\mathrm{g}/\omega)\mathcal{X}\mathcal{Y}\mathcal{T}$,	
	$\mathcal{X}\mathcal{Y}\mathcal{T} = \mathrm{e}^{\mathrm{i}(\pm\kappa x \pm \ell y + \omega t)}$	(2.76)	$\mathcal{X}\mathcal{Y}\mathcal{T} = \mathrm{e}^{\mathrm{i}(\pm\kappa x \pm \ell y + \omega t)}$	(2.83)

- Useful **textbooks** include Lighthill (1978), which takes a thorough applied-mathematics approach, while the **book** by Gill (1982) provides many useful results and data needed for ocean and atmosphere applications.

2.2 An example

The chief engineer of a small Pacific island is working in an emergency bunker with her staff and a few residents. A cyclone has impacted the island with winds gusting over 165 km h^{-1} for most of the day. Her estimates of wind speed and storm surge (chapter 7) have enabled the island's residents to manage the disaster well. The worst has passed.

A foreign backpacker is amongst those sheltering in the bunker. Ignoring warnings, he has gone outside several times to record the storm. He comes in to say the satellite dish on the roof - the island's only connection to the outside world - is close to being ripped off. Her phone rings - it is an international call, so the dish is still working. It is the International Tsunami Warning Center in Hawaii. A magnitude-nine earthquake has just occurred in the Cascadia Subduction Zone off the north-west coast of North America, about 6000 km from the island.

The chief engineer is responsible for distributing tsunami warnings. In a few hours, people would start to return to their waterfront homes. The caller assures her that simulations of the tsunami's arrival would be posted in a few minutes.

The message never arrives; the backpacker's warning was fulfilled and the satellite dish is gone, and with it all external information. The engineer knows there is a simple formula for tsunami propagation speed but cannot remember it. When will the tsunami hit?

Solution

The engineer makes several rough but appropriate approximations that can be formally justified in §2.3.11. The conservation of mass relation (1.15) for the tsunami can be understood from figure 2.13. A tsunami is an extremely long wave (measured from undisturbed water far ahead of it to undisturbed water far behind it) compared with the ocean depth. Therefore, it can be assumed there is no variation in the vertical direction, so the entire depth of the ocean, h, is a vertical 'column' of water that moves horizontally. Using the usual principles of calculus, the water column shown in figure 2.13 is considered to be infinitesimally thin in the x-direction, so its volume before the tsunami was $\Delta x Y h$ where Y is the width of the water column into the page. The temporary increase in water-column height above normal sea level is η, so that if η is increasing with time, the ocean surface at this location is rising.

The seawater is not being compressed significantly; for that to occur, variations in pressure similar to the differences between the sea surface and the ocean floor would have to be applied, and since the tsunami's height is a small fraction of the approximately 4000 m ocean depth, compressibility can be neglected. Thus, conservation of mass reduces to conservation of volume. With flow in the x-direction defined to be positive, the net volumetric rate with which water flows into the column is equal to the rate volume is flowing into its left-hand side, velocity u multiplied by the cross-sectional area $Y(h+\eta)$, minus the rate volume is flowing out of its right-hand side, $u + \partial u/\partial x$ multiplied by the area $Y(h+\eta)$. Conservation of volume requires that any net inflow must be equal to the rate at which the column is increasing in height. The continuity equation is thus

$$\left[u - \left(u + \frac{\partial u}{\partial x}\Delta x\right)\right] Y(h+\eta) = \Delta x Y \frac{\partial(h+\eta)}{\partial t}. \tag{2.1}$$

Since $\eta \ll h$, the $h+\eta$ in (2.1) is very well approximated by h, and since on the right-hand side, only η and not h varies with time, (2.1) becomes

$$-\frac{\partial u}{\partial x} = \frac{1}{h}\frac{\partial \eta}{\partial t}. \tag{2.2}$$

Considering only the conservation of momentum (1.20) in the x-direction, the direction in which the tsunami is travelling, and again applying the assumptions of incompressibility and that there is no variation in the vertical, the momentum equation becomes

$$\frac{\partial u}{\partial t} + u\frac{\partial u}{\partial x} = -\frac{1}{\rho_0}\frac{\partial p}{\partial x} + \nu\frac{\partial^2 u}{\partial x^2}. \tag{2.3}$$

Since variations in the vertical have already been assumed negligible, only variations in the horizontal are left. However, since the tsunami is extremely long, gradients in the horizontal are small and the gradient of the gradient is extremely small. Thus, neglecting horizontal gradients eliminates the second, viscous term on the right-hand side. Eliminating nonlinearity could be a problem because tsunamis are nonlinear waves (detailed in §12.3). The nonlinear term would be needed to predict how high the wave will be, but the urgent need is to predict when the tsunami will arrive. Assuming the velocity is small enough to eliminate the nonlinear term, (2.3) reduces to

$$\frac{\partial u}{\partial t} = -\frac{1}{\rho_0}\frac{\partial p}{\partial x}, \tag{2.4}$$

and remembering the hydrostatic relation between pressure and surface elevation, $p = \rho_0 g\eta$, (2.4) can also be written

$$\frac{\partial u}{\partial t} = -g\frac{\partial \eta}{\partial x}, \tag{2.5}$$

Now there are two equations, (2.2) and (2.5), in two unknowns, u and η. Either could be eliminated first. Choosing to eliminate u by differentiating (2.2) with respect to t and (2.5) with respect to x and adding the resulting two equations gives

$$\frac{1}{h}\frac{\partial^2 \eta}{\partial t^2} - g\frac{\partial^2 \eta}{\partial x^2} = 0, \tag{2.6}$$

or, slightly re-written,

$$\frac{\partial^2 \eta}{\partial t^2} - c^2\frac{\partial^2 \eta}{\partial x^2} = 0, \tag{2.7}$$

where $c^2 = gh$. This is the one-dimensional wave equation (§1.5.2). Any sum of sines and cosines of x multiplied by sines or cosines of t will satisfy (2.7). By writing a solution as, for example, $\eta = \hat{\eta}\cos[k(x - ct)]$ where k is some constant and $\hat{\eta}$ is the amplitude, it is clear that, while the height of the tsunami cannot be predicted - there is no information on the amplitude - time and space are related by the constant c.

Thus, c is the wave speed given by $c = \sqrt{gh}$. Knowing that the ocean between the island and the origin of the tsunami, like most of the Earth's ocean floor, is about 4 km down and that the acceleration due to gravity is roughly 10 m s^{-2}, a wave speed of $\sqrt{40,000} =$ 200 m s^{-1} is roughly estimated: that is 720 km h^{-1}. The tsunami will travel 6000 km in just over 8 hours.

That is just enough time to warn all the island's residents.

2.3 Linear water-wave theory

2.3.1 The waves we see

Surface gravity waves are the fluid waves that human beings are most used to thinking of as 'waves'. We see waves on the sea, on lakes and in rivers (figure 2.1), as well as in the swimming pool and bathtub. Thus, in one sense, it is helpful to begin with surface gravity waves, since our observations can immediately guide us. However, this approach is not without risk of complications, since surface gravity waves, even under the simplification of linear theory, have peculiarities not found in other linear wave theories, for example, the theory of sound waves.

The theories of this chapter directly inform the engineering design of ships and other water vessels, as well as the design and control of ports and harbours, coastal defences, and natural rivers and waterways. The installation of renewable-energy generators in the ocean (chapter 8), including offshore wind turbines, also requires careful calculation of the impact of surface waves. Naturally, the theories are also vital for understanding many oceanographic phenomena and for coupled ocean-atmosphere phenomena affecting climate change (chapter 10).

However, the same theories are also relevant to a great variety of other practical applications. For example, in pyrometallurgy, molten metals have a

FIGURE 2.1
Waves generated by the wind blowing over Cataract Gorge, Tasmania, Australia, reflect off different points on the lake shore, coming from top left and top right in this picture. These two sets of waves superpose without interference, illustrating how solutions of Laplace's equation $\nabla^2\phi = 0$ superpose. Photograph by Richard Manasseh.

free surface like the surface of water. Waves form on the surface of the molten steel in ladle steelmaking. In the minerals-processing, chemicals, petrochemicals and pharmaceuticals industries, enormous tanks, sometimes containing tens of thousands of litres of water or oil-based liquids, are stirred and mixed, and waves form on those surfaces, affecting the process. Transport of liquids in rail or road tankers can be destabilised by waves inside the tanks; furthermore, ships carrying bulk liquids, such as liquefied natural gas (or a carbon-free alternative such as ammonia), can suffer from a resonance between ocean waves and waves inside the tanks.

2.3.2 Potential flow

2.3.2.1 Physical assumptions that lead to potential flow

The properties of surface gravity waves can be determined by first deriving a wave equation as presented in §1.5.2 and then solving the wave equation for appropriate boundary conditions as discussed in §1.5.3. However, for surface gravity waves, it is common to bypass the wave-equation stage completely, because circumstances specific to surface gravity waves permit a simple mathematical approximation, that of *potential flow*. This permits variables of practical interest, such as the velocity in all three directions as well as the pressure, to be represented by a single variable. It also permits fast, efficient computational techniques, such as the *Boundary Element Method* (BEM) to be used for engineering problems, permitting relevant forces on ships, machines or structures in the ocean to be quickly calculated; calculations that would be prohibitively expensive otherwise.

In some ocean-wave texts, it is common to begin with a statement like 'assume potential flow ...'. In other branches of fluid dynamics, such as the aspect of aerodynamics that deals with the design of aircraft wings, it does seem to be necessary to jump straight to the assumption of potential flow. The disadvantage of assuming potential flow from the outset is that one makes a mathematical assumption from the outset, without that assumption being grounded in physics. However, with fluid waves, it is helpful to point out that potential flow is an automatic consequence of a sequence of four physical assumptions. At the end of this section, it will be emphasised that the reverse logic does not apply: potential flow does not imply the four assumptions.

The first assumption is that the flow is incompressible, so that the density ρ is a constant, ρ_0, and thus the momentum equation (1.20) takes the form,

$$\frac{\mathrm{D}\boldsymbol{u}}{\mathrm{D}t} = -\frac{1}{\rho_0}\boldsymbol{\nabla} p - \frac{1}{\rho_0}\boldsymbol{\nabla} P_{\emptyset} + \nu\nabla^2\boldsymbol{u} + \mathbf{g}, \tag{2.8}$$

where $\boldsymbol{u}$ is the velocity field, which is a vector $\boldsymbol{u} = (u, v, w)$, p is the dynamic pressure field and ν is the kinematic viscosity. Now, the dynamic pressure, p, is the pressure in excess of the ambient pressure, which, according to (1.2), includes atmospheric pressure, some further static pressure that may be applied by an engineering system, and the hydrostatic pressure due to the force of gravity. To emphasise the presence of gravity, two terms are shown on the right-hand side of (2.8), $-(1/\rho_0)\boldsymbol{\nabla} P_{\emptyset} + \mathbf{g}$, where $P_{\emptyset}$ is the ambient pressure and $\mathbf{g}$ is the gravitational acceleration vector (which points downwards). The atmospheric and engineering static pressures can be assumed to be effectively constant over short distances, so their gradients are zero, leaving only hydrostatic pressure. Hence, these two terms exactly cancel each other out, and after a few steps, they will be dropped. However, when the boundary conditions are considered in §2.3.4, it will be seen that gravity does come into play, since, of course, the water waves considered here are entirely due to gravity and are usually called *gravity waves* [1].

There are two further assumptions implicit in (2.8): that the effects of surface tension and of the flow of air above the water can be ignored. We will not count these in the 'four assumptions that lead to potential flow' for the following two reasons.

---

[1]Not to be confused with the *gravitational waves* in space-time predicted by relativity and only observed for the first time in 2016; our gravity waves have been observed as long as people have been watching the sea.

Firstly, including surface tension is still consistent with potential flow, and in the end, including it just modifies the final result without affecting the logic of the derivation; surface tension will be included in §2.3.5.4. Surface tension comes into play for waves that are very short, with crest-to-crest lengths less than about 0.1 m. These are called *ripples* and are irrelevant for most ocean-wave applications. However, ripples are seminal to gravity waves, since they roughen the surface of the water, permitting the wind to transfer greater momentum transfer to the surface, and, as detailed in §8.2.1, thus *create* the gravity waves.

Secondly, the air above the water is also a fluid, and if the water is to move, the air must move too. However, the density of air is only about 1.2 kg m^{-3}, while the density of the water is roughly 800 times higher. Thus, air pushed and pulled around by the water surface as it moves possesses only a fraction of a percent of the momentum of the water, so the modification that *still* air makes to the movement of the water can be safely neglected. However, once the air is moving relative to the water surface, so that there is a wind, the *shear stress* it applies to the water surface certainly matters. As just-noted, it is a wind that creates the waves we see most often on water bodies. To account for this wind shear-stress 'forcing' of the water surface, an extra force-per-unit-mass should be added to the right-hand side of (2.8). For now, following the classical approach to fluid-wave problems outlined in §1.5.3, we will study unforced waves, or, more mathematically speaking, the homogeneous momentum equation.

The second of the four assumptions is that the flow is inviscid. This means that energy losses due to fluid friction are ignored, reducing the momentum equation to its inviscid form (Euler's equation),

$$\frac{\mathrm{D}\boldsymbol{u}}{\mathrm{D}t} = -\frac{1}{\rho_0}\boldsymbol{\nabla} p - \frac{1}{\rho_0}\boldsymbol{\nabla} P_\emptyset + \mathbf{g}. \tag{2.9}$$

The third assumption is that the behaviour is linear. For example, if the water surface is not dead flat, this assumption means that any temporary water-surface slope that might exist due to water motion is extremely gentle. If the water is moving at all, it is moving with a very low velocity, and furthermore with a velocity that hardly varies at all with distance. Any disturbance to the water-surface level is infinitesimally small. A formal scaling analysis, such as that leading to (1.33), would show that nonlinearities can be neglected. Therefore, the nonlinear term in the material derivative (or 'total derivative') D/Dt disappears, reducing (2.9) to

$$\frac{\partial \boldsymbol{u}}{\partial t} = -\frac{1}{\rho_0}\boldsymbol{\nabla} p - \frac{1}{\rho_0}\boldsymbol{\nabla} P_\emptyset + \mathbf{g}, \tag{2.10}$$

and, as noted earlier, the hydrostatic and gravitational terms cancel out, so they can be dropped, giving

$$\frac{\partial \boldsymbol{u}}{\partial t} = -\frac{1}{\rho_0}\boldsymbol{\nabla} p. \tag{2.11}$$

The fourth assumption anticipates one part of the solution we expect to get, that the velocity and pressure vary sinusoidally with time. Thus, $\boldsymbol{u} = \boldsymbol{U}\mathrm{e}^{\mathrm{i}\,\omega t}$ where $\boldsymbol{U}$ is a spatially-varying velocity factor and ω is the radian frequency, and likewise with the pressure, consistently with the separation-of-variables concept introduced in §1.5.3.

Following this fourth assumption, (2.11) reduces to

$$\mathrm{i}\,\omega\boldsymbol{u} = -\frac{1}{\rho_0}\boldsymbol{\nabla}p, \tag{2.12}$$

which can be re-arranged by defining a *velocity potential*, ϕ, where

$$\phi = \frac{\mathrm{i}}{\rho_0\omega}p, \tag{2.13}$$

reducing (2.12) to

$$\boldsymbol{u} = \boldsymbol{\nabla}\phi, \tag{2.14}$$

Situations in physics in which some sort of vector field $\boldsymbol{u}$ is determined by the gradient of some sort of potential ϕ according to (2.14) are not limited to fluid dynamics. For example, an electric field in the absence of a time-varying magnetic field is given by the gradient of the electric potential; and the acceleration due to gravity that is vital to this chapter is not truly a constant but is given by the gradient of the gravitational potential.

Of the four assumptions, only the first assumption, that the water is incompressible, is a very good assumption. It is likely to be violated only if a shockwave from an explosion, a collapsing cavitation bubble, or a high-speed impact passed through the water. The assumption that the water is inviscid is fairly good, provided there is no solid surface nearby; but once we include the hull of a ship or a wave-energy converter, or indeed rocks or the sea bottom, the inviscid assumption runs into trouble, and indeed drag on solid surfaces is impossible in potential flow. Nevertheless, the inviscid assumption is widespread, since good estimations of the forces at right angles to solid surface can still be made. The third assumption, linearity, is surprisingly good; when waves get steep, modifications can still be made to linear potential-flow theory, and with these modifications, potential flow continues to be remarkably useful. Nonetheless, an important aspect of fluid dynamics, turbulence, is due to the existence of nonlinearities. The fourth assumption, sinusoidal behaviour with time, is likewise surprisingly reasonable.

Spelling out these four assumptions may place the use of potential-flow theory on a more physical basis, but it is important to understand that this logic only works one way. If one assumes potential flow from the outset, it need not imply all the four assumptions hold. For example, some viscous flows (for example a thin film of fluid between two parallel surfaces, as in lubrication) could be potential; and spherically-symmetric sound waves, despite being compressible, are also described by potential flow.

2.3.3 Laplace's equation

The assumption of incompressibility was used to write down (2.8) with a constant ρ, i.e. $\rho = \rho_0$. Now, the incompressibility assumption means that the Law of Conservation of Mass, (1.15), takes the form

$$\boldsymbol{\nabla} \cdot \boldsymbol{u} = 0. \tag{2.15}$$

Substituting (2.14) into (2.15) gives one of the most famous equations in mathematical physics,

$$\nabla^2 \phi = 0, \tag{2.16}$$

which is *Laplace's equation*, published by Pierre-Simon Laplace in 1783. (The Laplacian operator, ∇^2, was introduced in §1.2.3.2.) This equation is of immense usefulness, as it predicts the gravitational attraction of celestial bodies and electromagnetic fields as well as fluid flows including the tides, and the surface gravity waves that are our immediate interest. Thus, mathematicians in the 18th and 19th centuries were motivated to find many solutions to Laplace's equation, some of which we are about to use. Since Laplace's equation is linear (like the wave equation), its solutions can simply be super-imposed, and indeed this represents reality quite well; small waves can be observed passing across each other without interference. (See, for example, figure 2.1.) This is very helpful in predicting how waves reflected and refracted by rocks and reefs (see §2.3.12) interact with each other, and how ships, wave-energy converters and structures in the water interact with waves and each other.

2.3.4 Boundary conditions for water waves

Like any differential equation, Laplace's equation must be solved with knowledge of the *boundary conditions* - the values of ϕ or its derivatives at the boundaries of the domain over which the solution is valid. In this case, the boundaries are the bottom and top of the water body (the seabed and sea surface, for example). The first boundary condition (see figure 2.2), at the bottom, is simpler. Say the bottom of the water body is at $z = -h$, where z is defined to be pointing up from an origin at the undisturbed surface. It is only necessary to say that water cannot flow vertically through the bottom, so that the vertical component, w, of the velocity of the water is zero, i.e., $w = 0$ at $z = -h$. Away from the bottom, the water can move vertically as well as horizontally, and furthermore, the water can move with different horizontal speeds at different depths. From (2.14), the component of velocity in any direction is given by the derivative of ϕ in that direction, so the component of velocity in the z-direction is given by $w = \partial\phi/\partial z$; the bottom boundary condition is thus

$$\boxed{\frac{\partial \phi}{\partial z} = 0 \text{ at } z = -h.} \tag{2.17}$$

This is usually called the *kinematic boundary condition* at the bottom of the water column.

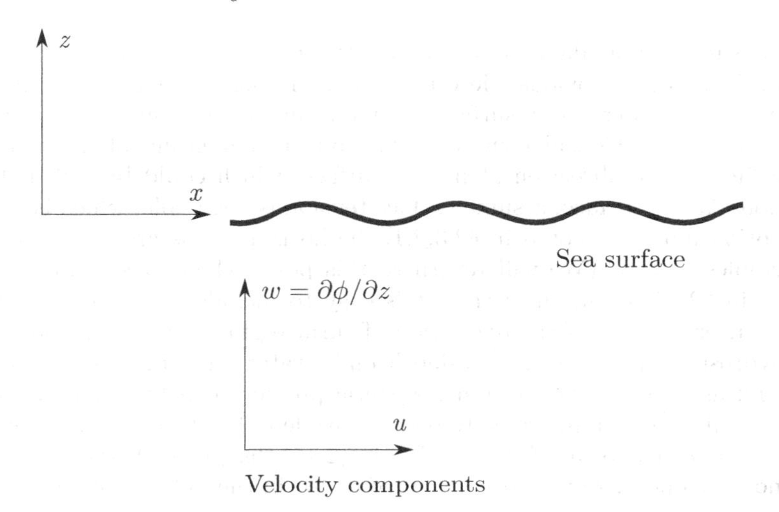

FIGURE 2.2
Boundary condition at the sea bottom (or the bottom of any body of water open to the atmosphere, such as a lake, river or channel) is $w = 0$ at $z = -h$.

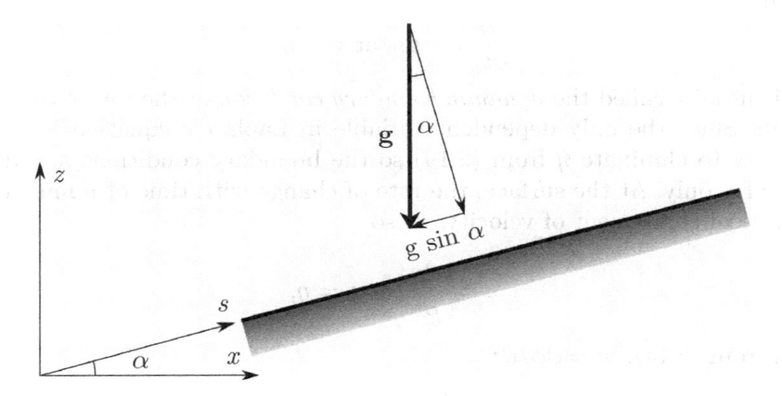

FIGURE 2.3
Definitions leading to the boundary condition at the sea surface (or the surface of any body of water open to the atmosphere, such as a lake, river or channel); recall that α must be a small angle for linear theory to be valid.

The surface boundary condition is trickier. Because the surface is in motion, it is not so obvious. Here is one explanation, referring to figure 2.3. Imagine the moving water surface is at one instant of time at an angle α to the horizontal. Consider the conservation of momentum of the water at the surface, in the direction along the surface, which could be called the s-direction. At the surface, assume surface tension is negligible, which is a good assumption unless as noted in §2.3.2.1, the horizontal distances are so short that ripples matter. (We will return to this point when considering surface tension in §2.3.5.4 and show that it is easy to include the effect of surface tension after the rest of the derivation of surface gravity waves is complete.) The hydrostatic pressure due to depth underwater must be zero at the surface, and as noted earlier, the atmospheric pressure can be assumed to not vary over distances comparable to ocean-wave lengths. The dynamic pressure p is zero. Thus the terms $\nabla P_\emptyset$ and ∇p in (2.10) disappear. However, gravity does not disappear, so the component of (2.10) in the s-direction becomes,

$$\frac{\partial}{\partial t}\left(\frac{\partial \phi}{\partial s}\right) = -\mathrm{g}\sin\alpha, \tag{2.18}$$

where (2.14) has again been used, this time to say that the component of velocity in the s-direction is $\partial\phi/\partial s$. The slope of the sea surface can also be written as $\partial\eta/\partial s$, where η is the *surface elevation*: the temporary vertical displacement of the sea surface from its equilibrium position at $z = 0$. From figure 2.3, $\partial\eta/\partial s = \sin\alpha$, therefore (2.18) can be integrated with respect to s, giving

$$\frac{\partial \phi}{\partial t} = -\mathrm{g}\eta \;\text{ at } z = \eta. \tag{2.19}$$

This is usually called the *dynamic boundary condition* at the top of the water column. Since the only dependent variable in Laplace's equation is ϕ, it is necessary to eliminate η from (2.19) so the boundary conditions are also in terms of ϕ only. At the surface, the rate of change with time of η must equal the vertical component of velocity, w, so

$$w = \frac{\partial \eta}{\partial t} \;\text{ at } z = \eta, \tag{2.20}$$

Since, from (2.14), $w = \partial\phi/\partial z$,

$$\frac{\partial \phi}{\partial z} = \frac{\partial \eta}{\partial t} \;\text{ at } z = \eta. \tag{2.21}$$

This is usually called the kinematic boundary condition at the top of the water column. Eliminate η by differentiating (2.19) with respect to t and using (2.21) to replace $\partial\eta/\partial t$, giving

$$\frac{\partial^2 \phi}{\partial t^2} + \mathrm{g}\frac{\partial \phi}{\partial z} = 0 \;\text{ at } z = \eta. \tag{2.22}$$

Some derivations quote (2.22) or an equivalent directly, calling it the 'Bernoulli equation at the surface', or similar. The value of g is often taken to be 9.81 ms^{-1} (a precise value is given on page 3) but varies with latitude and altitude. If calculations are done in which more than two significant figures of accuracy are desired, it is necessary to use a more precise relation for g that takes into account the local latitude (Moritz, 2000) and altitude (Li and Götze, 2001).

We could just as well have eliminated ϕ from (2.19) and (2.21) and arrived at an equivalent of (2.22) in terms of the sea-surface elevation η. Indeed, the boundary conditions in §2.3.4 could have been derived without using the velocity potential at all, just the velocity components and η, under three of the four assumptions: that the behaviour is incompressible, inviscid, and linear. However, as mentioned, using the velocity potential makes a powerful range of mathematical solutions to Laplace's equation available. It also permits a very convenient superposition of solutions using just one variable, ϕ, which unlike η is valid throughout the water depth and not just at the surface.

Finally, applying the fourth assumption of sinusoidal time dependence, the sea-surface boundary condition (2.22) becomes

$$\boxed{-\omega^2\phi + \mathrm{g}\frac{\partial \phi}{\partial z} = 0 \text{ at } z = \eta.} \tag{2.23}$$

2.3.5 Airy's solution for surface gravity waves

2.3.5.1 Separation of variables solution

Solutions to Laplace's equation (2.16) are a wide range of functions called harmonic functions, which includes the exponential function. The familiar sine and cosine functions that we expect to feature in descriptions of waves are harmonic functions, and they can be expressed as exponentials as well using Euler's formula, $\mathrm{e}^{\mathrm{i}x} = \cos(x) + \mathrm{i}\sin(x)$. Like the wave equation discussed in §1.5.2, Laplace's equation is known to be *separable*, so that its solution can be expressed as three functions of space and one function of time multiplied together, i.e.

$$\phi = A\mathcal{X}(x)\mathcal{Y}(y)\mathcal{Z}(z)\mathrm{e}^{\mathrm{i}\omega t}, \tag{2.24}$$

where, just as in (1.65) for the general wave equation, A is the *amplitude* and is a complex number.

Substituting (2.24) into the kinematic boundary condition at the bottom of the sea (2.17) gives

$$\frac{\partial \mathcal{Z}}{\partial z} = 0 \text{ at } z = -h, \tag{2.25}$$

while the sea-surface boundary condition (2.22) becomes

$$-\omega^2\mathcal{Z} + \mathrm{g}\frac{\partial \mathcal{Z}}{\partial z} = 0 \text{ at } z = \eta. \tag{2.26}$$

In general, $\mathcal{X}$ would be an exponential function of x and likewise $\mathcal{Y}$ and $\mathcal{Z}$ are exponential functions of y and z respectively. There are no boundary conditions in the horizontal directions x and y. However, there are still constraints on $\mathcal{X}$ and $\mathcal{Y}$ apparent from the physical reality: the velocity cannot increase to infinity with distance, nor can it decline to nothing with distance, since there is no viscosity or friction removing energy with distance. Thus the sine and cosine functions are good choices for $\mathcal{X}$ and $\mathcal{Y}$. By replacing $\mathcal{X}$ in (2.24) with either sine or cosine of positive or negative x and substituting into (2.16), it is clear that the sinusoidal functions are solutions of Laplace's equation; the equivalent applies to $\mathcal{Y}$. Thanks to Euler's formula, both sine and cosine are incorporated in a single exponential function of x where the exponent is imaginary. Thus $\mathcal{X} = \mathrm{e}^{\pm i\kappa x}$ where any real number κ will be valid and κ is the *wavenumber* in the x-direction. Both positive and negative signs are shown because either sign allows Laplace's equation to be satisfied. The different signs represent waves heading in opposite directions, since, in general, and just as in §1.5.3, wave propagation is possible in any direction.

2.3.5.2 Applying the boundary conditions

The vertical function, $\mathcal{Z}$, must be left as exponential with a real or complex exponent, not just a purely imaginary exponent, since all possible solutions must be permitted, not just sines or cosines, to have a chance of satisfying the boundary conditions. Thus,

$$\mathcal{Z} = a\mathrm{e}^{kz} + b\mathrm{e}^{-kz}, \tag{2.27}$$

where a, b and k are real constants that are arbitrary for now, though it will soon be seen that these constants have a physical meaning. To find these constants, the boundary conditions derived in §2.3.4 are now used. Substituting the vertical structure of the solution to Laplace's equation, (2.27), into the boundary conditions (2.25) and (2.26) converts them to

$$ak\mathrm{e}^{kz} - bk\mathrm{e}^{-kz} = 0 \text{ at } z = -h, \tag{2.28}$$

and

$$-\omega^2(a\mathrm{e}^{kz} + b\mathrm{e}^{-kz}) + \mathrm{g}(ak\mathrm{e}^{kz} - bk\mathrm{e}^{-kz}) = 0 \text{ at } z = \eta. \tag{2.29}$$

Inserting the values of $z = -h$ and $z = \eta$ into (2.28) and (2.29) respectively gives

$$ak\mathrm{e}^{-kh} - bk\mathrm{e}^{kh} = 0, \tag{2.30}$$

and

$$-\omega^2(a\mathrm{e}^{k\eta} + b\mathrm{e}^{-k\eta}) + \mathrm{g}(ak\mathrm{e}^{k\eta} - bk\mathrm{e}^{-k\eta}) = 0. \tag{2.31}$$

It is inconvenient that η, which had been eliminated in §2.3.4, has re-appeared in (2.31). Thus, an assumption will be made that $k\eta$ is very small, i.e. that

$k\eta \ll 1$. In §2.3.5.3, it will become clear that this is no more than the assumption of linearity already made, and thus is perfectly consistent. Then $e^{k\eta} \to 1$, reducing (2.31) to

$$-\omega^2(a+b) + gk(a-b) = 0. \tag{2.32}$$

There are two equations from the bottom and top boundary conditions, (2.30) and (2.32), and two unknowns, a and b; the following algebra eliminates the unknowns. Firstly, (2.30) becomes

$$b = a e^{-2kh}. \tag{2.33}$$

Substitution of (2.33) into (2.32) gives

$$\omega^2 = gk\left(\frac{1-e^{-2kh}}{1+e^{-2kh}}\right) = gk\left(\frac{e^{kh}-e^{-kh}}{e^{kh}+e^{-kh}}\right), \tag{2.34}$$

2.3.5.3 Dispersion relation

Now, $\frac{1}{2}(e^{kh} - e^{-kh})$ is the definition of the hyperbolic sine of kh, and similarly, the denominator of (2.34) is twice the hyperbolic cosine of kh, so (2.34) can be written

$$\boxed{\omega^2 = gk\tanh kh}. \tag{2.35}$$

This formula (2.35) was first published by George Airy in 1841 after several years during which various mathematicians and physicists came close, but did not quite get it right. We will now see why (2.35) is called a *dispersion relation.*

In (2.35), for a given water-body depth h, there is a direct relation between the frequency ω of the sinusoidal oscillations with time and a variable k. We will now show that k is the horizontal wavenumber. To understand this meaning of k, substitute the solution for the vertical structure (2.27) into the statement of separability (2.24) which then goes into Laplace's equation (2.16), noting from (2.27) that $\partial^2\mathcal{Z}/\partial^2 z = k^2\mathcal{Z}$, giving

$$\frac{\partial^2\mathcal{X}}{\partial x^2}\mathcal{Y}\mathcal{Z} + \mathcal{X}\frac{\partial^2\mathcal{Y}}{\partial y^2}\mathcal{Z} + \mathcal{X}\mathcal{Y}k^2\mathcal{Z} = 0. \tag{2.36}$$

It was already noted that purely sinusoidal functions were a perfect choice for $\mathcal{X}$ and $\mathcal{Y}$, so that

$$\begin{aligned} \mathcal{X} &= e^{\pm i\,\kappa x}, \\ \mathcal{Y} &= e^{\pm i\,\ell y}, \end{aligned} \tag{2.37}$$

where ℓ is the wavenumber in the y-direction. Unlike the exponent in the vertical function, the exponents in the horizontal functions $\mathcal{X}$ and $\mathcal{Y}$ are purely imaginary since they represent sinusoidal functions. Thus the second derivative of $\mathcal{X}$ is $-\kappa^2\mathcal{X}$ and similarly for $\mathcal{Y}$, so (2.36) becomes

$$-\kappa^2 - \ell^2 + k^2 = 0 \Rightarrow k = \pm\sqrt{(\kappa^2+\ell^2)}. \tag{2.38}$$

Thus, k is directly related to the wavenumbers κ and ℓ, so k can be called the horizontal wavenumber. If the waves are only propagating in one direction in the horizontal, say the x-direction, $\ell = 0$ and k is the same as κ. As discussed in §1.5.4, the wavenumber is effectively a 'frequency' in space. A relation between a frequency in time and a frequency in space, as in (2.35), is called a dispersion relation. As with any wave, the wavenumber is related to the *wavelength*, λ, which is the distance in metres from one crest to the next crest, measured at right angles to the crest, by (1.70),

$$k = \frac{2\pi}{\lambda}, \tag{2.39}$$

and similarly, the radian frequency with time is related to the frequency f in s^{-1}, usually reported in Hertz, by

$$\omega = 2\pi f. \tag{2.40}$$

Recall that in simplifying (2.31) to (2.32), it was assumed that $k\eta \ll 1$. It is now clear from (2.39) that this is the same as insisting $\eta/\lambda \ll 1$, i.e. that the height of the waves is very small compared with the wavelength and therefore that any temporary slope of the water surface must be very small. This is no more than the assumption of linear behaviour we made at the very start, so it is consistent.

It is worth noting that, just as the boundary conditions could have been derived without using a velocity potential, this dispersion relation could have been derived without using a potential. The three assumptions of incompressible, inviscid and linear behaviour would still have been made. An equation would have been found in terms of η or any velocity component, and it would have been the wave equation. However, we would not have the convenience of the solution expressed in a single variable, the potential ϕ, which completely describes all aspects of the motion, and, as noted earlier, permits fast and powerful computational techniques to be applied. We will soon see what this solution looks like.

2.3.5.4 Ripples

Ripples are waves with short enough wavelengths for surface tension to be significant, and, as mentioned in §2.3.2.1, they roughen the water surface, allowing the wind to transfer momentum to the water and create surface gravity waves (figure 8.1). Without tiny ripples, there would be no great surf.

At this point, the surface tension neglected earlier can be re-introduced without violating any of the preceding analyses. Provided the surface elevation, η, is sinusoidal, which was one of the four assumptions made in §2.3.2.1, the surface tension can be shown to cause an increase in the pressure in the water when the surface elevation is positive, and following the note on radii of curvature of the surface in §1.2.2.5, this increase in pressure is $\sigma k^2 \eta$. Noting that the value of σ is 0.0727 N m^{-1} at 22°C (Berry et al., 2015), consider waves of 1 m wavelength which therefore have a wavenumber k of 2π m^{-1}. Even if these waves have a surface elevation as large as 10% of their wavelength, i.e. $\eta = 0.1$ m(which, as we will see in §10.2.1, means the waves are likely to be breaking), the additional pressure under the wave crest would be 0.287 Pa, which is extremely small compared to the standard atmospheric pressure P_{atm} of 101 325 Pa. For ocean waves of 100 m wavelength with the same steepness as in the previous example, the additional pressure due to surface tension is a factor of 10 000 smaller still, since the surface-tension pressure is proportional to k^2. However, for waves of 0.01 m wavelength, the additional pressure under the wave crest due to surface tension rises to 2 870 Pa, a few percent of atmospheric pressure. For shorter wavelengths still, the surface tension will begin to be quite significant.

The formal inclusion of the surface-tension force in the derivations beginning in §2.3.2.1 would lead to the mathematics becoming rather more complicated, since the dynamic pressure would have to include a term that depends on η. Fortunately, returning to the point where the surface boundary condition was considered, just before (2.18), the dynamic pressure term, $-(1/\rho_0)\boldsymbol{\nabla}p$, now no longer negligible, becomes $-(1/\rho_0)\sigma k^2(\partial\eta/\partial s)$. Since, just as in §2.3.4, $\partial\eta/\partial s = \sin\alpha$, the effect of surface tension can be included at the point (2.18) was derived, simply by replacing the fixed acceleration due to gravity, g, with a new variable, say g_σ, such that

$$\boxed{g_\sigma = \mathrm{g} + \frac{\sigma k^2}{\rho_0}}. \tag{2.41}$$

Since (2.18) in the derivation marks the once and only instance where the all-important acceleration due to gravity enters the equations, it is possible to consistently replace g with g_σ everywhere from §2.3.4 up to here, and onwards.

An example of a dispersion relation is plotted in figure 2.4.

2.3.5.5 Phase speed

The dispersion relation (2.35) has an interesting consequence. As in all waves, the frequency and wavelength multiplied together give the wave speed, c, so, repeating (1.72),

$$c = f\lambda, \tag{2.42}$$

or equivalently, noting (2.39) and (2.40),

$$c = \omega/k. \tag{2.43}$$

There are many 'speeds' associated with wave motion, and for the precision of meaning, the wave speed is usually called the *phase speed* or sometimes the wave *celerity*; it is the speed with that wave crests travel. Thus, it is the speed of propagation of the waves; the speed with which a disturbance made in the surface of the water will travel, and thus the speed with which *information*

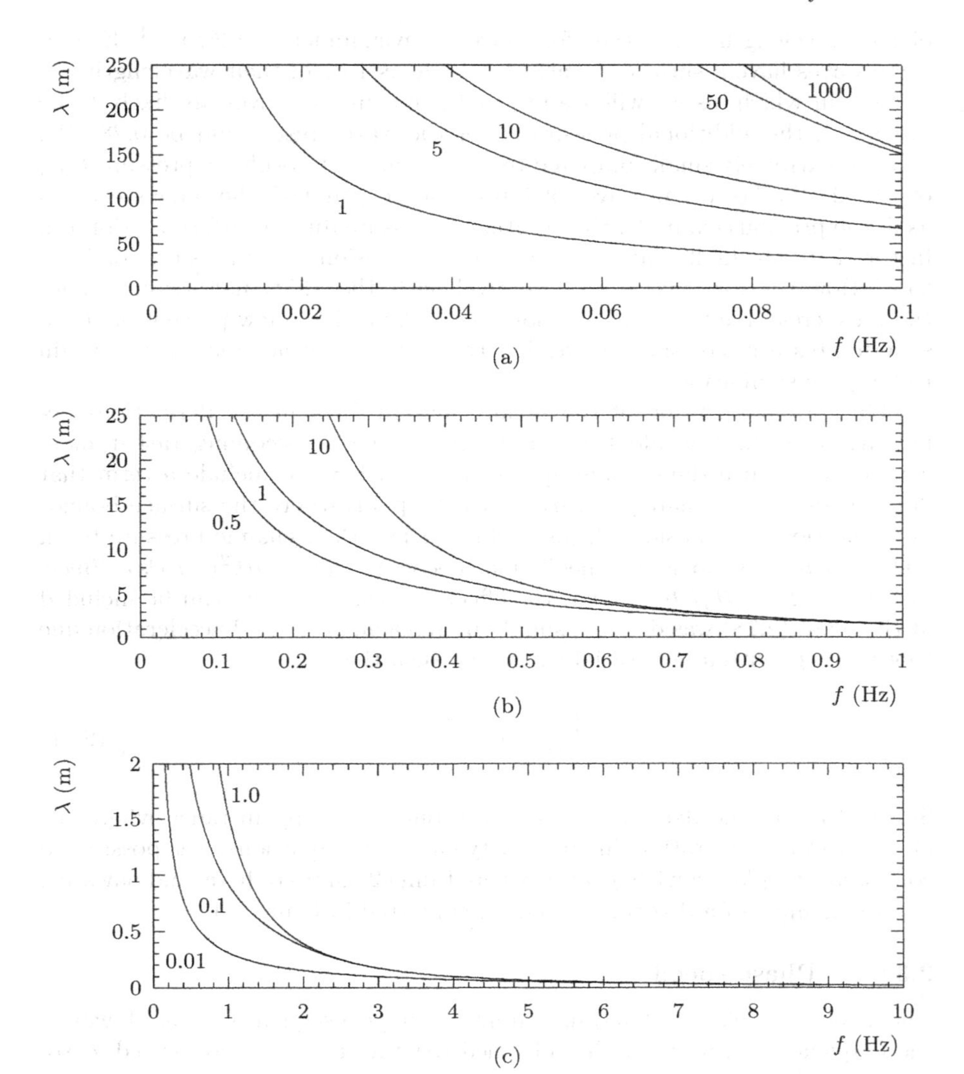

FIGURE 2.4
The dispersion relation (2.35) with g replaced by g_σ given by (2.41), so it is valid for waves ranging from ripples to ocean swell. (a) Long wavelengths appropriate to the ocean for water depths $h = 1, 5, 10, 50$ and 1000 m. Curves for $h = 50$ m and 1000 m are very close for frequencies more than ~ 0.1 Hz. (b) Medium wavelengths appropriate to small lakes, rivers, etc, for $h = 0.5, 1$ and 10 m. Curves for $h > 10$ m are very close to the $h = 10$ m curve. (c) Short wavelengths appropriate to a laboratory, for $h = 0.01, 0.1$ and 1 m. Surface tension is only significant for $\lambda < 0.1$ m.

travels - information about the disturbance that created the waves. Using (2.35), (2.39), (2.40) and (2.42) gives

$$\boxed{c = \sqrt{\left(\frac{g}{k}\tanh kh\right)}}, \tag{2.44}$$

from which it can be seen that the speed with which wave crests travel increases with wavelength λ. Waves that have a phase speed that varies with wavelength (and thus with frequency) are called *dispersive* waves. This phenomenon is not unique to water waves.

2.3.5.6 Velocity field

For many purposes, whether calculating wave impacts on an engineering structure or determining how melting ice interacts with waves, it is necessary to know the actual velocity with which the water is set into motion and the corresponding pressure. Thus, substituting (2.33) into (2.27) gives

$$\mathcal{Z} = a\mathrm{e}^{-kh}\left[\mathrm{e}^{(kz+kh)} + \mathrm{e}^{(-kz-kh)}\right], \tag{2.45}$$

which can be written using the definition of cosh noted earlier as

$$\mathcal{Z} = a\mathrm{e}^{-kh}\, 2\cosh(kh + kz). \tag{2.46}$$

Substitution of (2.46) back into the separable solution for ϕ, (2.24), gives

$$\phi = A\mathcal{X}(x)\mathcal{Y}(y)a\mathrm{e}^{-kh}\, 2\cosh(kh + kz)\mathrm{e}^{\mathrm{i}\,\omega t}. \tag{2.47}$$

It might seem that this is the required relation for ϕ enabling the velocity and pressure to be calculated if it were not for a subtle point. This becomes apparent on using (2.47) to write down the solution for the surface displacement, η. Using (2.19), plus the fact that the time dependence is sinusoidal, gives η as

$$\eta = -\frac{\mathrm{i}\omega}{g}\phi. \tag{2.48}$$

Now, using the newly-found solution for ϕ, (2.47), at the surface $z = 0$, the relation giving η becomes

$$\eta = -\mathrm{i}A\frac{\omega}{g}\mathcal{X}(x)\mathcal{Y}(y)a\mathrm{e}^{-kh}\, 2\cosh(kh)\mathrm{e}^{\mathrm{i}\,\omega t}, \tag{2.49}$$

recalling that the amplitude A introduced at the stage of (2.24) is a complex number. Now, the subtle point is that η is the surface displacement, and thus, unlike the velocity and pressure, η can only be some constant amplitude, A, multiplied by functions that vary in the horizontal (in x and y) and in time. However, (2.49) appears to depend on the depth, h, as well, not just because of the $\mathrm{e}^{-kh}\, 2\cosh(kh)$ factor, but also because ω is now known to vary with

h. This issue is resolved on realising the functions of h must be absorbed into the remaining unknown co-efficient a, so that a is given by

$$a = \frac{g}{\omega}\left[\frac{e^{kh}}{2\cosh(kh)}\right], \tag{2.50}$$

so that the vertical structure function (2.46) becomes

$$\mathcal{Z}(z) = \frac{g}{\omega}\frac{\cosh(kz + kh)}{\cosh(kh)}, \tag{2.51}$$

and now the solution (2.47) can take the form,

$$\phi = A\frac{g}{\omega}\mathcal{X}(x)\mathcal{Y}(y)\frac{\cosh(kh + kz)}{\cosh(kh)}e^{i\omega t}. \tag{2.52}$$

Replacing $\mathcal{X}$ and $\mathcal{Y}$ with their solutions (2.37), the final solution for the potential is

$$\boxed{\phi = A\frac{g}{\omega}\frac{\cosh(kh + kz)}{\cosh(kh)}e^{i(\pm\kappa x \pm \ell y + \omega t)}}. \tag{2.53}$$

Using this final relation (2.53) and (2.14), $\boldsymbol{u} = \boldsymbol{\nabla}\phi$, all three components of the velocity, at any horizontal location and any depth of interest in the water body, be it a puddle or the deep ocean, are immediately determined, and the pressure is likewise determined using (2.13), $p = -i\rho_0\omega\phi$. Simplified, real-number expressions for the water velocity and pressure may be preferred to the general expression (2.53), in particular, a formula that can be related to the easily-observed surface-elevation amplitude may be desirable. This will be given in §2.3.7, but first, the surface elevation must be clearly related to the amplitude in §2.3.6 below.

2.3.6 Surface elevation

The surface elevation η was last given by (2.49), and is the feature of surface water waves most obvious to the observer, so it is useful to have an explicit formula for η. Now that the unknown constant a has been determined, (2.49) can be written as

$$\boxed{\eta = -iAe^{i(\pm\kappa x \pm \ell y + \omega t)}}. \tag{2.54}$$

This accords with observations that the surface elevation varies sinusoidally with horizontal space and with time.

Imagine the direction in which the waves are propagating is defined to be the x-direction. (If the waves are not interacting with any fixed obstacle such as a coast, we are free define x to be the same as the propagation direction; this just makes the mathematics that follows simpler.) Thus, as noted when discussing (2.38), $\kappa = k$ and $\ell = 0$. Then (2.54) becomes

$$\eta = -iAe^{i(\pm kx + \omega t)}, \tag{2.55}$$

It is easy to illustrate some key properties of waves by drawing some graphs, and in order to draw those, it will be necessary to move from the mathematical convenience of complex numbers to the reality of real numbers. Writing $-\mathrm{i}A = a_1 + \mathrm{i}a_2$ where a_1 and a_2 are real numbers, and expanding the complex exponential gives

$$\begin{aligned}\eta &= a_1 \cos(\pm kx + \omega t) - a_2 \sin(\pm kx + \omega t) \\ &\quad + \mathrm{i}a_2 \cos(\pm kx + \omega t) + \mathrm{i}a_1 \sin(\pm kx + \omega t).\end{aligned} \tag{2.56}$$

Now it is time to enforce the fact that the real world is, well, real and not imaginary or complex. Thus, the water-surface elevation we would like to draw a picture of is given by the real part of (2.56). It is also necessary to think about the signs of the kx terms, since both possibilities are present. As mentioned in §2.3.5, the two sign possibilities, which arose because the square root in (2.38), simply represent the possibility that the waves could propagate in either the positive or the negative direction. Since the following discussion would be exactly the same irrespective of which way the waves propagate, choose the negative sign; this gives the most common representation, in which waves are travelling in the positive x-direction. Then (2.56) becomes

$$\eta_R = a_1 \cos(kx - \omega t) + a_2 \sin(kx - \omega t), \tag{2.57}$$

in which the symbol η_R is used to emphasise that the real part of (2.56) has been taken, i.e. $\eta_R = \Re(\eta)$. (In many texts, the same symbol would be used for the complex variable as well as for its real part, and taking the real part when the solution is finally compared with reality is expected to be 'understood' by the reader.) The choice of signs that leads to (2.57) can be simply understood by noting that increases in t in (2.57) are equivalent to decreases in x; if x points from left to right, at a later time, one sees a part of the wave that was farther to the left at an earlier time.

In order to plot the surface elevation, first, consider the initial conditions of the water surface. If η_R is zero at $t = 0$ and at $x = \lambda/4$, a_2 must be zero, so that $\eta_R = a_1 \cos(kx - \omega t)$. Conversely, if η_R is zero at $t = 0$ and at $x = 0$, a_1 would have to be zero. Any other initial conditions in between could exist and would be represented by both a_1 and a_2 being non-zero. This general situation is most conveniently represented by switching from a_1 and a_2 to another two variables, $\hat{\eta}$ and Φ, such that

$$a_1 = \hat{\eta} \cos \Phi \quad \text{and} \quad a_2 = \hat{\eta} \sin \Phi, \tag{2.58}$$

where $\hat{\eta}$ is the amplitude of the surface elevation and Φ (not to be confused with the symbol ϕ used for the potential) is the *phase* of the wave, and recalling $\sin^2 \alpha + \cos^2 \alpha = 1$, it is clear that

$$|A| = \hat{\eta}. \tag{2.59}$$

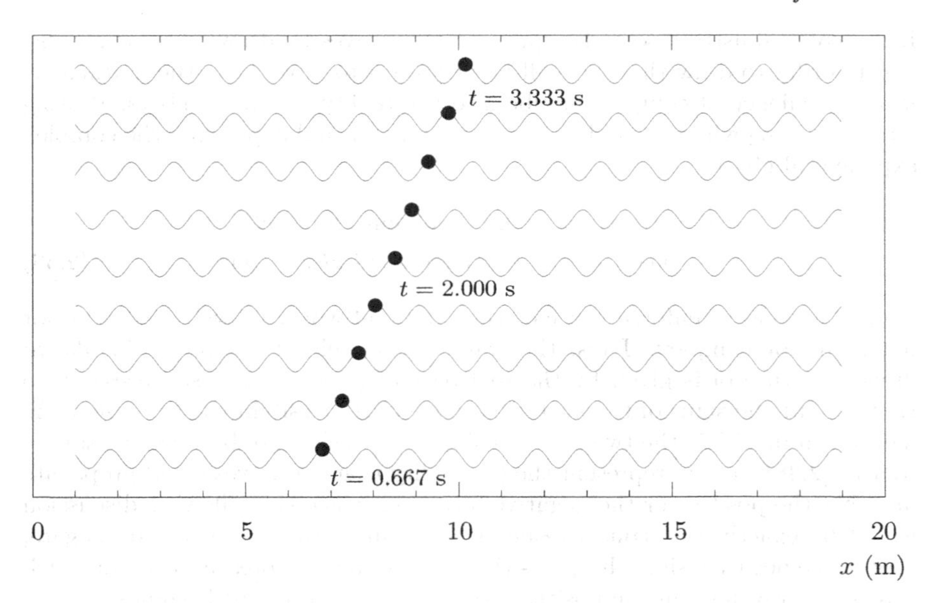

FIGURE 2.5
Surface elevation of a wave in water 4000 m deep; in this example, λ has been set at 1 m, so $k = 2\pi$ m^{-1}; therefore, equation (2.35) gives $\omega = 7.84$ rad s^{-1}, so $f = 1.25$ s^{-1}. Curves are for 9 different times about 0.33 s apart. The wave amplitude has been greatly exaggerated, so that it is visible on the plot; recall that for linear theory to be valid, the amplitude must be very small relative to the wavelength. Dots mark the location of one of the crests; the crest travels about 3.3 m in about 2.67 s, giving a phase speed of about 1.25 m s^{-1} as expected from equation (2.42) The dots show the characteristic line of the wave equation.

The amplitude is half of the *wave height* and the phase simply indicates the part of the wave's cycle that was occurring at $t = 0$ and $x = 0$. With these general initial conditions, and noting the standard formula for the cosine of a sum, $\cos(\alpha \pm \beta) = \cos\alpha\cos\beta \mp \sin\alpha\sin\beta$, the form of (2.57) is

$$\eta_R = \hat{\eta}\cos(kx - \omega t - \Phi), \tag{2.60}$$

which for zero phase Φ is

$$\boxed{\eta_R = \hat{\eta}\cos(kx - \omega t)}, \tag{2.61}$$

and an example is shown in figure 2.5.

Noting from the definition of phase speed given by (2.43), $c = \omega/k$, an alternative 'wavenumber-speed form' expression for (2.60), as in d'Alembert's solution, (1.63), is

$$\eta_R = \hat{\eta}\cos\left[k(x - ct) - \Phi\right], \tag{2.62}$$

which emphasises that for a travelling wave, time and the space through which it travels are interchangeable; this applies to any travelling wave, not just water waves.

Exactly the same manipulation in this section 2.3.6 could be applied to all the other variables in Airy's solution: the three components of velocity given by the potential, as well as the pressure, since they all share the same horizontal-space and time dependence.

2.3.7 Particle trajectories

If we float in the ocean, the passing waves cause us to 'bob' up and down and 'surge' to and fro. It is interesting to plot what this motion actually looks like, not just at the surface, but at all depths. Such plots are also the starting point for a study of how particles, including living organisms, are transported by waves, which will be further detailed in §6.4. Firstly, as in §2.3.6, define the direction in which the waves are propagating to be the x-direction, so $\kappa = k$ and $\ell = 0$. Now, using (2.53) and (2.14), $\boldsymbol{u} = \boldsymbol{\nabla}\phi$, the velocity components in the x- and z-directions, u and w respectively, are given by

$$\begin{aligned} u &= -\mathrm{i}kA\frac{\mathrm{g}}{\omega}\frac{\cosh(kh+kz)}{\cosh(kh)}\mathrm{e}^{\mathrm{i}(-kx+\omega t)} \\ \text{and} \qquad w &= kA\frac{\mathrm{g}}{\omega}\frac{\sinh(kh+kz)}{\cosh(kh)}\mathrm{e}^{\mathrm{i}(-kx+\omega t)}, \end{aligned} \tag{2.63}$$

where, just as for (2.57), the negative sign was chosen for the $\pm kx$ term to illustrate waves travelling in the positive x-direction.

As an aside, it was mentioned at the end of §2.3.5.6 that it may be convenient to see a simplified, real-number expression for the velocity field and pressure, rather than the general expression for the velocity potential. For a phase Φ of zero, and recalling (2.59), $|A| = \hat{\eta}$, the real part of the velocity vector, with horizontal and vertical components u_R and w_R respectively - the water velocity one would actually experience - can be extracted from (2.63), and the real part of the pressure, p_R, can likewise be found by applying (2.13) to (2.53) giving all together

$$\begin{aligned} u_R &= \hat{\eta}k\frac{\mathrm{g}}{\omega}\frac{\cosh(kh+kz)}{\cosh(kh)}\cos(kx-\omega t), \\ w_R &= \hat{\eta}k\frac{\mathrm{g}}{\omega}\frac{\sinh(kh+kz)}{\cosh(kh)}\sin(kx-\omega t), \\ p_R &= \hat{\eta}\rho_0\mathrm{g}\frac{\cosh(kh+kz)}{\cosh(kh)}\cos(kx-\omega t). \end{aligned} \tag{2.64}$$

Next, imagining $\tilde{x}$ and $\tilde{z}$ are the x and z co-ordinates of a particle that moves exactly with the water around it,

$$u = \frac{\mathrm{d}\tilde{x}}{\mathrm{d}t} \quad \text{and} \quad w = \frac{\mathrm{d}\tilde{z}}{\mathrm{d}t}. \tag{2.65}$$

Now, substituting (2.65) into (2.63) and integrating with respect to time gives

$$\tilde{x} = -kA\frac{g}{\omega^2}\frac{\cosh(kh+kz)}{\cosh(kh)}e^{i(-kx+\omega t)}$$

$$\text{and} \qquad \tilde{z} = \frac{k}{i}A\frac{g}{\omega^2}\frac{\sinh(kh+kz)}{\cosh(kh)}e^{i(-kx+\omega t)}. \tag{2.66}$$

Take the real parts of $\tilde{x}$ and $\tilde{z}$, and recalling $\sin^2\alpha + \cos^2\alpha = 1$ and that $|A| = \hat{\eta}$, squaring these real parts and adding them gives a relation between the real parts of $\tilde{x}$ and $\tilde{z}$,

$$\boxed{\left(\frac{\tilde{x}_R}{\hat{x}}\right)^2 + \left(\frac{\tilde{z}_R}{\hat{z}}\right)^2 = 1,} \tag{2.67}$$

where as before the subscript R denotes the real part, and

$$\hat{x} = \frac{gk}{\omega^2}\frac{\cosh(kh+kz)}{\cosh(kh)}\hat{\eta}$$

$$\text{and} \qquad \hat{z} = \frac{gk}{\omega^2}\frac{\sinh(kh+kz)}{\cosh(kh)}\hat{\eta}. \tag{2.68}$$

The relation (2.67) is the equation of an ellipse and (2.68) gives the lengths of the semi-major axis $\hat{x}$ and semi-minor axis $\hat{z}$. Particles in the water trace out elliptical paths as waves go past (figure 2.6). Thus, the water (and anything floating in it) bobs up and down and surges to and fro, and returns to its starting point with the passage of each wave. It is clear from figure 2.6 that water-surface waves are not purely transverse waves, since motion occurs in the same direction as the wave propagates as well as at right angles to it. Thus water-surface waves are a combination of transverse and longitudinal waves.

Since the sinh function is zero when its argument is zero, it is clear that at the bottom of the sea, $z = -h$, the minor-axis length becomes zero, so the particles move horizontally only. This is, of course, because of the boundary condition (2.17) that ensures no vertical flow occurs through the bottom boundary. Meanwhile, the absence of friction owing to the inviscid assumption in §2.3.2.1 permits water to move freely parallel to the bottom boundary. Meanwhile, at the surface, $z = 0$, the trajectories look close to circular, but are still 'flattened'.

2.3.8 Group velocity

The dispersive nature of water-surface waves has an interesting consequence. This will turn out to be essential to our understanding of how energy is transported by ocean waves. This is renewable energy, which we might want to use (chapter 8), or possibly destructive energy, which we might want to avoid (§12.2), or energy affecting our calculations of climate change (§10.3.1), or of ship design (§2.3.10.2 below).

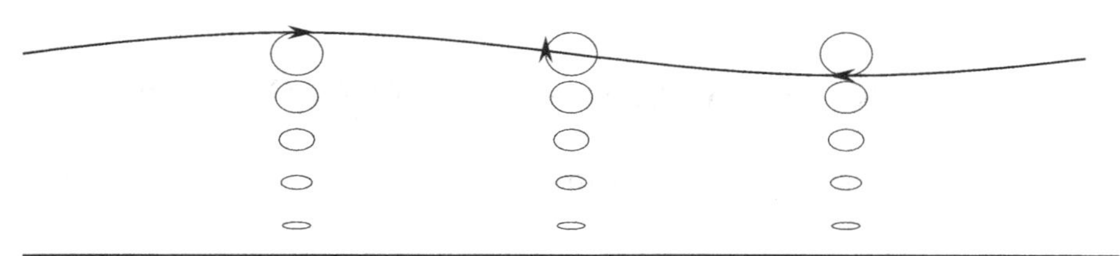

FIGURE 2.6
Trajectories of particles suspended in the water as waves go past (travelling from left to right). In this example, waves with wavelength $\lambda = 50$ m and surface-elevation amplitude $\hat{\eta} = 1$ m (a wave height of 2 m) travel across a sea of depth $h = 10$ m. At the surface, $z = 0$, elliptical trajectories with major- and minor-axis lengths given by (2.68) are close to circular, but towards the bottom, the motion becomes horizontal only. Arrows show the direction of water motion at the surface.

Imagine there are two waves that are both travelling in the x-direction, but have different frequencies ω_1 and ω_2 and therefore different wavelengths k_1 and k_2. Since they are both solutions to a linear differential equation (Laplace's equation) they can be linearly superposed, just as the waves travelling in different directions in figure 2.1 superpose. Using (2.60), the water-surface elevation at any point in space and time would be given by

$$\eta_R = \hat{\eta}\cos(k_1 x - \omega_1 t) + \hat{\eta}\cos(k_2 x - \omega_2 t), \tag{2.69}$$

Both waves have been given the same amplitude $\hat{\eta}$ and the same phase of $\Phi = 0$ to keep the mathematics uncluttered. Giving them different amplitudes would not alter the following analysis; the analysis would simply be applied to all of the smaller-height wave and a matching height of the larger-height wave, while the excess height of the larger wave would simply be linearly superposed. Also, giving the waves different initial conditions and hence different phases would simply shift their waveforms relative to each other, but would not alter following analysis. Exactly the same addition could be done for all the other variables in Airy's solution.

Using the standard formulas for the cosine of a sum, in particular in the form $\cos(2\alpha) = 2\cos^2\alpha - 1 = 1 - 2\sin^2\alpha$, (2.69) can be re-written, following a few lines of algebra, as

$$\eta_R = 2\hat{\eta}\cos(k'x - \omega' t)\cos(\bar{k}x - \bar{\omega}t), \tag{2.70}$$

where $\bar{k} = \frac{1}{2}(k_1 + k_2)$ and $\bar{\omega} = \frac{1}{2}(\omega_1 + \omega_2)$ are clearly the average wavenumber and frequency of the two waves, while $k' = \frac{1}{2}(k_1 - k_2)$ and $\omega' = \frac{1}{2}(\omega_1 - \omega_2)$ are half of the difference between them. When the two waves are close in frequency (and hence wavenumber), the result, shown in figure 2.7, is the phenomenon of *beats*. A base frequency $\bar{\omega}$ that is the average of the two waves is modulated by a beat frequency ω' that is much lower, giving an 'envelope'

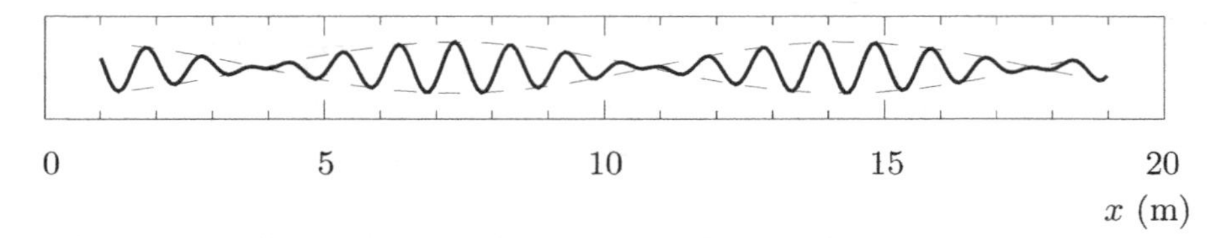

FIGURE 2.7
Surface elevation of two linearly superposed waves in water 4000 m deep, given by (2.70). In this example, the average of the two waves has the same wavenumber and frequency as the single wave in figure 2.5; whereas the single wave in figure 2.5 had a wavelength of exactly 1 m, here one wave has a wavelength of about 0.93 m and the other about 1.07 m. As in figure 2.5, the wave amplitude has been greatly exaggerated. The envelope function connecting each crest and each trough is shown by the dashed line.

function joining each crest and joining each trough. Since figure 2.7 is plotted as a function of x at a single instant of time t, it shows a base wavenumber $\bar{k}$ modulated by a beat wavenumber k' forming the envelope function; plotting as a function of t at a single point x would show the same sort of pattern. Although the present section, §2.3.8, is in the context of travelling water waves, the algebra giving (2.70) is of course perfectly general and applies to all waves. For example, sound waves are often generated with slightly different frequencies, either intentionally in music or unintentionally in equipment like ventilation fans, resulting in a 'wowing' noise: a sound with a base frequency modulated by a much lower beating sound.

Most significantly, it can be seen from figure 2.8 that the modulation, representing the envelope of the base waveform, travels at a different speed to the wave crests that travel at phase speed c. The speed with which the envelope travels is given by $c_g = \omega'/k'$ and in figure 2.8 this speed is lower than c. The peak of the envelope is where the *energy* of the motion created by the combination of the two waves is a maximum, and therefore the speed c_g with which the envelope travels is the *speed with which the energy is travelling*. Meanwhile, phase speed c is the speed with which information is travelling - information about the initial conditions that created the waves. (§8.2.1).

If the frequencies (and hence wavenumbers) of the two waves are infinitesimally close, the two frequencies could be written as $\omega_1 = \omega - \Delta\omega$ and $\omega_2 = \omega + \Delta\omega$ and the corresponding wavenumbers as $k_1 = k - \Delta k$ and $k_2 = k + \Delta k$, where $\Delta\omega$ and Δk are infinitesimally small. Substituting these new definitions of the two waves into (2.70) gives

$$\eta_R = 2\hat{\eta}\cos(\Delta kx - \Delta\omega t)\cos(kx - \omega t), \tag{2.71}$$

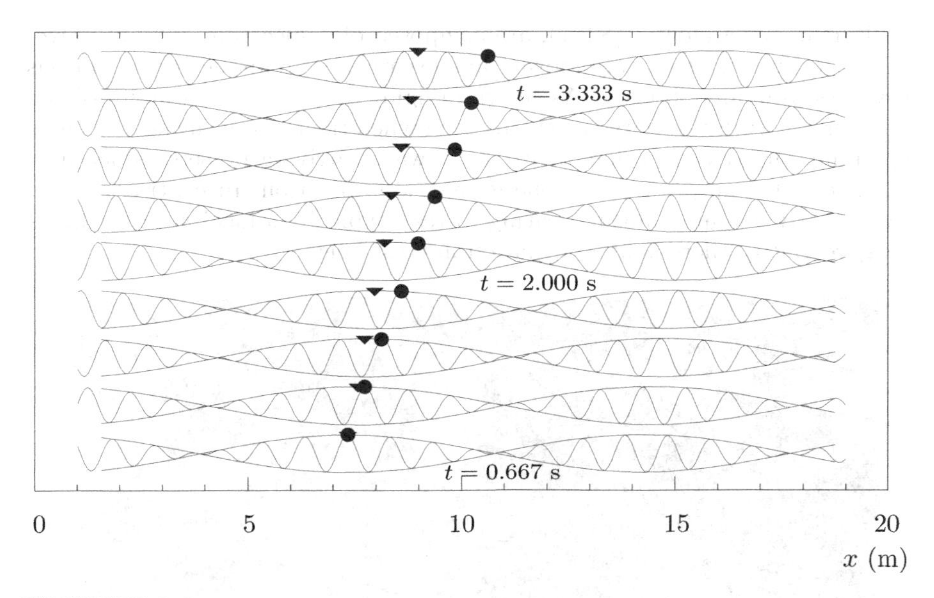

FIGURE 2.8
Surface elevation of two linearly superposed waves in water 4000 m deep, given by (2.70). The two waves are those shown in figure 2.7. As in figure 2.5, curves are for 9 different times about 0.33 s apart, with the first curve at $t = 0.667$ s and the last at $t = 3.333$ s. The envelope function connecting each peak and each trough is shown by a solid line. Similarly to figure 2.5, dots mark the location of one of the crests; the crest travels at the phase speed of about 1.24 m s^{-1}. Triangles mark the peak of the envelope function, showing the energy propagates more slowly than the crests, in this case at about 0.64 m s^{-1}.

and switching to the wavenumber-speed form, (2.71) becomes

$$\eta_R = 2\hat{\eta}\cos\{\Delta k[x - (\Delta\omega/\Delta k)t]\}\cos[k(x - ct)]. \tag{2.72}$$

Now, on taking the limit as $\Delta\omega$ and Δk tend to zero and using the usual definitions of a derivative, (2.72) then applies not just to two distinct waves but to any group of waves, and indeed to any continuously-varying distribution of wave frequencies, and the speed with which the energy of this group of waves propagates, c_g, is given by

$$\boxed{c_g = \frac{\mathrm{d}\omega}{\mathrm{d}k}.} \tag{2.73}$$

The dispersive nature of water-surface waves creates phenomena often noticed in nature. Ocean waves, particularly in the energetic *mid-latitudes* (the regions between 30° and 60° in either northern or southern hemispheres of the Earth, as detailed in §8.2.1), are composed of a spectrum of many wavelengths owing to the many storms and weather systems active at various distances from any observer. As shown in figure 2.9(a), wave 'groupiness' can be observed, with very large crests arriving when the peak of the energy envelope function arrives. Surfers report large waves arriving in 'sets' which may correspond to the peak of the energy envelope function. In contrast, where there is only one mechanism creating waves - typically a local mechanism, as in figure 2.9(b), there is little variation in crest height.

(a) (b)

FIGURE 2.9
(a) Southern Ocean waves approaching the coast at Wongarra, Victoria, Australia. From this vantage point, the horizon (approximately at the top of the image) is about 46 km away while the edge of the land is about 1 km away. The large crests, appearing as the darkest ridges in the image, are in the order of kilometres apart, whereas the wavelengths typical of this region are about 250 m; hence the large crests are likely to be due to wave groups. (b) A gentle breeze blowing over the Daintree River, Queensland, Australia, creates short waves with approximately a single wavelength of roughly 0.1 m; in this case, there is no significant dispersion and thus no significant grouping of waves. Photographs by Richard Manasseh.

The same argument applies to waves travelling in different directions, for example, with some angle between them as in figure 2.1. Once the x-direction is defined, a group speed could be calculated based on the components of the waves travelling in the x-direction. Now, however, the components in the y-direction would also need to be calculated, and the result would be a *group velocity*: a two-dimensional vector representing the direction in which energy travels over the water surface as well as the speed of energy propagation. Since velocity is a vector, the group-velocity vector is calculated by recognising that the wavenumber is a vector too, so the equivalent of (2.73) has exactly the same form, but with k and c_g replaced by their vector equivalents, $\boldsymbol{k}$ and $\boldsymbol{c}_g$.

2.3.9 Deep-water approximation

The dispersion relation (2.35) is comprehensive. Subject only to the four assumptions listed in §2.3.2 (and, as noted in §2.3.5.4, the replacement of g with g_σ for wavelengths less than about a tenth of a metre), it predicts the frequency and speed of waves in water of any depth. However, the tanh function is inconvenient; in particular, if the frequency ω is known and the wavenumber k is unknown, finding the root of the resulting transcendental equation requires iteration. Traditionally, this computational issue was circumvented by two approximations, one for deep and one for shallow water; however, in the present day, it is easy to solve (2.35) iteratively. Nevertheless, very interesting consequences that help our understanding of the physics, and thus of engineering and scientific applications, emerge when approximations to (2.35) are made.

Thus, consider what happens if the sea is very deep compared to the wavelength, so that kh is very large. When the argument of the tanh function is very large, it tends to unity (as can be seen from (2.34) with $kh \to \infty$), so (2.35) reduces to

$$\boxed{\omega^2 = \mathrm{g}k}, \tag{2.74}$$

and the phase speed relation (2.44) likewise reduces to

$$\boxed{c = \sqrt{\frac{\mathrm{g}}{k}}}. \tag{2.75}$$

These *deep-water* waves are a fair approximation if $h > \lambda/2$, and therefore $kh > \pi$; so the water need not be very deep relative to the wavelength for this convenient approximation to hold. Deep-water waves are still dispersive: c is still a function of k. However, the dependence on the depth h has disappeared, because the waves no longer 'feel' the bottom. Indeed, one could derive (2.74) by realising the constant b in (2.27) must be zero, because as the water becomes very deep, so that as $z \to -\infty$, the velocity and pressure due to the wave motion must disappear, not become infinite. This represents reality very well. Imagine a storm at the surface creates waves a hundred metres in wavelength (so $k = 2\pi/100\ \mathrm{m}^{-1} \simeq 0.06\ \mathrm{m}^{-1}$). The bottom of the ocean may be four kilometres down, and therefore $kh \simeq 250$ which is certainly much

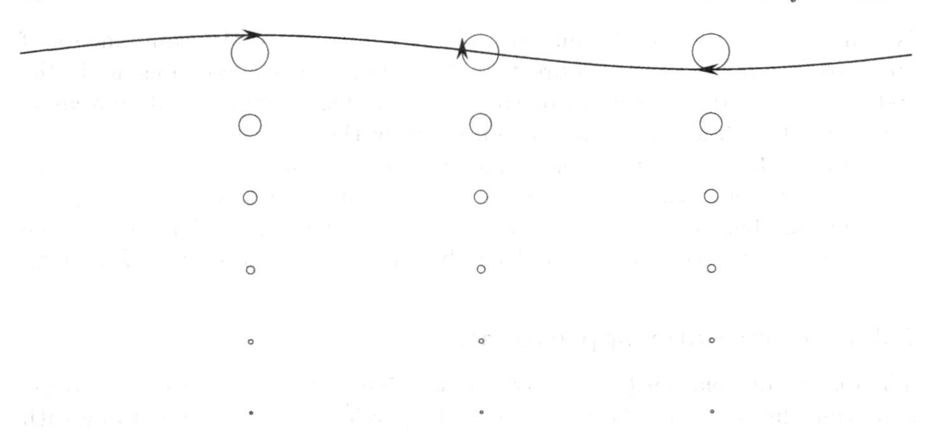

FIGURE 2.10
Trajectories of particles suspended in water as deep-water waves go past (travelling from left to right). In this example, waves with wavelength $\lambda = 50$ m and surface-elevation amplitude $\hat{\eta} = 1$ m (a wave height of 2 m) travel across a sea of depth $h = 4000$ m, of which the top 20 m is shown. Arrows show the direction of water motion at the surface.

larger than π and thus large enough for the tanh function to be considered unity. Thus these waves do not 'feel' the bottom. For $kh \to \infty$, the potential function, (2.53), becomes

$$\boxed{\phi = A\frac{g}{\omega}e^{kz}e^{i(\pm\kappa x \pm \ell y + \omega t)}}, \tag{2.76}$$

showing that the velocity and pressure fluctuations due to the surface waves decrease exponentially with depth (recall that z is negative as one goes down). Considering the particle trajectories, (2.68) reduces to

$$\hat{x} = \hat{z} = \hat{\eta}\frac{gk}{\omega^2}\,e^{kz}, \tag{2.77}$$

Thus, the ellipse axes are equal: the particle-trajectory ellipse has become a circle. As a deep-water wave passes, the water moves in a circular orbit in a vertical plane. The circles decrease exponentially with depth (figure 2.10). If a bridge or ocean platform were built in water where the waves tend to be deep-water waves, its piers would experience much higher horizontal forces near the surface, and virtually zero at the bottom, creating bending loads. Wave-energy converters built in deep-water wave regimes must have their energy-extraction mechanism near the surface.

By applying (2.73) to (2.74), the group velocity of deep-water waves turns out to be

$$\boxed{c_g = \frac{1}{2}\sqrt{\frac{g}{k}}}, \tag{2.78}$$

and comparing (2.78) to (2.75) shows that $c_g = \frac{1}{2}c$; the energy of a group of deep-water waves travels at exactly half the speed of the information. Thus, in theory, we receive 'news' that some dispersive waves have been created before we are hit with their maximum energy. In fact, even though figure 2.8, which was for very deep water, showed a combination of only two waves, the approximation was still good enough to show that the energy propagated at about half the speed of the information.

2.3.10 Consequences of deep water

2.3.10.1 Maximum wavelength of ocean swell

The overwhelming majority of waves on the surface of the water are created by winds. As mentioned in §2.3.2.1 and §2.3.5.4, and as will be detailed in §8.2.1, winds initially create ripples, increasing the roughness of the surface and thus permitting more momentum to be transferred from the wind to the water. This creates short-wavelength, 'choppy' waves. As they travel, these short waves transfer their energy via nonlinear processes to increasingly longer waves. There is a brief outline of these nonlinear processes at the end of §7.4, and photographs of the initial and advanced stages of wave generation are in figure 8.1. As waves travel, the nonlinear processes make them longer and longer, unless, of course, they reach a shore. Once they reach a shore, the wavelength is said to be limited by the *fetch* or maximum size of the water body in the wind direction.

The longer the waves in deep water, the faster they travel, since from (2.75), $c = \sqrt{g\lambda/(2\pi)}$. When the waves become so long that they are travelling as fast as the wind, the transfer of energy from the wind to the waves may be limited. This allows some estimates to be made of the maximum wavelength of waves, which will be further discussed in §8.2.1, and together with some knowledge of *wave breaking* over the open ocean, which will be discussed in §10.2.1, the energy of ocean waves can begin to be estimated. This has relevance to the extraction of renewable energy from ocean waves, to the impact of waves on engineering systems in the ocean, to the impact of waves coastal ecosystems and infrastructure, and to the role of waves in the breakup of ice in the Arctic and Antarctic.

2.3.10.2 V-shaped wakes in deep water

The fact that in deep water, $c_g = \frac{1}{2}c$, has an interesting consequence that was first elucidated by Kelvin. It had always been noticed that an object travelling along the surface of deep water creates a V-shaped wake with the same half-angle of about 19.5°, irrespective of how large or what shape the object is. The wakes of vessels are important not only to the design of ships but to the design of harbours and waterways impacted by vessel wakes. Figure 2.11 shows examples. It is remarkable that the angle of what is now known as the

FIGURE 2.11
Objects create a V-shaped wake as long as the disturbance wavelength satisfies $\lambda < 2h$, irrespective of the size of the object. Photographs by Richard Manasseh.

Kelvin wake pattern can be derived solely from the fact that $c_g = \frac{1}{2}c$, using only trigonometry.

Figure 2.12 shows a graphical derivation. Firstly, imagine the object is moving with speed U along the surface. It will create a spectrum of gravity waves, from the shortest ripples one could measure to waves even longer than the object's size. That is because waves will arise in order to match the boundary conditions of the shape of the object. As mentioned at the end of §1.5.6, a spectrum can be in space instead of time. In general, a spectrum of waves matching an arbitrary shape contains an infinite number of wavelengths. Consider only one of these wavelengths, which has a phase speed c such that $c < U$. As the object moves along, it continually makes waves of this wavelength that propagate radially outwards from each point along the object's track where they were created. The line connecting all the farthermost crests of these waves has an angle of $\tan^{-1}(c/U)$. This angle is an analogue of the *Mach angle* when an object moves faster than the speed of sound. However, because a continuous spectrum of deep-water waves of many wavelengths is being produced, there are many values of c, ranging from $c \simeq 0$ to $c = U$, and therefore wake angles due to these phase crests range from 0° to 90° without any particular angle standing out. (Values of $c > U$ simply represent waves sent ahead of the object, which form no angle.)

Consider an object that had travelled over some time T at speed U from point 'A' to point 'B' (figure 2.12(a)). At time T, the locus of all possible points of tangency of lines from the object back to the circles of radius cT forms a semi-circle of radius $UT/2$, centred a distance $UT/2$ behind the object. That is merely a graphical way of representing the fact that the wake angles formed by the phase crests could take any value between 0° and 90°.

Secondly, recall that it is the group velocity and not the phase velocity that defines where there is energy propagating through the water, and thus where waves are actually visible. The fastest-propagating wave that could create a wake, which has $c = U$, therefore has $c_g = U/2$. Just as with the phase crests, at some time T, the locus of all possible points of tangency of lines from the

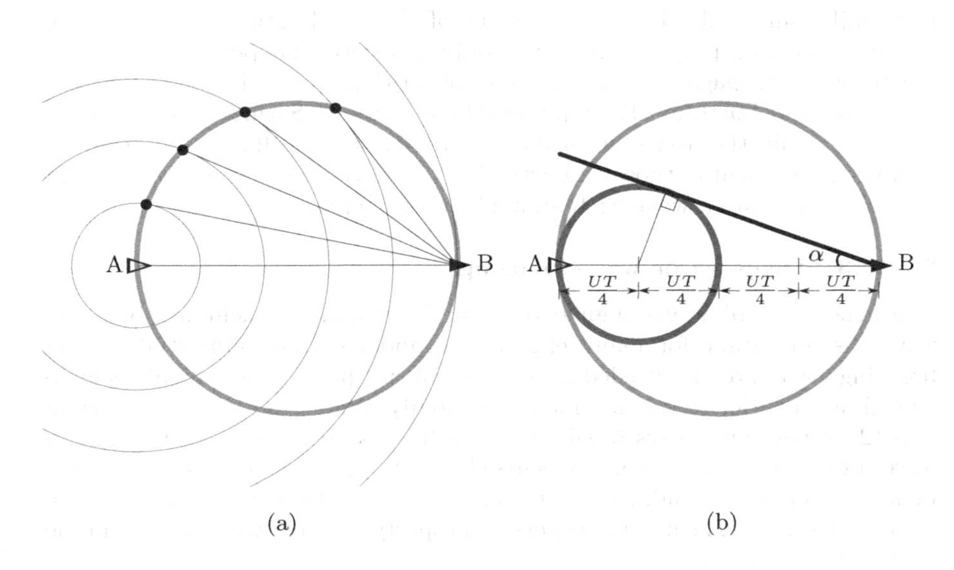

FIGURE 2.12
Graphical derivation of the V-shaped wake half-angle. (a) Tangent lines from an object now at point 'B' to wave crests emitted from the object when it was at 'A', at various phase speeds $c \leq U$. Distance from 'A' to 'B' is UT. Black dots mark tangent points and the locus of all tangent points forms a light-grey circle of radius $UT/2$. (b) Dark-grey circle of radius $UT/4$ is the equivalent locus of tangent points, but for the group speed $c_g \leq U/2$; line tangent to it is at angle α that encloses all observed wake energy.

object back to the circles of radius c_gT forms a semi-circle. Now, however, the semi-circle of group-crest tangency points is half the size of the phase-crest semi-circle; it has a radius $UT/4$, and is centred a distance $3UT/4$ behind the object.

It is now clear that when visible wakes are considered, and thus where the energy is significant, there is a maximum wake angle much less than 90°. The line drawn from the object back to the semi-circle of all possible group-velocity tangency points forms one side of a right-angled triangle with a hypotenuse of length $3UT/4$ and shortest side of length $UT/4$. The angle α is thus

$$\alpha = \sin^{-1}\left(\frac{UT/4}{3UT/4}\right) = \sin^{-1}(1/3) \simeq 19.4712°, \tag{2.79}$$

so that the two arms of the 'V' form an angle of $2\sin^{-1}(1/3) \simeq 38.9424°$ between them.

The full details of the Kelvin wake pattern, which includes the pattern formed within the 'V', require more detailed analysis, but the angle $\sin^{-1}(1/3)$

is literally universal - it is independent of the acceleration due to gravity or any other fluid property, requiring only a surface that permits horizontal gravity-wave propagation. At the time of writing, the only liquid surfaces known beyond Earth are the liquid-methane lakes on Saturn's moon, Titan, where not only the acceleration due to gravity but the fluid properties are completely different to those on Earth. A vessel navigating Titan's lakes would nevertheless create a wake with an angle of $\sin^{-1}(1/3)$.

2.3.10.3 Deep-water wave focusing

The generation of waves high enough to be nonlinear, including breaking waves, is important for many engineering and environmental studies, and breaking waves are also needed in commercial surf pools, where breakers large enough for a surfer to ride are made repeatedly. While, as discussed below in §2.3.12, waves will always break on a beach, if studies are to be relevant to waves breaking far from shore, (discussed in §10.2) it is necessary to have them break in deep water, and, ideally, break at a predictable location far from the wave-maker mechanism. The dispersive property of deep-water waves can be exploited for this goal.

Since, from (2.75), $c = \sqrt{(g/k)} = \sqrt{g\lambda/(2\pi)}$, longer waves travel faster. Therefore, it is possible to begin by generating a set of short-wavelength waves for a few periods of those short waves, by driving the wave-maker with a high frequency defined by (2.74). Then, a set of longer waves is generated for a few periods, and then if desired still-longer waves, and so on. Provided the wavelength of each generated wavelength is calculated correctly using (2.75) and (2.74) and the relative phase of each generated wavelength is also set correctly, the longer waves will 'catch up' to the shorter waves at the same time and location, such that the crests are superposed. Even if the individual sets of waves had small amplitudes, so that the behaviour of each individual set is well approximated by linear theory, when enough crests superpose, the result will be a wave that is high enough to behave nonlinearly, including breaking. Furthermore, by altering the relative wavelength and phase of the generated waves, it can be made to break in any desired location.

2.3.11 Shallow-water approximation

Now consider the other extreme, water that is very shallow compared with the wavelength. Of course, (2.35) shows that ω^2 also tends to zero, but it is interesting to see how it tends to zero. This can be found using the Taylor's series expansion for $\tanh(\epsilon)$, for some argument ϵ that is small,

$$\tanh(\epsilon) \simeq \epsilon - \frac{1}{3}\epsilon^3 + \dots ,$$

therefore $\tanh(kh) \simeq kh$ as $kh \to 0$, and thus (2.34) becomes

$$\boxed{\omega^2 = ghk^2}, \tag{2.80}$$

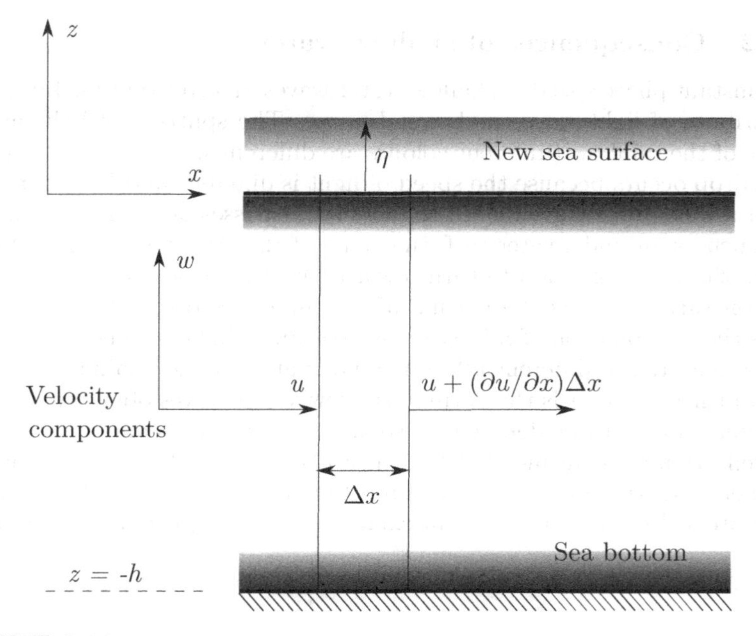

FIGURE 2.13
Conservation of mass in a body of water where the entire depth h is set into horizontal motion with no vertical variation.

and the phase speed relation (2.44) likewise becomes

$$\boxed{c = \sqrt{gh}}. \tag{2.81}$$

This *shallow-water* approximation (also called the *long-wave approximation*) is valid in the situation shown in figure 2.13. The relation (2.81) was derived in the introductory example in §2.2, where greatly simplifying assumptions of horizontal motion only with no variation in the vertical were made, as shown in figure 2.13. In §2.3.12 below, it will become clear that these physical assumptions are consistent with the mathematical assumption $kh \to 0$.

Shallow-water waves are a fair approximation if $h < \lambda/20$ and therefore $kh < \pi/10$. This shallow-water result has some interesting and useful consequences. The phase speed in shallow water no longer depends on k, so shallow-water waves are no longer dispersive; thus

$$\boxed{c_g = c}. \tag{2.82}$$

Shallow-water waves travel at a constant speed as long as the depth is constant. In other words, as long as the 'medium' (the water-column depth) is constant, the phase speed is constant.

2.3.12 Consequences of shallow water

The constant phase speed of shallow-water waves in a constant medium is also an attribute of light waves and sound waves. The splitting of light into the colours of the rainbow (different colours are different wavelengths) by a prism or raindrop occurs because the speed of light is different in different media. If part of a wave front travelling in shallow water passes over an even shallower zone, such as an underwater reef, that part of the wave will be slowed down, causing the wave direction to turn towards the reef. More precisely, the wave direction turns towards the normal of the interface to the shallower region. This is the phenomenon of *refraction* which causes light to bend if it enters or leaves a material of different refractive index at an angle, making lenses possible. In turning towards the normal, shallow-water waves obey the refraction law (Snell's Law), to be detailed for sound waves in §3.3.5 and figure 3.5, and obey reflection laws outlined in §3.3.7, just like rays of light entering a medium of higher refractive index. This is important for the design of harbours and breakwaters. Figure 2.14 shows an example. Another phenomenon occurring

FIGURE 2.14
Waves refract as they enter shallower water at an angle. See also figure 3.5. Photograph by Richard Manasseh.

in light and sound waves is the shift in frequency owing to movement of a source or reflector of waves, or (in the case of sound waves) owing to movement of the medium, usually called the *Doppler shift* and covered in §3.3.10. Surface gravity waves in a strong current may be subjected to this effect.

While, as mentioned earlier, the Earth's ocean basins have a typical depth of four kilometres, the continental shelves are not more than about 150 m deep, and, of course, close to the shore, the depth usually becomes much less, down to tens of metres, and at a beach, down to zero. The ocean swell may have a wavelength of order 100 m, and almost inevitably finds itself becoming a shallow-water wave as it progresses towards the coast. Thus, the continental shelf itself acts as a giant reef, turning the arriving ocean swell towards the normal to the coast and therefore towards the shore. This is why waves always end up coming towards shore, no matter which way they were heading far out to sea.

For $kh \to 0$, the potential function, (2.53), becomes

$$\boxed{\phi = A\frac{\mathrm{g}}{\omega}\mathrm{e}^{\mathrm{i}(\pm\kappa x \pm \ell y + \omega t)}}, \tag{2.83}$$

It is clear that the potential no longer varies in z: in shallow-water waves, the entire water column is set equally into motion. Moreover, since $\partial\phi/\partial z = 0$, there is no vertical component to the velocity; the velocity is purely horizontal. As a shallow-water wave passes, the water (and anything floating in it) only surge to and fro. This may not accord with our experience of bathing in the ocean. We definitely feel ourselves getting lifted up and dropped down as well as moved horizontally. However, remember that the shallow-water approximation is only good when the depth is less than 1/20 of the wavelength; at greater depths, there will be some vertical motion too.

Using (2.63), and assuming the wave has no structure in the y-direction, i.e. $\ell = 0$ so that $\kappa = k$ as discussed in §2.3.5.3, the horizontal component of water velocity, u, is given by

$$u = A\sqrt{\frac{\mathrm{g}}{h}}\,\mathrm{e}^{\mathrm{i}(-kx+\omega t)}, \tag{2.84}$$

in which, as for (2.63), the negative sign was chosen for the kx in the exponent to represent only the waves propagating towards the shore. Noting (2.59), $|A| = \hat{\eta}$, it becomes clear that the magnitude of the horizontal velocity in shallow water, $\hat{u}$, is related to the surface elevation amplitude $\hat{\eta}$ by

$$\hat{u} = \sqrt{\frac{\mathrm{g}}{h}}\hat{\eta}. \tag{2.85}$$

If waves are heading towards a beach, a process called *shoaling*, the depth h continues to decrease as the waves travel towards the shore. Thus their phase speed c becomes slower and slower according to (2.81). Meanwhile, (2.84) predicts the actual water moves faster and faster! Recall one of the assumptions that led to potential flow was that the water is inviscid, so that no energy loss due to friction (for example, friction with the seabed) is possible. And another assumption was that the behaviour was linear, so no energy loss due to turbulence is possible. Thus, a faster water motion closer to shore is consistent. The same kinetic energy is getting squeezed into a shallower and shallower depth, and the only way for this to occur in a frictionless sea is for the water velocity to get faster and faster.

At some point that depends on the wave amplitude, the phase speed c becomes so slow that the horizontal component of water velocity u equals c. The water 'catches up' with the wave crest, steepening it. Once u exceeds c,

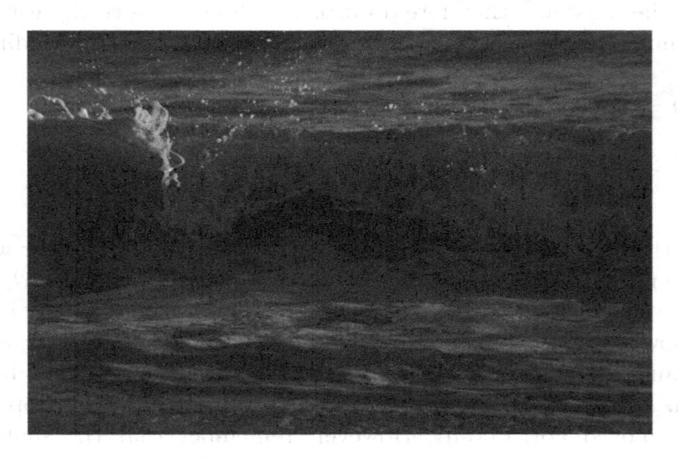

FIGURE 2.15
A small wave breaks on the beach in Port Campbell, Australia, illustrating a situation where the shallow-water phase speed c has dropped below the horizontal component of water speed u. Photograph by Richard Manasseh. See also figure 10.1, page 230.

the water overshoots the crest, forming a breaking wave: surf (figure 2.15). It is clear from (2.85) and (2.81) that the magnitude of u equals c when $(\sqrt{g/h})\hat{\eta} = \sqrt{gh}$, so the amplitude is equal to the depth; in other words, when the trough of the wave reveals bare sand, the wave's life is over. In reality, many other factors can cause a wave to break well before it reaches this point. Moreover, it is evident that waves in deep water or intermediate-depth water also break. Some further details on wave-breaking are given in §10.2.1.

The ratio of the magnitude of some *steady* velocity of water flow (U, say) to the speed of shallow-water waves is very important in civil engineering and maritime engineering, where, thanks to William Froude's findings published in 1861, it is usually called the *Froude number*, or Fr, which was introduced in §1.2.4 as (1.27), reproduced here as

$$\mathrm{Fr} = \frac{U}{\sqrt{gh}}. \tag{2.86}$$

The Froude number is usually applied in situations where something is moving in shallow water (or shallow water is moving relative to some fixed walls). For example, the motion of a ship in shallow water, or of a river or channel flow, is described with a Froude number. When the Froude number is less than unity, the motion is *subcritical*; waves affecting the entire water depth travel faster than the ship, or faster than water flows relative to the fixed river bank. Therefore, the waves carry information about the object's motion ahead of it, or upstream in the river. When the Froude number is more than unity, the motion is *supercritical*; information about the object's motion cannot go ahead of it to warn of its arrival, and the object forms a V-shaped pattern of waves behind it. In the supercritical river or channel, V-shaped waves come off every rock or fixed protuberance.

It is important to note that while the V-shaped pattern referred to above occurs in the context of shallow-water waves, as explained in §2.3.10, a V-shaped pattern is also formed in deep water, but unlike the V-shaped pattern in shallow water, in deep water there is always the same angle of $\sin^{-1}(1/3)$.

The denominator $\sqrt{gh}$ in the Froude number does not just apply to an ocean or river depth. This scaling crops up in any situation in which gravity-dominated motion is integrated over some depth H, such as the up-and-down or rocking motion of a ship or wave-energy converter of *draught* H, where here we are using a capital H to emphasise we are now referring to an engineered object. Integration over some draught H merely means that an object is a solid object, but it has the same consequence of removing variation in the vertical in a gravity-dominated motion, which also occurs in shallow-water wave theory. So it is not surprising that a $\sqrt{gH}$ also occurs. If engineers want to build a model ship and obtain the correct size of waves relative to the size of the ship, they must ensure the Froude number is the same in their model as in the real ship. Thus, if their model is 1/100 the size, they need to test it, not at 1/100 of the speed, but at 1/10 of the speed.

The Froude number is the analogue of the *Mach number* (1.31) which measures the speed of motion relative to the speed of sound, and the analogues of subcritical and supercritical flow are subsonic and supersonic flow respectively.

Many real-world aspects of surface gravity waves are not predicted by the linear theory of this chapter. Drifts are described in §6.4. The steepening of waves, so that they no longer have a sinusoidal shape, is outlined in §10.2.1. Phenomena like tsunamis are also nonlinear and are discussed in §6.5.1 and §12.3.

Finally, the same incompressible, inviscid and linear assumptions used in this chapter, plus the inclusion of the Coriolis force (§5.3.1) and the force due to the gravity of celestial bodies like the Moon and Sun, permits the combination of the continuity and momentum equations into a set of equations that predict the *astronomical tides*. Variations in the vertical can be neglected, as in the shallow-water approximation. This was first done by Laplace in 1776, and despite their simplifications, these assumptions permit the prediction of tides with great precision.

2.4 Problems

1. Waves travelling at a phase speed of 1 m s^{-1} have a wavenumber of 0.5 m^{-1}. What is their frequency in radians per second?
2. Waves travelling at a phase speed of 10.5 m s^{-1} have a frequency of 1.5 Hz (1.5 s^{-1}). What is their wavelength?
3. Travelling waves have a wavelength of 20 m and their frequency is calculated to be 6.28 rad s^{-1}. What is their phase speed?
4. Waves measured to travel at 2.0 m s^{-1} are theoretically predicted to have a wavenumber of 1.571 m^{-1}. What should their frequency be in Hertz (cycles per second)?
5. A cargo barge is known to suffer from vibration-fatigue problems. If it operates in water 3 m deep where strong winds typically create waves of wavelength 12 m, what frequency of vibrations (in Hz) will impact the barge?
6. A dock in a harbour suffers from a serious sloshing problem, preventing cargo unloading if waves of approximately 15 m wavelength enter the dock. The water is 20 m deep. A piledriver is proposed to operate nearby, generating waves. Approximately what frequency (In Hertz) of pile-driving must the operator be told to avoid?
7. A pharmaceuticals-industry tank 7 m deep is agitated by the injection of bubbles that create surface disturbances with a spectral peak around 1 Hz. What is the wavelength of the surface gravity waves created?
8. Storm waves of wavelength 100 m enter a harbour of uniform of depth of 4 m. What is the approximate frequency (in Hertz) with which the dock at the end of the harbour is shaken?
9. A disused harbour 4 m deep with a single narrow entrance is proposed for testing full-scale wave-energy converters. Serious sloshing would occur with a wavelength four times the length L. If the wave-energy converter models oscillate with a frequency of 0.08 Hz, what length of the harbour would be problematic?
10. The theory of linear surface gravity waves is sufficient for

    A Stabilising spacecraft in Earth orbit containing liquid fuels

    B Design of high-speed ferries that use hydrofoils

    C Predicting the speed of water flooding from a collapsed dam

    D Measuring viscosity of sugar solutions in the food industry

11. Which of the following assumptions would lead to the assumption of potential flow for water-surface gravity waves?

    A compressible, inviscid, linear, sinusoidal in space

    B incompressible, inviscid, linear, sinusoidal in time

    C incompressible, viscous, linear, sinusoidal in time

    D compressible, inviscid, nonlinear, sinusoidal in space

12. Surface gravity waves are sinusoidal in the horizontal but hyperbolic in the vertical because

    A The boundary conditions in the horizontal are sinusoidal in space

    B The horizontal direction extends to infinity

    C The vertical direction is subject to the force of gravity

D The boundary conditions in the vertical are different from those in the horizontal

13. Waves are calculated to have a frequency of 10 rad s^{-1} and a wavenumber of 2.5 m^{-1}. What is their phase speed?

14. Travelling waves have a wavelength of 7 m and a frequency of 0.8 Hz (0.8 s^{-1}). What is their phase speed?

15. Linear inviscid surface gravity waves in water of finite depth create elliptical particle trajectories because

A There is horizontal velocity plus zero vertical velocity at the bottom and nonzero vertical velocity at the surface

B There is horizontal velocity in the centre of the water column and nonzero vertical velocity at the surface

C There is nonzero vertical velocity and nonzero horizontal velocity at the surface

D There is vertical velocity in the centre of the water column and nonzero horizontal velocity throughout the depth

16. A proposed high-speed river ferry would make waves of approximately 3 m wavelength in a river dredged to a uniform depth of 10 m. Passengers would wait on a floating terminal quay. At approximately what frequency (in Hertz) would the waves make the quay rock?

17. An Olympic swimming pool is 50 m long and 1.5 m deep. What is the approximate frequency (in Hertz) of waves created in this pool with a wavelength of twice the pool length?

18. Reciprocating baths 20 cm long and 3.5 cm deep are used to gently agitate samples in laboratories of a pathology-testing business. The drive is a variable-frequency motor. Operators report liquid splashing out when sloshing waves four times the bath length arise. What motor frequency should **not** be used?

19. When the Froude number is equal to 1, it indicates

A The deep-water wave speed equals the water speed

B The shallow-water wave speed equals the water speed

C The shallow-water wave speed is less than the water speed

D The deep-water wave speed is greater than the water speed

20. Travelling waves with a phase speed of 2 m s^{-1} have a wavenumber of 0.25 m^{-1}. What is their frequency in radians per second?

21. Waves with a frequency of 0.8 Hz (0.8 s^{-1}) travel with a phase speed of 3.6 m s^{-1}. What is their wavelength?

22. Travelling waves have a wavelength of 40 m and their frequency is calculated to be 3.14 rad s^{-1}. What is their phase speed?

23. Waves measured to travel at 1.0 m s^{-1} are theoretically predicted to have a wavenumber of 3.142 m^{-1}. What should their frequency be in Hertz (cycles per second)?

24. The structure of a new catamaran vessel has a low-frequency vibration mode. If it operates in a bay 5 m deep impacted by ocean waves 20 m long, what frequency of vibrations (in Hz) will the ship suffer?

25. The design of a proposed offshore renewable-energy platform requires vibrations to be less than 1 s^{-1} (1 Hz) for safety. It is known that passing ships typically make waves 12 m long. The water is 100 m deep. Determine by calculation if the design is safe.

26. An experimental carbon-free steelmaking ladle 2 m deep has hydrogen gas injected, creating surface waves on the molten steel with a peak frequency of approximately 2 Hz. What is the wavelength of the surface gravity waves created?

27. A small-vessel harbour is 150 m long from the ocean end to the land end and has a depth of 5 m. What is the approximate frequency (in Hertz) of sloshing-mode waves with a wavelength equal to the harbour length?

28. An exercise swimming pool of length $L = 25$ m is being designed for a colony on Mars where the acceleration due to gravity, g, is 3.72 m s^{-2}. Serious sloshing would occur with a wavelength four times the pool length. Since the pool is only wide enough for one swimmer at a time, swimmers with speed V create a frequency of $V/(2L)$ Hz as they do laps. If $V = 1.0$ m s^{-1}, what design pool depth h should be avoided?

29. Discuss which of the deep or shallow-water approximations of surface gravity waves is likely to be relevant to

(a) Preventing destabilising waves inside a bulk-liquids transporter

(b) Predicting the spectrum of ocean waves

(c) Estimating where basin-scale resonances could occur in a port design

(d) Calculating whether a floating offshore wind turbine would rock unacceptably

30. Discuss whether potential flow could be a valid assumption for the prediction of surface gravity waves under the following circumstances.

(a) The liquid-methane lakes of Saturn's moon, Titan

(b) Flood-water on a city street under cyclonic winds

(c) Flow of low-viscosity lava from a volcanic eruption

(d) Water released from a dam and flowing over its spillway

31. What is the maximum length of waves created as a typhoon with wind speeds of 200 km h^{-1} passes over a sea of depth 1000 m?

32. A proposed new design of wave-energy converter has two segments with a joint in between. As the segments tilt relative to each other, electricity is generated at the joint. For this to work, the length of each segment must be 1/4 of the ocean wavelength. The machine is to operate in water 60 m deep and its natural frequency of 0.2 Hz is designed to match the known ocean-wave frequency at that location. How long should the segments be?

33. A large open pond is used for the cultivation of algae for bio-fuels from sequestered carbon dioxide. It is a 'racetrack' design with an inner radius of 10 m and an outer radius of 15 m and is 1 m deep. Approximately how long would it take a large disturbance to circle the pond? If the aim is to generate a periodic flow to mix the carbon dioxide into the water, discuss the frequency with which you would pulse the gas injection and the number and location of injection points.

34. A river is flowing over a rocky shelf, so the depth of the water is reduced from h_1 to h_2, where $h_2 < h_1$, while the width of the river is unchanged. Upstream of the shelf, the river flows at speed u_1. On the rocky shelf and distance L from the start of the shelf, a rock reaches the surface, presenting a hazard to boats. Derive a formula for the time that waves warning about the rock's existence take to reach a boat drifting towards the shelf and a distance X from the shelf edge.

3

Sound waves

3.1 Summary of key points

- The **speed of sound**, c, as a function of the properties of the ambient medium, which may be a solid, liquid or gas, is given by (3.19) on page 97,

$$\boxed{c = \sqrt{\frac{E_v}{\rho_0}}},$$

where ρ_0 is the density of the medium and E_v is the bulk modulus (which can be calculated from density and speed-of-sound measurements in Fine and Millero (1973) or Gill (1982)); for a gas, c reduces to (3.20) on page 97,

$$\boxed{c = \sqrt{\frac{\gamma P_0}{\rho_0}}},$$

where γ is the adiabatic index (or ratio of specific heats, C_P/C_V) and P_0 is the steady-state total pressure (ambient pressure in the absence of steady flows).

- The **sound pressure** for plane waves in Cartesian coordinates is given by (3.21) on page 98,

$$\boxed{p = A\mathrm{e}^{\mathrm{i}(\pm\kappa x \pm \ell y \pm kz + \omega t)}},$$

where A is a complex amplitude and κ, ℓ and k are the wavenumbers in the x, y and z directions respectively; for plane waves propagating in the x-direction only and a phase of zero, the **real pressure** is given by (3.23) on page 100,

$$\boxed{p_R = \hat{p}\cos(\kappa x - \omega t)},$$

where $\hat{p}$ is the sound-pressure amplitude and in spherical polar coordinates the real pressure is given by (3.24),

$$\boxed{p_R = \frac{1}{r}\,\hat{p}\cos(\kappa r - \omega t)}.$$

DOI: 10.1201/9780429295263-3

- The **acoustic impedance relation** for plane waves is given by (3.28) on page 103,

$$\boxed{\hat{p} = \hat{Z}\hat{u},}.$$

where the characteristic specific acoustic impedance is given by $\hat{Z} = \rho_0 c$, and $\hat{u}$ is the amplitude of the fluid particle velocity.

- The **refraction** angle though which rays of any kind of wave are bent as they pass from a medium with one phase speed into a medium with a different phase speed is given by Snell's Law, (3.25) on page 102,

$$\boxed{\frac{\sin\theta_1}{c_1} = \frac{\sin\theta_2}{c_2}}.$$

- The **reflection** fraction of the pressure amplitude of a plane sound wave as it passes through a boundary between two fluids, $\hat{p}_r/\hat{p}_1$, is given by (3.29) on 104,

$$\boxed{\frac{\hat{p}_r}{\hat{p}_1} = \frac{\hat{Z}_2\cos\theta_1 - \hat{Z}_1\cos\theta_2}{\hat{Z}_2\cos\theta_1 + \hat{Z}_1\cos\theta_2}};$$

see Kinsler and Frey (1962) for reflection formulae for other situations such as a boundary with a solid, or spherical waves.

- The **sound pressure level**, L_p, is given by (3.34) on page 106,

$$\boxed{L_p = 20\log_{10}\left(\frac{\hat{p}}{p_\emptyset}\right) \quad \text{dB}},$$

where $\hat{p}$ is the pressure amplitude in Pa and $p_\emptyset$ is the reference pressure, usually 20 μPa in air and 1 μPa in water.

- The **sound intensity level**, L_I, is given by (3.32) on page 106,

$$L_I = 10\log_{10}\left(\frac{I}{I_0}\right) \quad \text{dB},$$

where I is the sound intensity in W m^{-2} and I_0 is the reference intensity, usually 10^{-12} W m^{-2}.

- The **Inverse Square Law** for spherically-symmetric sound radiation from a spherical source is given by (3.38) on page 111,

$$\boxed{\frac{I(r)}{I(R_0)} = \left(\frac{R_0}{r}\right)^2},$$

where $I(R_0)$ is the sound intensity at the source of radius R_0 and $I(r)$ is the sound intensity at any radius r.

- The **general Doppler-shifted frequency**, f_{12}, for a transmitter generating sound with frequency f_0 while moving with speed U_t, and a receiver moving with speed U_r in the same direction as sound propagation, is given by (3.39) on page 112,

$$\boxed{f_{12} = \frac{c - U_r}{c - U_t} f_0}.$$

- The **Doppler shift**, f_d, for the special case relevant to many measurement instruments, where a stationary transmitter generates sound with frequency f_0 that is reflected from an object and returns to the transmitter, is given by (3.44) on page 113.

$$\boxed{f_d = \frac{2U_r}{c + U_r} f_0}.$$

  Where the object speed $U_r \ll c$ and the line from the transmitter to the object is at an angle θ to the object's line of travel, f_d is given by (3.46) on page 113,

$$\boxed{f_d = \frac{2U_r \cos\theta}{c} f_0}.$$

- The **reverberation time**, T_{60}, is given by (3.47) on page 115,

$$\boxed{T_{60} = \frac{24\ln(10)}{c} \frac{V}{a_a S}},$$

  where c is the speed of sound in air at 20°C, V is the volume of the room, a_a is the acoustic absorption coefficient that must be measured for each thickness of each type of wall material, and S is the surface area of the room.

- Useful **textbooks** include Lighthill (1978) for a thorough applied-mathematical approach to sound and Kinsler and Frey (1962) for many practical formulae in acoustics.

3.2 An example

Some weeks after the cyclone and tsunami in §2.2, several dolphins beach themselves on the island and cannot be saved. The backpacking tourist is curious about a mining survey ship that arrived a week ago, its long-awaited arrival delayed by the cyclone. He posts an opinion that it caused the dolphin deaths. The chief engineer confronts him, pointing out that a mining lease would be valuable to the island's people and that he has no evidence. He says it is known that the ship uses a standard seismic air gun, which creates an approximately spherical bubble of radius 0.5 m and pressure amplitude 300 kPa. The bubbles collapse

(§9.2.1) to create a shockwave used to probe the seabed geology. The engineer challenges him to demonstrate what that might do to a dolphin 1 km away. He surprised her by doing a crude acoustical calculation.

Solution

Assuming the seabed is far away, apply the inverse square law, (3.38), $I/I(R_1) = (R_1/r)^2$, where I is the sound intensity, $R_1 = 0.5$ m and r is radial distance, and knowing that $I = \hat{p}^2/Z$ from (3.30), the problem reduces to a simple ratio, $p/p(R_1) = R_1/r$, so that $p = 0.5/1000 \times 300\ 000$, and hence $p = 150$ Pa.

He claims this sound pressure would cause sufficient trauma to cause the beachings. The calculation appeared correct, but the chief engineer was not convinced; there were too many other uncertainties.

3.3 Linear sound-wave theory

3.3.1 Use and control of sound

FIGURE 3.1
Music and even the instruments on which it is played may be centuries old, but its modern recording and reproduction is a complex technological process demanding an understanding of sound propagation and transduction. Image courtesy of the Australian Chamber Orchestra.

Sound is obvious to all humans excepting those with complete hearing loss, and even those individuals can perceive sound using cybernetic implants connected to microphones. Almost all animal life can detect sound and very many creatures can make a sound. Whales can produce sound frequencies well below 1 cycle per second (1 Hz), while some insects can produce sound frequencies well above 100 000 cycles per second.

In addition to our use of sound for speech, sound is also valued as music. Music has been present in all human societies since ancient times, and its creation, recording, performance and reproduction - and all the promotion and publicity associated with music - gives rise to a vast set of global industries (figure 3.1).

Unwanted sounds - noises - are a problem in cities and factories, and must be blocked or attenuated.

Sound often replaces light for detection and ranging underwater, so that ships and submarines utilise sound, as do marine animals like dolphins. Sound too high for humans to hear - conventionally assumed to be above 20 000 Hz - is called *ultrasound.* Ultrasonic sensors tell us when our cars are too close to obstacles and report the levels of liquid in industrial tanks. Ultrasound is used in aerospace and mechanical industries for non-destructive testing, in biological and chemical industries to promote reactions, and in electronic industries for cleaning. Ultrasound is used in medicine to image inside the human body, a procedure so routine that in advanced societies virtually everyone born after 1990 experienced ultrasonic waves before birth. In some cases, ultrasound is used to treat as well as to detect disease.

Despite the diversity of applications of sound, all sound waves share the same attributes. Sound waves exist because of the compressibility of the fluid. In this respect, they are completely different to the water-surface waves considered in chapter 2, where variations in the density of the fluid could safely be neglected in order to derive the basic properties of water-surface waves. As a sound wave travels, the fluid is compressed at the crest of the pressure wave, increasing its density, and expanded (or *rarefied*) at the trough of the pressure wave, decreasing its density. The fluid velocity is parallel to the direction of travel of the wave. Sound waves are therefore purely *longitudinal waves*, whereas water-surface waves are a combination of transverse and longitudinal waves.

The fluid motions created by most sound waves are microscopic, and even if they were not so small, the oscillations are far to rapid for human eyes to perceive. The corresponding changes in density are likewise extremely small relative to the undisturbed density of the fluid. Nonetheless, the wave-like nature of sound and analogies with water-surface waves could be deduced by early scientists and engineers. The architect and engineer Vitruvius wrote well over 2000 years ago, "Voice ... moves in an endless number of circular rounds, like the innumerably increasing circular waves which appear when a stone is thrown into smooth water".[1]

[1]Vitruvius Pollio, *Ten books on architecture*, Book V, Chapter III, article 6 (translated by Morris Hicky Morgan, 1914, Harvard University Press.)

3.3.2 The wave equation for sound waves

Owing to the importance of compressibility, it is helpful to first consider the Law of Conservation of Mass, (1.15), reproduced here as

$$\frac{\partial \rho}{\partial t} = -\boldsymbol{\nabla} \cdot (\rho \boldsymbol{u}), \tag{3.1}$$

noting that for a fluid assumed to be incompressible, as in chapter 2, the left-hand side of (3.1) would be zero, but for the compressible flow created by sound waves, it must be nonzero. Meanwhile, the momentum (or Navier-Stokes) equation, (1.19), reproduced here, is

$$\frac{\mathrm{D}(\rho \boldsymbol{u})}{\mathrm{D}t} = \boldsymbol{\nabla} \cdot \boldsymbol{\tau} + \rho \mathrm{g}. \tag{3.2}$$

The first simplifying assumption is that the fluid is inviscid, as in chapter 2. This permits the neglect of the shear-stress elements from the stress tensor $\boldsymbol{\tau}$, which then only has nonzero elements on its diagonal, which are the normal stresses due to pressure, P. Of course, as with all fluid waves, friction (as well as possible diffusion of heat from the tiny changes in temperature during compressions and expansions) will eventually cause the sound waves to peter out. However, in many cases, scattering (§3.3.7) and geometrical spreading (§3.3.9) of the waves will cause them to weaken much more rapidly with distance than the weakening due to viscous or thermal losses.

As in all fluid wave problems, the nonlinearities in (3.1) and (3.2), which arise from the advective operator $\boldsymbol{u} \cdot \boldsymbol{\nabla}$ in the material derivative, make solution difficult. However, as noted in §3.3.1 above, the changes in density are extremely small relative to the density of the undisturbed fluid, ρ_0, in most cases - and certainly for speech, music and ultrasonic imaging. The corresponding fluid velocities are also small. This means the density can be written

$$\rho = \rho_0 + \varrho \tag{3.3}$$

where ϱ is that small variation in space and time due to sound, and ρ_0 is the much larger density of the undisturbed fluid that is assumed to be constant. For the derivation of sound waves, the total pressure, P, can be expressed as

$$P = P_0 + p, \tag{3.4}$$

where P_0 is the constant part of the total pressure, the constant pressure of the undisturbed fluid due to atmospheric pressure, plus, if the fluid is a liquid, hydrostatic pressure due to depth, plus any additional static pressure applied by engineered systems, and any additional pressure due to steady fluid flows. As discussed in §1.2.2.3, if there is no steady fluid flow in the 'background', P_0 equals the ambient pressure $P_\emptyset$, which will be assumed for the rest of this chapter. Even if there is a 'background' steady flow, the rest of this chapter is still valid provided any such flow has only a small spatial variation. It is worth noting that just as ϱ is small compared with ρ_0, p is small compared with P_0. Because the velocities are also small, any term where ϱ and $\boldsymbol{u}$ appear multiplied together disappears relatively to the other terms. Thus, the left-hand side of the equation of conservation of mass (3.1) is linearised by neglecting the term with the advective factor, and the right-hand side is linearised by neglecting the term where ϱ multiplies $\boldsymbol{\nabla} \cdot \boldsymbol{u}$. The same linearising assumption must be applied to the momentum equation: the product of ϱ and $\boldsymbol{u}$ on the left-hand side of (3.2) disappears, reducing it to $\rho_0\, \partial \boldsymbol{u}/\partial t$.

Thus, these assumptions of small density variations and small velocities, and of an inviscid fluid, reduce the two conservation laws, (3.1) and (3.2), to

$$\frac{\partial \varrho}{\partial t} = -\rho_0 \boldsymbol{\nabla} \cdot \boldsymbol{u}, \tag{3.5}$$

and

$$\rho_0 \frac{\partial \boldsymbol{u}}{\partial t} = -\boldsymbol{\nabla} P + (\rho_0 + \varrho)\mathbf{g}. \tag{3.6}$$

The two conservation laws are now two equations, but they involve three variables, ϱ and p and the vector $\boldsymbol{u}$. Clearly, a third equation will be needed. First, however, a few more steps can be taken. Eliminate one of the variables by combining these equations. Differentiating (3.5) with respect to time, the mass-conservation law becomes

$$\frac{\partial^2 \varrho}{\partial t^2} = -\rho_0 \boldsymbol{\nabla} \cdot \frac{\partial \boldsymbol{u}}{\partial t}. \tag{3.7}$$

Now, calculate the divergence of (3.6). The last, gravitational term in (3.6) becomes $\boldsymbol{\nabla} \cdot [(\rho_0 + \varrho)\mathbf{g}] = \rho_0 \boldsymbol{\nabla} \cdot \mathbf{g} - (\partial \varrho/\partial z)\mathrm{g}$, the second term only involving a derivative in z because $\mathbf{g}$ points in the vertical only. Note that $\boldsymbol{\nabla} \cdot \mathbf{g} = 0$, because there is no significant variation in the acceleration due to gravity with distance; as mentioned in §2.3.4, $\mathbf{g}$ does vary slightly, but over a substantial fraction of the size of the planet, not over the scale of sound waves. Thus the momentum-conservation law becomes

$$\rho_0 \boldsymbol{\nabla} \cdot \frac{\partial \boldsymbol{u}}{\partial t} = -\nabla^2 P - \frac{\partial \varrho}{\partial z}\mathrm{g}, \tag{3.8}$$

where the operator ∇^2 is the scalar operator equal to $\boldsymbol{\nabla} \cdot \boldsymbol{\nabla}$, reflecting the fact that (3.8) is a scalar equation just like (3.7). The operator ∇^2 is the Laplacian operator introduced in §1.2.3.2. Combining (3.7) and (3.8) gives

$$\frac{\partial^2 \varrho}{\partial t^2} - \nabla^2 P - \frac{\partial \varrho}{\partial z}\mathrm{g} = 0. \tag{3.9}$$

Recalling (3.4), $P = P_0 + p$, and recalling (1.2) for negligible 'background' steady flow, that $P_0 = P_\emptyset = P_{\text{atm}} + P_h + P_s$, a scaling analysis would show that variations in the atmospheric pressure P_{atm} that is one part of P_0 are very small over lengths corresponding to the wavelengths of sound, so the derivatives of P_0 due to variations in atmospheric pressure can safely be neglected. Any engineering static pressure P_s is by definition constant so its derivatives are zero. Meanwhile, the remaining variation in P_0 is due to hydrostatic pressure, P_h. That does vary in the vertical, but linearly, according to (1.1), so the second derivative in the vertical of P_h caused by the ∇^2 operator is also zero. Even if there were some 'background' steady flow (so that $P_0 \neq P_\emptyset$), as noted earlier, the following still applies, provided derivatives of the steady flow are negligible. Thus the derivatives of P_0 are all zero, and $\nabla^2 P = \nabla^2 p$. At this point a scaling analysis would also show that the timescale over which the oscillations of sound waves occurs is very small, typically a fraction of a second, the length scale is relatively large, typically the order of a metre, and the magnitude of the acceleration due to gravity, g, is about 10. Thus the last, gravitational term in (3.9) is negligible compared to the first term. Thus (3.9) becomes

$$\frac{\partial^2 \varrho}{\partial t^2} - \nabla^2 p = 0. \tag{3.10}$$

This looks almost like one of the forms of the wave equation listed in §1.5.2, but it is not; as noted earlier, it involves two variables, ϱ and p. In order to solve (3.10), a further equation is needed, provided by a Constitutive Law. In this case, the Constitutive Law required is simply the relation between pressure and density. In general - and indeed for solids as well as fluids - this relation is given by the expression for the bulk modulus, (1.4), here reproduced as

$$E_v = \rho \frac{\partial P}{\partial \rho}, \tag{3.11}$$

$$\Rightarrow E_v \frac{1}{\rho} = \frac{\partial P}{\partial \rho}. \tag{3.12}$$

Integrating (3.12) with respect to ρ gives

$$E_v \ln \rho = P + \text{const.} \tag{3.13}$$

When the fluid is undisturbed by any waves (or any kind of flow), $P = P_0 = P_\emptyset$ and $\rho = \rho_0$, so the constant is $E_v \ln \rho_0 - P_0$. Thus, (3.13) becomes

$$E_v \ln\left(\frac{\rho}{\rho_0}\right) = P - P_0, \tag{3.14}$$

and since the variation in pressure, p, is defined by (1.3) as $P - P_0$ in the absence of 'background' steady flows, the right-hand side is simply equal to p. However, the left-hand side of (3.14) is nonlinear owing to the natural log function. For consistency, the same assumption of small density variations has to be applied to the left-hand side of (3.14). Thus, noting as before that $\rho = \rho_0 + \varrho$,

$$\ln\left(\frac{\rho}{\rho_0}\right) = \ln\left(1 + \frac{\varrho}{\rho_0}\right). \tag{3.15}$$

Now, the Taylor series expansion of the natural log function for any small variation ϵ about unity is

$$\ln(1+\epsilon) \simeq \epsilon - \frac{1}{2}\epsilon^2 + \dots. \tag{3.16}$$

Thus, identifying ϵ as ϱ/ρ_0, the linearised version of (3.14) is simply

$$E_v \frac{\varrho}{\rho_0} = p. \tag{3.17}$$

This linearised Constitutive-Law relation (3.17) between ϱ and p provides the third equation along with the conservation of mass and momentum equations already combined into (3.10). One of ϱ and p should be eliminated. In this derivation of sound waves, we can recognise that it is the pressure that is measured by microphones in the air (and by hydrophones in the water, as detailed in §3.3.8.2), so a solution in terms of p would be more useful. We can choose to eliminate ϱ from (3.17) in favour of p. Differentiating (3.17) twice with respect to time and substituting into (3.10) gives the familiar wave equation,

$$\frac{\partial^2 p}{\partial t^2} - c^2 \nabla^2 p = 0, \tag{3.18}$$

which is exactly (1.50), one of the standard forms of the wave equation, where the wave speed c is given by

$$\boxed{c = \sqrt{\frac{E_v}{\rho_0}}}. \tag{3.19}$$

This wave speed is the *speed of sound.*

The bulk modulus E_v, as defined in §1.2.2.3, is a measure of the resistance to compression of any substance, solid, liquid or gas. The bulk modulus of water is about 2.2×10^9 Pa (Fine and Millero, 1973); like the density of water, the bulk modulus of water does vary slightly with temperature (and, in seawater, with salinity), causing a variation in the speed of sound in the ocean, as shown in figure 3.2. For a gas undergoing the rapid expansions and compressions of small-amplitude sound waves, the transfer of heat can be neglected, so the bulk modulus is well approximated by $E_v = \gamma P_0$, where γ is the adiabatic index detailed in §1.2.2.3. Thus, for a gas, c is given by

$$\boxed{c = \sqrt{\frac{\gamma P_0}{\rho_0}}}. \tag{3.20}$$

If the gas is air, the value of γ is 1.4 and if the air is at the standard atmospheric pressure and is motionless apart from the sound waves, so that $P_0 = P_{\text{atm}} =$ 101 325 Pa, the bulk modulus, E_v, is 141 855 Pa, or four orders of magnitude

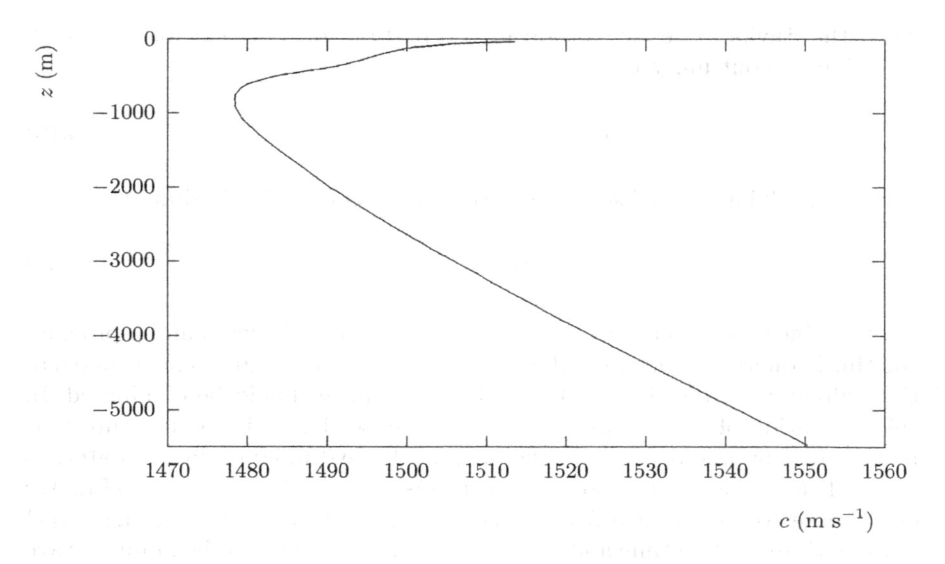

FIGURE 3.2
Speed of sound c depends slightly on depth z in the ocean because water is slightly compressible. Data adapted from `https://commons.wikimedia.org/wiki/Underwater_speed_of_sound.svg`.

less than that of water, illustrating the dramatic difference in compressibility between liquids and gases. At the Earth's average ground-and-sea-level air temperature of about 15° C, so that $\rho_0 = 1.225$ kg m^{-3}, (3.20) predicts that the speed of sound should be about 340 m s^{-1} and indeed it is. A useful rule-of-thumb is that sound at the surface of the Earth travels roughly one kilometre in three seconds. Once a lightning flash is seen, counting the seconds until the thunderclap is heard and dividing by three gives the distance of the lightning strike in kilometres. The speed of sound drops with temperature; assuming the atmospheric pressure is still the same, it would be about 331 m s^{-1} at 0° C, whereas it would be about 358 m s^{-1} at 45° C (figure 3.3). Since temperature and pressure vary with altitude in the atmosphere, so the speed of sound varies with altitude too (figure 3.4).

3.3.3 Solution of the wave equation

The wave equation for sound waves (3.18) is solved just as with any wave equation, using the methods in 1.5.3, giving in Cartesian coordinates x, y and z

$$\boxed{p = A e^{\mathrm{i}(\pm\kappa x \pm \ell y \pm kz + \omega t)}}, \tag{3.21}$$

where, as with all waves, A is a complex constant that depends on the initial conditions in space and time, and κ, ℓ and k are the wavenumbers in the x,

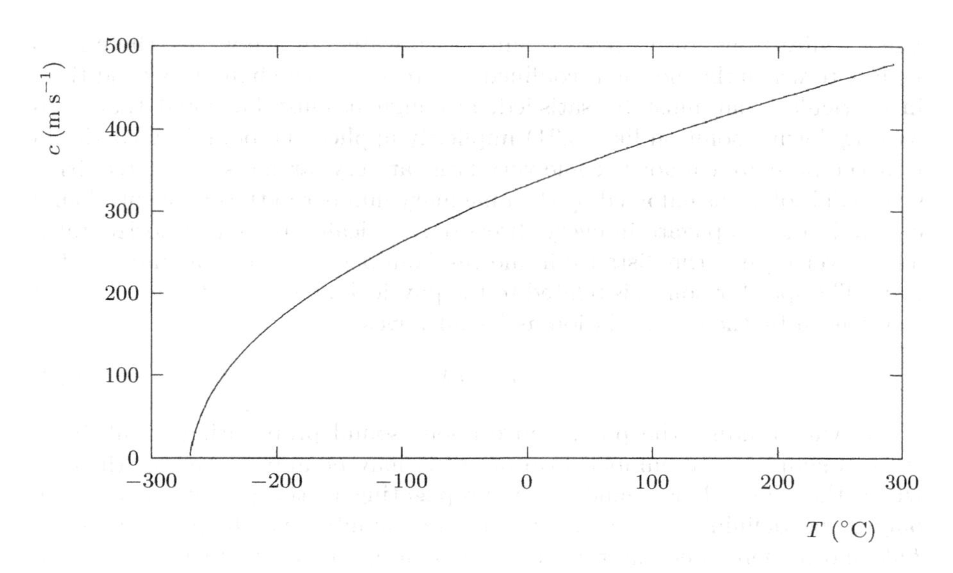

FIGURE 3.3
Speed of sound c as a function of dry-air temperature T. Data adapted from `https://commons.wikimedia.org/wiki/File:Speed_of_sound_in_dry_air.svg`

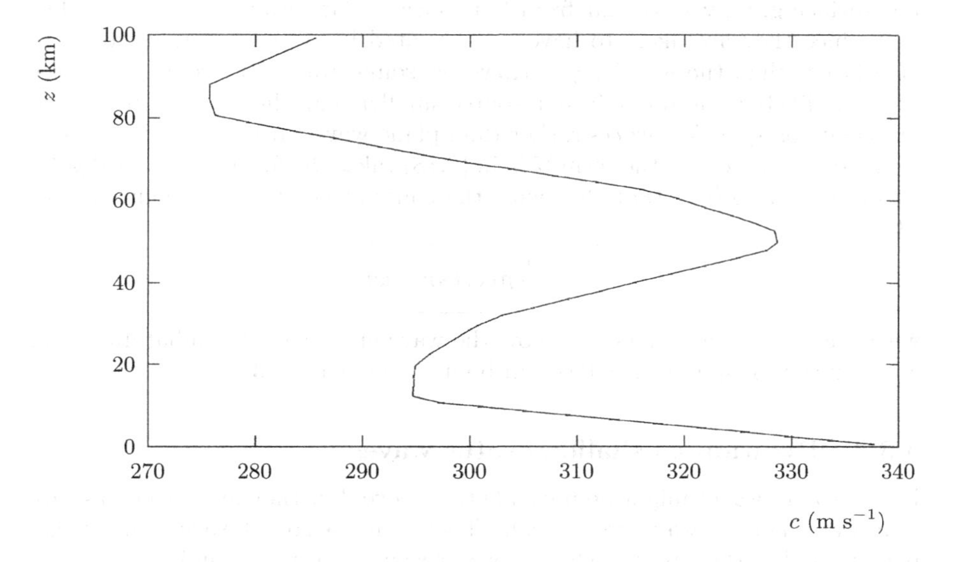

FIGURE 3.4
Speed of sound c in the atmosphere varies with altitude z. Data from `https://commons.wikimedia.org/wiki/File:Comparison_US_standard_atmosphere_1962.svg`

y and z directions, respectively. Unlike with water-surface waves, there is no surface to which the motion is confined. There are also no boundary conditions in particular that must be satisfied, although it must be noted that even writing down a solution like (3.21) implicitly applied the boundary condition that the pressure cannot become very large at very large distances, requiring sinusoidal solutions enforced by the imaginary number in the exponent. Sound can and does propagate in every direction, vertically as well as horizontally. The wavelength is the distance in metres from one pressure maximum to the next. The speed of sound is related to the physical quantities of frequency and wavelength by the same relation as for all waves,

$$c = f\lambda \tag{3.22}$$

Notwithstanding the point above about sound propagating in all directions, a simpler, real-number form of (3.21) may be appreciated for the case where there are plane sound waves propagating in the positive x-direction only. Thus, defining $A = \hat{p}\,\mathrm{e}^{\mathrm{i}\Phi}$, where the real number $\hat{p}$ is the pressure *amplitude* and if there is no external reference phase, setting the phase $\Phi = 0$ and taking the real part of the pressure, p_R, reduces (3.21) to

$$\boxed{p_R = \hat{p}\cos(\kappa x - \omega t)}. \tag{3.23}$$

For surface gravity waves far from boundaries, plane waves represent reality well, since they are likely to have been created by a source (winds) over an area larger than the wavelength. However, sound waves far from boundaries are more likely to originate from a source smaller than the wavelength, and so propagate as *spherical waves* rather than plane waves. If the only variation is in the radial direction, the term $\nabla^2 p$ in (3.18) takes the form in *spherical polar coordinates* of $(1/r)\partial^2(rp)/\partial r^2$, with the consequence that the equivalent of (3.23) is

$$\boxed{p_R = \frac{1}{r}\,\hat{p}\cos(\kappa r - \omega t)}, \tag{3.24}$$

where, for spherical polars, κ is now the wavenumber in the radial direction. The solution in spherical polars will be used further in §3.3.9.

3.3.4 Relation to shallow-water waves

Linear sound waves might be invisible to the eye, but they are easier to study than water-surface waves in general. This is because sound waves are not dispersive, and so they are like shallow-water waves that were studied in §2.3.11. In this respect, the analogy of Vitruvius mentioned in §3.3.2, while brilliantly perceptive, was not perfectly correct, since the ripples he referred to are dispersive waves. The sound-wave speed varies with pressure and temperature, but there is no dependence on the depth of the medium: no equivalent of 'shoaling', where shallow-water waves slow down as they approach a beach.

Moreover, water-surface waves occur only *on* the boundary between water and air. In contrast, sound can pass *across* any of the boundaries between solids, liquids or gases, with the consequent variation in c in different media giving rise to refraction, reflection and transmission phenomena. Water-surface waves do experience refraction, reflection and transmission phenomena, but mostly when they travel over changes in depth.

As with all refraction phenomena, as described in §3.3.5 below, and reflection phenomena in §3.3.7, it is helpful to introduce the concept of a *ray*. It is simply a line that is perpendicular to the wave crests and troughs, and points in the direction that the wave is travelling.

In addition to changes in c due to passing into completely different substances, sound waves can also experience changes in c owing to temperature and density changes within the same substance, again giving rise to refraction, reflection and transmission phenomena. It is sometimes observed that sounds travel far on a cold, clear night or early morning. During such weather, there is sometimes a layer of cold air immediately above the ground, called a *temperature inversion*. This is because the ground has cooled off faster than the atmosphere in general, owing to the solid ground being better at radiating heat into space than the air (a point we will return to in §10.3.2 when discussing climate change). As rays of sound waves made at ground level travel upwards as well as horizontally, when they approach the boundary with the warmer atmosphere above, they start to travel slightly faster. The ray of sound waves is turned towards the medium with lower c, and for rays propagating at a sufficiently shallow angle, the temperature inversion causes them to be completely reflected back towards the ground. This means that not all of the sound energy created at a point on the ground spreads out in a perfect sphere; some energy is trapped close to the ground, allowing it to travel farther than normal. This is the *waveguide* phenomenon that occurs for all waves, including light as well as sound. (The optical fibres that are the basis of the fastest internet links are waveguides.) In §12.4.3 waveguides will be mentioned in the context of internal gravity waves in the atmosphere. In principle, shallow-water waves could experience waveguide effects too, but they would need to be propagating from shallow towards deeper water, which is not so common in the sea.

Sound waves also refract and reflect when they pass through tissues of slightly different densities in the human body. Modern ultrasound scanners can thus distinguish slight variations in the density of the soft tissues in the body better than X-rays.

3.3.5 Refraction

A ray of waves is 'bent' as it travels through a medium with speed c_1, hits the boundary at angle θ_1 to the normal, and crosses the boundary into another medium with speed c_2, where it makes an angle of θ_2 to the normal (figure 3.5). The angle through which the ray turns is derived using geometry and the

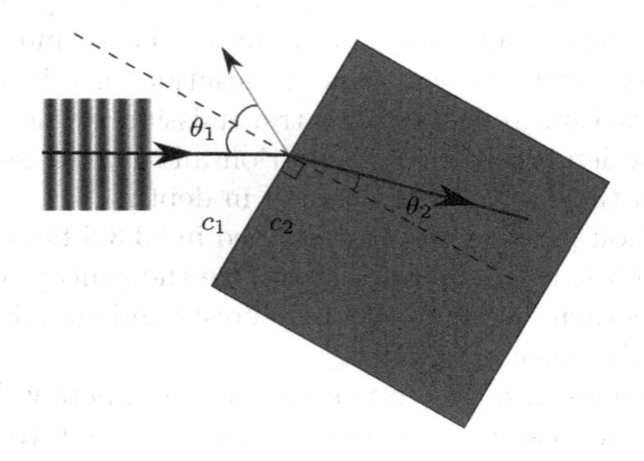

FIGURE 3.5
The angle of refraction of an incident ray of waves, shown by the black arrow at left, depends on the phase speed c in each medium. Snell's Law, (3.25), applies to electromagnetic waves such a light, and also to fluid waves such as sound or surface gravity waves. Here, $c_2 < c_1$, so that $\theta_2 < \theta_1$; for sound waves, the region with lower speed c_2 could be colder air (see figure 3.3) and for surface gravity waves it could be a zone of shallower water (e.g. figure 2.14). A reflection from the boundary (§3.3.7) is also shown, at an equal angle θ_1 to the normal.

principle that the total time to travel from one medium to another must be minimised. These considerations give

$$\boxed{\frac{\sin\theta_1}{c_1} = \frac{\sin\theta_2}{c_2}}, \tag{3.25}$$

which is traditionally attributed to Willebrord Snellius in 1621, and thus is called *Snell's Law*. However, owing to its importance in optics, it has a far more ancient provenance, and its first correct derivation is now ascribed to Ibn Sahl in 984. The relation (3.25) applies to any waves, be they electromagnetic waves or fluid waves, as long as the wave speed c is different in different media. The *refractive index*, n, that is commonly quoted for light is defined as $n \equiv c_0/c$, where c_0 is the speed of light in a vacuum, and c is the speed of light in some medium such as air, water, or glass. For fluid waves, there is no equivalent to c_0, which in any case would cancel out of the optical version of Snell's Law based on refractive index, $n_1 \sin\theta_1 = n_2 \sin\theta_2$, to give (3.25).

3.3.6 Acoustic impedance

The relation (3.25) allows the calculation of the angle of refraction. However, some wave power is typically refracted, and some is reflected. In order to calculate the proportions of wave power subjected to refraction and reflection phenomena, and, ultimately, to estimate how much power gets through - how much is transmitted - it is necessary to relate the variations in pressure to the variations in velocity, $\boldsymbol{u}$. This is the velocity with which the fluid particles move as sound waves pass. Typically, the magnitude of $\boldsymbol{u}$ is much smaller than the sound-wave speed c. Returning to (3.6), assume for simplicity one-dimensional plane waves. Then, (3.6) becomes

$$\rho_0 \frac{\partial u}{\partial t} = -\frac{\partial p}{\partial x}. \tag{3.26}$$

Substitution of the general solution (3.21) into (3.26), and choosing constants of integration such that waves are propagating in the positive x-direction gives

$$u = \frac{1}{\rho_0 c} p, \tag{3.27}$$

or, using the real numbers $\hat{p}$ and $\hat{u}$ to represent the amplitude of the pressure and velocity respectively,

$$\boxed{\hat{p} = \hat{Z}\hat{u}}, \tag{3.28}$$

where $\hat{Z} = \rho_0 c$. The same relation would occur if the constants were chosen so the wave propagates in the $-x$ direction.

It is clear that $\hat{Z}$ is purely a property of the fluid medium, called the *characteristic specific acoustic impedance.* (In cases where the propagation is as a spherical wave, or where one of the media is solid, the equivalent of $\hat{Z}$ in the equivalent of (3.28) would be a complex number, and the term 'characteristic' refers to the magnitude of that complex number, and the term 'specific' means it is per unit volume.) In fact, recalling the relation between velocity and pressure for water-surface waves, differentiating (2.13) with respect to x would give an equation looking just like (3.27), only with c being the speed of water waves; and if the water waves were shallow-water waves, the equivalent product $\rho_0 c$ would also be purely a property of the fluid medium: the depth of the water and its density.

If the relation (3.27) is written as $\hat{u} = \hat{p}/\hat{Z}$, an analogy with electric circuits is evident, with $\hat{u}$ the analogue of electric current and $\hat{p}$ the analogue of voltage. The characteristic specific acoustic impedance represents the alternating velocity produced in the fluid in reaction to the application of an alternating pressure, analogous to electric impedance represents the alternating current produced in a wire in reaction to an alternating voltage.

If wave power is to be transmitted from one medium to another with minimum power losses, the impedances should be matched; this is a perfectly general requirement in any physical system that supports waves or oscillations and will be derived in §8.5.2.

3.3.7 Reflection, scattering and transmission

As a wave crosses a boundary, both the pressure and the normal component of the velocity of particles must be the same on the other side. The consequence may be *reflection*, in which a fraction of the wave's energy must bounce back from the boundary. The angle with which it bounces back, according to the *Law of Reflection*, is the same angle to the normal to the boundary, θ_1, with which the incident ray arrived (as shown in figure 3.5). If the boundary is between a fluid and another fluid (such as an air-water boundary) and if the waves are plane waves, the fraction of the wave's energy that is reflected can be determined in a dozen or so lines of algebra by matching the pressures and the components of velocity normal to the boundary on either side (see Kinsler and Frey, 1962). This creates another relation involving the angles θ_1 and θ_2 and the pressure amplitude of the reflected wave, $\hat{p}_r$, relative to the pressure amplitude of the incident wave, $\hat{p}_1$, given by

$$\boxed{\frac{\hat{p}_r}{\hat{p}_1} = \frac{\hat{Z}_2 \cos\theta_1 - \hat{Z}_1 \cos\theta_2}{\hat{Z}_2 \cos\theta_1 + \hat{Z}_1 \cos\theta_2}}. \tag{3.29}$$

If $\theta_1 = \theta_2 = 0$, it is clear that the sign of $\hat{p}_r/\hat{p}_1$ will be positive if $\hat{Z}_2 > \hat{Z}_1$, so the reflected wave is in phase with the incident wave (it has a phase shift of 0°), re-inforcing it; for example, sound passing from air to water will undergo this sort of reflection. Meanwhile, if $\hat{Z}_2 < \hat{Z}_1$, the reflected pressure is negative (it has a phase shift of 180°), cancelling some of the incident wave; this would happen if sound passes from water to air. However, it is important to remember that spherical wave propagation, or a solid boundary, generate complex impedances, so that a complex version of (3.29) is required (given in Kinsler and Frey, 1962, for various combinations of boundaries and wave geometry); the complex ratio of pressures represents some phase shift that is neither 0 nor 180 degrees. The relation (3.29) could be applied to plane water-surface waves with $\hat{Z}$ replaced by $\rho_0\sqrt{gh}$.

Both refraction and reflection lead to *scattering*. Any kind of wave suffers scattering as it travels through media with varying properties and hence varying values of $\hat{Z}$. Wave energy is refracted or reflected at each boundary, wherever small (or not so small) variations in the medium occur (figure 3.6). For sound waves in a fluid, the variation could be due to zones with different temperatures. The phenomenon of scattering, like the refraction and reflection that cause it, is perfectly general and occurs for sound waves travelling through solids as well, where materials of different internal compositions present internal boundaries with different values of $\hat{Z}$. Thus, less wave energy gets through in the direction the wave was originally travelling, and hence there is less *transmission*.

It is important to note that there are many classes of scattering, and the situation in the cartoon of figure 3.6 is specific to only one of them, in which the wavelength is small compared to the size of the scattering objects. The processes of refraction and reflection discussed above are *elastic processes*, in which no energy is lost to the various media through which the wave is travelling, and their consequence is *elastic scattering*, which in a fluid-wave context means that there is no viscosity causing conversion of the wave energy

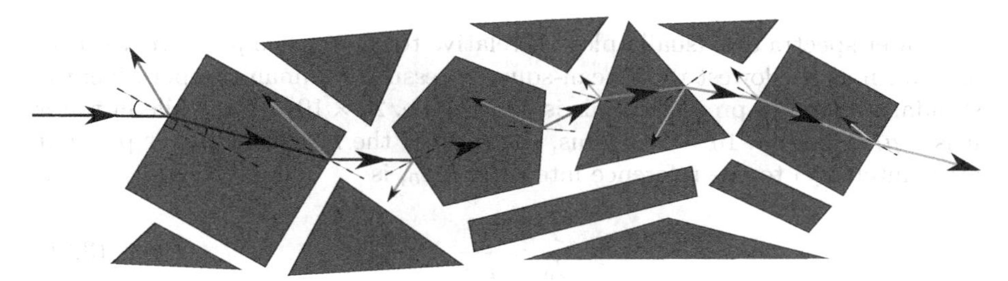

FIGURE 3.6
Scattering due to refraction and reflection. As the ray of wave energy passes through each boundary, it is refracted, causing changes of direction, and some of the energy is also reflected. The result is that only a fraction of the incident energy passes through.

to heat. As discussed below in §3.4, for sound waves, the practical outcome is that all the refractions and refractions, even if elastic, would cause a loss of transmitted energy. Moreover, in reality, viscosity does exist, and the great increase in the number and lengths of the paths taken by the energy owing to scattering hastens the transformation of wave energy to heat. In §9.6.2, one of many forms of *inelastic scattering* is described, that due to microbubbles interacting with ultrasonic waves.

3.3.8 Representation and measurement of sound

3.3.8.1 Spectral representation of sound

Because absorption can only be determined by energy loss, when we measure sound and select materials to control sound, it is necessary to measure the rate of transmission of sound energy. Furthermore, the vast majority of practical applications require such measurements to be made as a function of frequency, so that sound should be represented spectrally, as detailed in §1.5.6. The rate of transmission of energy is power. Now, Power = Force × Velocity, or Pressure × Volume Flux. It is necessary to multiply two sinusoidal quantities, p and u, and, as pointed out in §1.5.1, the real part of each quantity must first be taken. Since from (3.28), $\hat{p} = \hat{Z}\hat{u}$, pressure and velocity amplitudes are directly proportional, and, for plane waves (but not spherical waves) the time-varying pressure and velocity are also directly proportional. Now, from (3.21), $p = A\exp[\mathrm{i}(\pm\kappa x \pm \ell y \pm mz + \omega t)]$, so pressure and velocity amplitudes can be written $\hat{p} = |A|$ and $\hat{u} = |A|/\hat{Z}$. Taking the time average of the power per unit area, pu, and recalling that the time-averaging operation (1.47) in §1.5.1 generates a factor of 1/2, the *sound intensity*, I, is given by

$$\boxed{I = \frac{\hat{p}^2}{2\hat{Z}}}, \tag{3.30}$$

where I is measured in Watts per square metre.

Power spectra are usually plotted relative to a *reference pressure*, $p_\emptyset$, usually taken as the lowest root-mean-square pressure a human can perceive: the standard reference pressure in air is 20 μPa, i.e. 20×10^{-6} Pa, while in water it is 1 μPa, i.e. 1×10^{-6} Pa. Thus, the ratio of the measured power per unit area (intensity) to the reference intensity, I/I_0, is

$$\frac{I}{I_0} = \frac{\hat{p}^2}{{p_\emptyset}^2}, \tag{3.31}$$

where the reference intensity for air is usually taken as $I_0 = 10^{-12}$ W m^{-2}. In fact, it is this number, I_0, that is the standard estimated from the human-hearing threshold, and the value of $p_\emptyset = 20 \times 10^{-6}$ Pa for $p_\emptyset$ is derived from I_0, using (3.30), with $\hat{Z}$ under conditions of standard atmospheric pressure and 20°C. When the log to the base 10 of the ratio on the left-hand side of (3.31) is taken, the result is usually called the *sound intensity level*, L_I, and although the ratio has no dimensions, it is conventionally given the unit of 'bel' with abbreviation B, derived from the name of Alexander Graham Bell, inventor of the telephone. Thus, L_I is given by

$$\begin{aligned} L_I &= \log_{10}\left(\frac{I}{I_0}\right) \quad \text{B} \\ \text{or} \quad L_I &= 10\log_{10}\left(\frac{I}{I_0}\right) \quad \text{dB}, \end{aligned} \tag{3.32}$$

where dB stands for decibel, since a decibel is one-tenth of a bel. Alternatively, applying the log to the base 10 to the right-hand side of (3.31) gives what is usually called the *sound pressure level*, L_p, given by

$$L_p = 2\log_{10}\left(\frac{\hat{p}}{p_\emptyset}\right) \quad \text{B}, \tag{3.33}$$

or, as is much more commonly used, in decibel,

$$\boxed{L_p = 20\log_{10}\left(\frac{\hat{p}}{p_\emptyset}\right) \quad \text{dB}}. \tag{3.34}$$

Great care must be taken in interpreting the output of measurement devices and software, or of reports quoting 'dB'. Many practitioners become familiar with what constitutes typical 'dB' values in their field of work and blithely continue to quote values in dB, without ever relating those values to the physical quantity of pressure, p, and without mentioning the reference pressure, $p_\emptyset$. Thus, it is necessary to determine if the 'dB' scale is based on sound pressure level and not some other quantity, such as voltage inside an instrument, and to confirm the reference value, before attempting to relate measurements to predictions, or using measurements in calculations.

Of course, provided the device has a linear relationship between voltage and pressure and provided all that is of interest is a *relative* comparison of measured dB values using exactly the *same* device with exactly the *same* amplification, it does not matter; the problem arises when a translation to physical quantities is needed. The general public is often unaware that the decibel is a logarithmic measure of sound; a 100 dB sound may be considered loud, but a 140 dB sound causes pain (Kinsler and Frey, 1962).

3.3.8.2 Sound-measurement instruments

It is estimated that the number of mobile telephones, both the simple telephones that just make calls and the smartphones that permit programs ('apps') to be installed, exceeded the global population in about 2016[2]. To this total can be added an estimated 1 billion tablets and similar small computing devices[3] and even more laptop computers. All of these devices contain a microphone. While the number of telephones and computing devices is ever-changing and can only be roughly estimated, it is clear that there is an enormous number of microphones on Earth, making it one of the most ubiquitous electro-mechanical technologies.

All microphones must *transduce* (convert) the pressure fluctuations due to sound waves into an electrical quantity that can be transmitted and recorded. Microphones contain a *diaphragm* which is a very thin plate of flexible material such as paper, aluminium or plastic. Behind the diaphragm is a cavity and then a backplate. The cavity between the diaphragm and the backplate is open the atmosphere via tiny holes, and it must be, so that gradual changes in atmospheric pressure owing to changes in the weather or altitude are equalised on both sides of the diaphragm. The wavelength of audible sound waves is at least an order of magnitude greater than the size of the cavity between the diaphragm and the backplate, so one might have thought that there would be equal pressure on both sides of the diaphragm at any instant. However, the pressure fluctuations due to sound are so rapid that air is not able to flow quickly enough into and out of the tiny holes to equalise the pressure on both sides of the diaphragm, so that there is an instantaneous difference in pressure across the diaphragm. Thus, the pressure fluctuations due to sound waves impinging on the diaphragm cause it to move.

In fact, the preceding description is also valid for the human ear, in which the diaphragm is called the *eardrum*. The study of human hearing and the treatment of its disorders is a considerable branch of acoustics called *psychoacoustics*, which overlaps with medical neuroscience, psychology and electrical and biomedical engineering; an introduction is in Kinsler and Frey (1962) and an aspect of psychoacoustics will be briefly covered in §3.4.1.

---

[2]Murphy, M., 2019, Cellphones now outnumber the world's population `https://qz.com/1608103/there-are-now-more-cellphones-than-people-in-the-world/`

[3]Statista, 2020, `https://www.statista.com/statistics/377977/tablet-users-worldwide-forecast/`

In a *condenser microphone* (figure 3.7(a)), the diaphragm movement relative to the backplate changes the electrical capacitance of this gap. A power source applies a *polarizing voltage* between the diaphragm and a backplate. The diaphragm movement causes a small change in voltage which is amplified. The best-quality condenser microphones, used professionally, require more power and are usually large - and expensive. They can be damaged if moisture gets in the cavity.

The *electret microphone* (figure 3.7(b)) is a type of condenser microphone. A permanent static-electric-charged diaphragm or backplate eliminates the requirement for a polarizing voltage. Electret microphones can be very small - the vast majority of inbuilt microphones are electrets. They are usually much cheaper than polarized condenser microphones, so that the term 'condenser microphone' is often taken to mean the more expensive polarized condenser microphones. Consistently with their lower cost, electret microphones are often lower in quality than polarized condenser microphones; however, where the small size of an electret microphone is valuable, very high-quality electret microphones are also available, for a matching cost. Like polarized condenser microphones, electret microphones are damaged by moisture.

Dynamic microphones (figure 3.7(c)) work rather like loudspeakers in reverse. A coil of wire is moved by a membrane relative to a magnet. They are comparatively robust and moisture-resistant. They have a narrower frequency-response bandwidth than a good condenser microphone, which may facilitate their use in specialised applications where multiple microphones are available. Thus, they tend to be used in professional performance and studio situations, in conjunction with other microphones. The *shotgun* microphone (figure 3.7(d)) only transducers sound from the source to which it is directly pointing. Shotgun microphones are highly directional, and thus reduce noise from sources in other directions; for example, they may be used to pick up the words from a single speaker in a noisy crowd.

Underwater sound is transduced by a *hydrophone*; an example is shown in figure 3.8. Its operating principle is completely different to that of a microphone. Clearly, a cavity filled with water that, unlike air, conducts electricity, would render the condenser microphone principle useless. The dynamic-microphone principle would also be problematic underwater. Thus, the hydrophone does not have a delicate flexible membrane with an air cavity within it; rather, it is entirely made of solid materials. A *piezoelectric crystal* is embedded in a special rubber with acoustic impedance $\hat{Z} = \rho_0 c$ designed to be the same as the impedance of water. Pressure fluctuations cause tiny compressions and rarefactions of this crystal, which in turn causes tiny movements of charge into and out of the crystal. The hydrophone requires a specialised electronic *charge amplifier* that is usually kept above water or in a secure dry place. Hydrophones are very robust, but the hydrophone is connected to its charge amplifier by a long, sensitive coaxial cable. In practice, this cable participates in the sensing of pressure fluctuations, so the entire hydrophone and

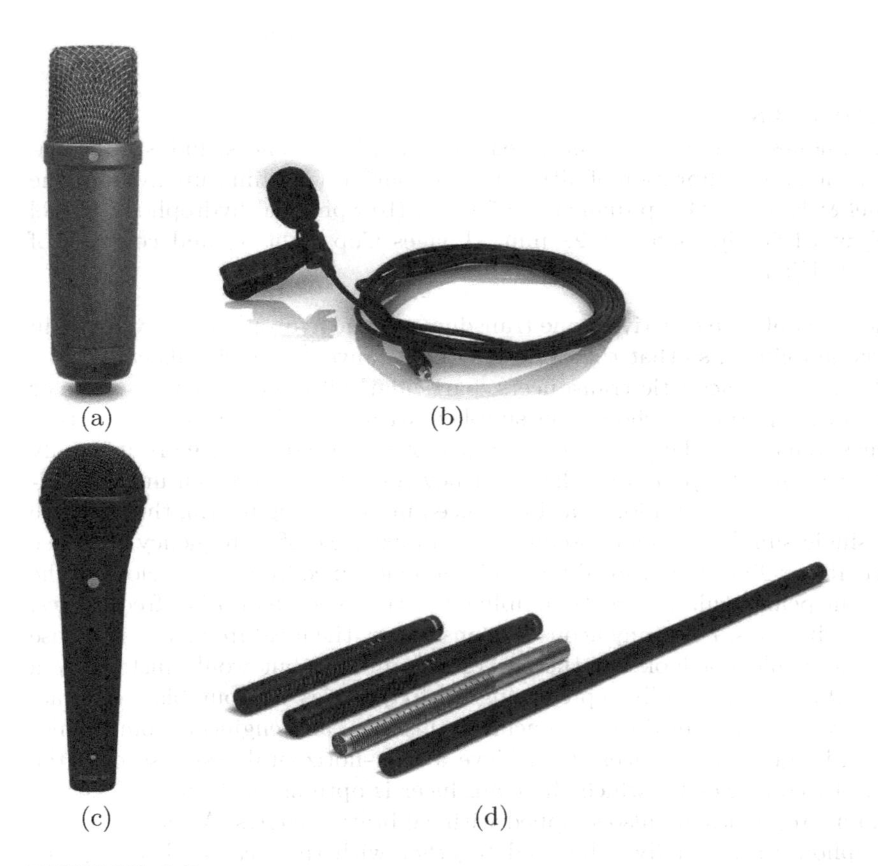

FIGURE 3.7
A variety of professional microphones intended for the musical performance and recording, and journalism industries. (a) Studio condenser microphone, length 190 mm (Røde NT1-A); (b) Electret microphone, mass 1 g (Røde Lavalier); (c) Dynamic microphone, length 171 mm (Røde M1); (d) Shotgun microphones, diameters 19-22 mm (Røde NGT range). Images copyright by and courtesy of Røde Microphones.

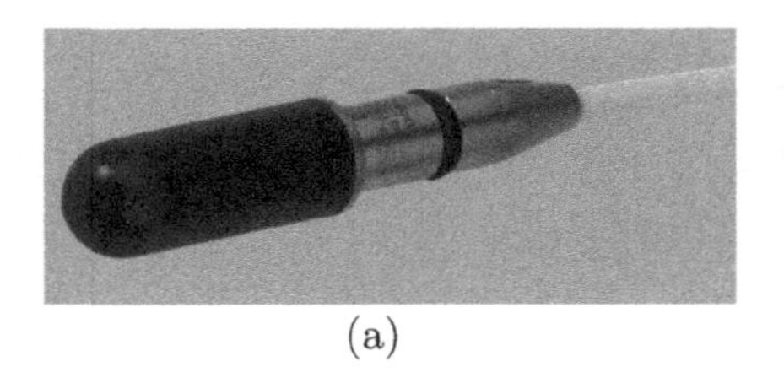

(a)

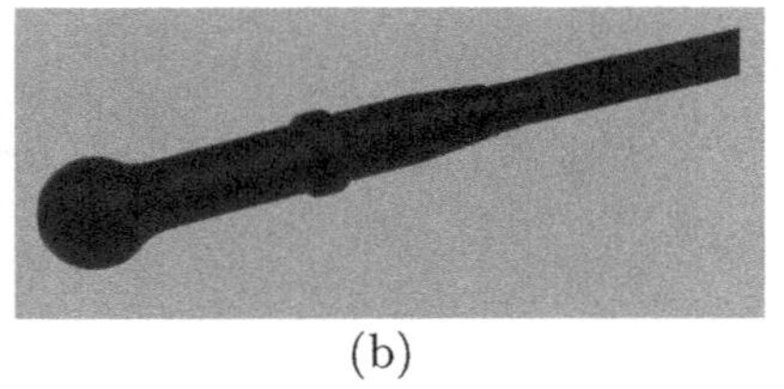

(b)

FIGURE 3.8
Hydrophones used in scientific research in marine sciences, industrial measurements, and calibration of ultrasonic equipment. (a) Miniature hydrophone (Brüel & Kjær 8103, tip diameter 9.5 mm); (b) Spherical hydrophone (Brüel & Kjær 8105, tip diameter 22 mm). Images Copyright © and courtesy of Brüel & Kjær.

cable assembly are effectively one transducer. Some hydrophones have built-in charge amplifiers, so that the cable may be a conventional signals cable.

Professional acoustic transducers, particularly if intended for scientific or engineering purposes, should be supplied with a *calibration* chart. This is a measurement of the *frequency response* of the instrument, more generally called the *transfer function.* The frequency response requires an understanding of sound in spectral form, as introduced in §1.5.6. Figure 1.6, the response of a single simple-harmonic oscillator, is an example of a frequency response chart. It is definitely not ideal for a microphone, since frequencies close to the resonant peak would be greatly amplified at the expense of other frequencies, introducing bias. For many acoustic transducers, the ideal frequency response function would not look like the curves in figure 1.6 but would instead be a horizontal line, faithfully reproducing all frequencies without bias. In practice, however, no transducer can achieve this ideal and engineers compromise in the design of transducers to achieve a near-horizontal response over the band of frequencies for which the transducer is optimised. Some professional audio microphones are also supplied with calibration charts. As noted earlier, hydrophones are usually calibrated together with their cable. The highest-quality transducers may be supplied with individualised calibration charts, which take into account the inevitable tiny differences that may occur in the manufacture of each device.

3.3.9 Geometrical spreading

Geometrical spreading applies to all waves, but since sound waves are more likely to begin from a source that is small compared to the distances over which the sound is heard, and the distances can vary from quite close to the source to very far away, geometrical spreading can be very marked for sound waves. Consider the conservation of mass given by (3.1) in a spherical coordinate system that is symmetric, so that the only spatial variation is in the radius, r. The radial factor in the solution of the wave equation when written in a spherically-symmetric coordinate system is given by the *spherical Bessel functions* of order zero,

$$\frac{1}{r}e^{\pm i\kappa r},$$

(the real part of which was used in (3.24)) so that the solution is

$$p = \frac{A}{r}e^{i(\pm\kappa r+\omega t)}, \qquad (3.35)$$

with amplitude

$$\hat{p}(r) = \frac{|A|}{r} \qquad (3.36)$$

A sound source of radius R_0 will therefore give a pressure amplitude at a radius r given by

$$\frac{\hat{p}(r)}{R_0} = \frac{\hat{p}(R_0)}{r} \qquad (3.37)$$

Since, from (3.30), $I = \hat{p}^2/(2\hat{Z})$,

$$\boxed{\frac{I(r)}{I(R_0)} = \left(\frac{R_0}{r}\right)^2}, \qquad (3.38)$$

which is the *Inverse Square Law* familiar in many areas of physics, including the spreading of electromagnetic radiation.

3.3.10 Doppler effect

As with any waves, if sound waves are emitted by an object moving towards an observer, the observed frequency is higher than the emitted frequency; and if the object is moving away, the observed frequency is lower. This has many applications, such as fisheries or naval sonar and medical ultrasound. A common derivation of this effect considers first a moving transmitter and stationary receiver, and secondly a stationary transmitter and a moving receiver.

In the first case, experienced if standing by a road while an ambulance sounding its siren approaches, a transmitter of sound is moving with speed U_t towards a stationary receiver. It is making sound with frequency f_0 and the speed of sound is c. The wavelength of sound is, as usual, given by $\lambda_0 = c/f_0$. Say that a wave crest was emitted by the transmitter one period ago, i.e., T_0 seconds where $T_0 = 1/f_0$; it will have travelled a distance λ_0 over that time. Now, the transmitter is just about to emit another wave crest, but it has moved a distance U_tT_0 towards the receiver. Therefore, the apparent wavelength, called λ_1, say, has been reduced from λ_0 to $\lambda_1 = \lambda_0 - U_tT_0$. As always, the crests and troughs travel *relative to the medium*, and therefore relative to the stationary receiver, at the phase speed c. The frequency f_1 that the receiver experiences is therefore, $f_1 = c/\lambda_1$, i.e.

$$f_1 = \frac{c}{\lambda_0 - U_t T_0} = \frac{1}{(1 - U_t/c)}\frac{c}{\lambda_0} = \frac{c}{c - U_t} f_0.$$

When the ambulance is approaching, $U_t > 0$ so a higher-frequency sound is heard; once it passes, $U_t < 0$ so a lower-frequency sound is heard.

In the second case, a receiver of sound is moving away at speed U_r from a stationary transmitter. This is different to the moving transmitter, because now, like the sound, the *receiver is moving relative to the medium*. Thus the speed of the wave relative to the receiver is $c - U_r$ and the frequency f_2 that the receiver experiences is therefore,

$$f_2 = \frac{c - U_r}{\lambda_0} = \frac{c - U_r}{c} f_0.$$

The general case when both are moving is obtained by replacing f_0 with f_1 in the expression for f_2, giving

$$\boxed{f_{12} = \frac{c - U_r}{c - U_t} f_0}. \tag{3.39}$$

Now, in a situation typical of most measurement systems such as marine sonar or medical ultrasound, imagine an object moving at speed U_r is receiving sound waves emitted with frequency f_0 from a stationary transmitter ($U_t = 0$). According to (3.39), it will receive them at frequency f_2 given by

$$f_2 = \frac{c - U_r}{c} f_0, \tag{3.40}$$

Some of this received wave energy will be reflected back to the source, and this reflection is effectively a new transmission from a moving transmitter travelling at speed U_r. The fact that it is a reflection can be represented by a negative sign of c. Thus the frequency reflected back, f_1' is

$$f_1' = \frac{-c}{-c - U_r} f_2 = \frac{c}{c + U_r} f_2, \tag{3.41}$$

so that the frequency received back at the source is

$$f_1' = \frac{c - U_r}{c + U_r} f_0. \tag{3.42}$$

The difference in frequency, f_d, between the received reflection, and the original transmitted frequency is

$$f_d = f_0 - f_1' = \left(1 - \frac{c - U_r}{c + U_r}\right) f_0, \tag{3.43}$$

so that this difference (or 'shift') in frequency is seen to be

$$\boxed{f_d = \frac{2U_r}{c + U_r} f_0}. \tag{3.44}$$

This result is generally known as the *Doppler shift* after Christian Doppler, who studied frequency changes due to motion in 1842 in an astrophysical context.

If the object's speed is small compared with the speed of sound,

$$f_d \simeq \frac{2U_r}{c} f_0. \tag{3.45}$$

If the sound waves are travelling at an angle θ to the direction of the moving object, then the component of the object's velocity in the direction of motion is what causes the Doppler shift, so (3.45) is written

$$\boxed{f_d = \frac{2U_r \cos\theta}{c} f_0}. \tag{3.46}$$

The relation (3.46) is used in ultrasound measurements of blood flow, where θ is the angle between the direction the probe is pointing and the centreline of a blood vessel. The ultrasound-scanner software typically overlays a colour representing the magnitude of U_r - the speed of blood flow - on the image of the blood vessel and the surrounding tissue. The doctor or technician holding the probe usually has to interactively enter a line identifying the blood vessel and hence θ.

3.4 Building acoustics

3.4.1 Reverberation

Sound generated in a room will reflect from the walls, ceiling and floor unless these are perfect absorbers. Each reflection will in turn reflect off the opposite surface, and return to reflect off the first surface, and so on. In general, the sound will become weaker with each reflection, though it does so by some combination of geometrical spreading and loss of sound energy on each reflection that clearly depends on the size and shape of the room as well as the material and structure of the walls.

The fraction of the sound pressure amplitude that reflects from a surface of an extremely thick, pure and uniform substance of known c and ρ_0, and hence known impedance $\hat{Z}$, could be predicted from the reflection equation, (3.29). In practice, however, walls, ceilings and floors are not extremely thick, nor are they made of pure substances; common building materials such as wood, plaster, bricks and concrete are complex materials made up of different grains and particles with different values of $\hat{Z}$. As discussed in §3.3.7, scattering due to internal reflection and refraction within the wall-coating material occurs, and *absorption* of energy within the material occurs, in which the viscosity of the medium ultimately converts the energy to heat. From a theoretical perspective, one could try to distinguish energy that is elastically scattered, and in principle is still available as sound, from the energy that is absorbed, and thence converted to heat. The surviving sound that is transmitted through the material can reflect off whatever surface is behind the wall coating. It then undergoes more scattering and absorption on the way back through the coating. If there is no reflective surface behind the wall, the sound is transmitted into the space beyond and is therefore 'lost' from the room.

The internally scattered and internally absorbed energy is generally too complex to predict, and some energy initially scattered within the material is almost inevitably also absorbed and converted to heat. Furthermore, as just-noted, some sound could survive the processes within the wall and be transmitted through the wall into whatever space is beyond the wall, thence being unavailable for reflection back into the room. Thus, in practice, all these losses are dealt with empirically. The sound energy reflected back from the material, whatever the cause of the loss, is measured for various thicknesses of each material. All the scattering, energy-absorption and transmission phenomena are therefore lumped together into an *acoustic absorption co-efficient*, a_a.

The fraction of the incident sound intensity 'absorbed' (i.e. not reflected) by a surface is, therefore, a_a, and the fraction of the incident sound intensity reflected is $1 - a_a$. A door open to the outside world has complete absorption and hence $a_a = 1$; even though there is nothing there, nothing is reflected, so 100% of the energy is defined as 'lost'.

The time it takes the reflections to die down to a level imperceptible to humans is called the *reverberation time*. Reverberation is in essence a practical concept that tries to encapsulate the aural experience of a room or other internal space in numbers that can be scientifically measured. Thus, in order to understand the importance of reverberation time in engineering and architectural design, we need to briefly study an aspect of psychoacoustics, which, as noted in §3.3.8.2, is a broad multi-disciplinary field.

Humans perceive sound frequencies between about 20 Hz and 20 kHz, and we perceive sound pressure amplitudes between about 20 μPa and 100 kPa; the lower value is defined as the reference pressure used in (3.31). However, humans perceive amplitudes nonlinearly, explaining the benefit of the logarithmic decibel scale introduced in §3.3.8.1. Our brains can discriminate between tiny differences in frequency: less than 4 Hz (0.4%) between 1000 and 2000 Hz. The difference between musical notes C_4 and $C\sharp_4$, which are adjacent notes on the piano keyboard, is (from table 1.1) 15 Hz and thus is easily perceived.

Humans perceive reflections returning after more than about 50 milliseconds as *echoes*. Reverberations are therefore defined as reflections returning in less than 50 milliseconds. We are not consciously aware of the individual reflections occurring during reverberation. But the brain definitely notices reverberation in processing sound, which is easily demonstrated by listening to standard signals with varying amounts of reverberation added.

Reverberation time, T_{60}, is defined as the time the sound pressure level takes to fall by 60 dB. The measurement of T_{60} is an engineering standard defined in the International Standard Organisation (ISO) standard ISO 3382 and is measured in seconds.

Reverberation time is given by the semi-empirical relation

$$\boxed{T_{60} = \frac{24\ln(10)}{c}\frac{V}{a_a S}}, \tag{3.47}$$

where c is the speed of sound in air at 20°C, V is the volume of the room, a_a is the acoustic absorption coefficient and S is the surface area of the room. The relation (3.47) can be derived from considerations of the energy lost over time but is essentially an empirical relation determined by the experiments on lecture-hall acoustics undertaken by Wallace Clement Sabine in 1895. It is clear that (3.47) takes into account the geometry of the room as well as the properties of the walls.

If T_{60} is too long, speech may be distorted by the reflected sound returning before the speaker has finished speaking a syllable. Furthermore, noise from other speakers will linger, making it difficult and tiring to make out the words of one's interlocutor. If T_{60} is too short, sound may 'disappear' too quickly for listeners to detect, unless they are directly facing the speaker, which may also create miscommunications. Thus, an appropriate reverberation time is important in any space in which humans are to communicate with each other verbally, and moreover, the nature of that communication affects the desired reverberation time (table 3.1). Appropriate reverberation times are also vital for spaces in which music is performed. An extreme example of a very short

TABLE 3.1
Desirable reverberation times for various spaces. From Rakerd et al. (2018).

Interior space	T_{60} (s)
Anechoic chamber	0.05
Kindergarten	0.4
Bedroom	0.5
Office	0.6
Music room	1.0
Large performance hall	2.0

T_{60} is an *anechoic chamber*, which is designed to almost completely eliminate reflected sound and is used for testing instruments and products for audible sound. Some people in anechoic chambers report feeling strange or uncomfortable, indicating the importance of reverberant sound in human sensing of the environment around them.

3.5 Problems

1. A specialised medical ultrasound device is being designed to measure the speed of blood flow in small arteries, using the Doppler shift. Its operating frequency will be 10 MHz (1×10^7 s^{-1}). What change in frequency must it detect to measure a blood-flow speed of 1 cm s^{-1}? The speed of sound in the blood is 1500 ms $^{-1}$.

2. A recording studio acquires wall material with pyramid-shaped wedges that project 0.15 m inwards. Assuming the wedges are quarter-wavelength absorbers, what frequency will have the minimum reflection from the walls? The speed of sound in air is 340 ms $^{-1}$.

3. A new design of an underwater inspection system using ultrasound imaging is being tested in a tank 10 m long, 2 m deep and 4 m wide. The system is meant to operate in the open ocean, imaging objects up to 1 m away, so reverberations from the test-pool walls with acoustic absorption co-efficient 0.5 should have fallen by 60 dB before returning to the detector. The time between ultrasound pulses is 0.04 s so the reverberation time must be less than this. The speed of sound in water is 1480 m s^{-1}. What is the largest distance a false wall (of the same material as the other walls) could be placed to reduce the pool length from 10 m?

4. A medical ultrasound Doppler blood-flow measurement is fundamentally a measurement of

   A The acoustic impedance of the patient's blood

   B The component of blood-flow velocity at right angles to the centreline of the ultrasound probe

   C The blood-flow speed irrespective of the orientation of the ultrasound probe

   D The component of blood-flow velocity in the direction of the centreline of the ultrasound probe

5. It is proposed to use fixed sound recorders to estimate the speed of passing vehicles. As a vehicle approaches, a frequency of 1000 Hz is recorded and as it recedes, a frequency of 916 Hz is recorded. How fast is it travelling? The speed of sound in air is 340 ms $^{-1}$.

6. The theory of linear sound waves is sufficient for

   A Calculation of the sound volume during a musical performance

   B Design of medical devices for acoustically destroying kidney stones

   C Estimation of the time for a dolphin's sonar echo to return

   D Predicting the speed of the shock wave from a supersonic aircraft

7. An ultrasonic level sensor in a minerals-industry tank operates in the air with $c =$ 340 m s^{-1} above a liquid surface. It measures a time of 23.5 milliseconds between each pulse and the returning echo. The distance to the surface is

A 3.4 m

B 4.0 m

C 6.9 m

D 8.0 m

8. Traffic noise typically has a peak frequency of 1 kHz. The speed of sound in air is 340 ms $^{-1}$. Assuming a quarter-wavelength thick sound barrier can be constructed of suitable material, how thick should it be in metres?

9. The characteristic specific acoustic impedance is

A The fluid particle velocity divided by the acoustic pressure

B The acoustic pressure divided by the fluid particle velocity

C The ratio of the fluid particle velocity to the acoustic pressure

D The ratio of the fluid density to the fluid particle velocity

10. As sound waves cross a boundary into a medium with a higher speed of sound they will be

A Always be totally internally reflected

B Bent towards the normal to the boundary

C Partially reflected and continue at the same angle to the normal to the boundary

D Bent away from the normal to the boundary

11. A pneumatic tool at a construction site creates a bang in the surrounding air with a pressure amplitude of 20 Pa. What is the sound pressure level in dB?

A 60 dB

B 100 dB

C 120 dB

D 240 dB

12. Factory renovations removed a wall between workers and a machine, increasing their noise exposure by 18 dB. By what factor does the sound intensity experienced by workers increase?

13. Trains on a busy line create noise with a sound pressure level measured at 60 dB (re 20 μPa) in a nearby residence over the 20 s trains take to pass. Trains pass every 120 s in the morning rush hour. What is the average sound pressure level during rush hour?

14. A firework detonation creates an approximately spherical sound wave measured from a test tower 10 m away to have a pressure amplitude of 860 Pa. What would be the pressure amplitude experienced by observers on an apartment balcony 50 m away?

15. Hydrophones measure underwater sound based on

A A piezoceramic crystal coupled to a charge amplifier

B A piezoceramic crystal with a polarizing voltage applied

C A thin flexible membrane coupled to a charge amplifier

D A thin flexible membrane with a polarizing voltage applied

16. The time taken for the sound pressure level in a room to fall by 60 dB after a single musical note is struck is

A The musical note duration

B The musical beat period
C The reverberation time
D The echo time

17. If a room is nearly anechoic the reverberation time is

A Extremely short
B Extremely long
C Greater than 50 ms
D Between 50 ms and 100 ms

18. The ratio of the frequencies of the musical notes C and D is

A $2^{1/12}$
B $2^{1/6}$
C $2^{1/4}$
D $2^{1/3}$

19. An open-plan office space is 10 m long, 6 m wide and 3.5 m high. What should be the absorption coefficient of material lining the walls, floor and ceiling to achieve a reverberation time of 0.5 second? The speed of sound in air is 340 ms $^{-1}$.

20. A musical performance is scheduled to occur in a rectilinear room 10 m long, 8 m wide and 3.5 m high. The walls are plasterboard with an acoustic absorption coefficient of 0.15 and the back wall can be moved to make the room less than 10 m long. What should be the location of the back wall to achieve a reverberation time of 1.0 s? The speed of sound in air is 340 m s^{-1}.

21. A navy sonar system is being designed to automatically measure the speed of objects moving towards it, using the Doppler shift. Its operating frequency will be 100 kHz (1×10^5 s^{-1}). What change in frequency must it detect to measure a target speed of 0.5 m s^{-1}? The speed of sound in seawater is 1500 ms $^{-1}$.

22. A submarine is coated with material 0.2 m thick intended to absorb sonar signals. Assuming the material is a quarter-wavelength absorber, what sonar frequency will be optimally absorbed? The speed of sound in seawater at the submarine depth is 1500 ms $^{-1}$.

23. An aircraft travelling at twice the speed of sound emits in sequence three frequencies: 50 Hz for 1 second, then 75 Hz for 1 second, then 100 Hz for one second, and then the cycle of three frequencies is repeated. After a stationary observer has experienced the 'sonic boom' owing to the shockwave, what does the observer hear?

24. Discuss the limitations of linear inviscid sound-wave theory for

(a) The absorption of noise by building walls
(b) Designing the shape of an underwater vehicle to minimise flow noise
(c) Collecting data on thunderstorms using thunderclap sounds
(d) Predicting volcanic eruption intensity from pre-eruption noises

25. A soprano practising at two octaves above middle C finds her voice creates an annoying reverberation in a chamber-music hall. The management proposes to install quarter-wavelength sound-absorbing material. How thick should it be? The speed of sound in air is 340 ms $^{-1}$.

26. Discuss the appropriateness of an analogy between linear sound waves and

(a) The transmission of an alternating electrical current in wires
(b) The propagation of electromagnetic waves
(c) The propagation of seismic waves in the Earth's crust
(d) The propagation of waves in shallow water

4

Internal gravity waves

4.1 Summary of key points

- The **reduced gravity** in a two-layer system of two fluids of different density is given by (4.7) on page 124,

$$\boxed{g' = \left[\frac{\rho_2 - \rho_1}{\rho_1 + \rho_2}\right] \mathrm{g}},$$

where g is the acceleration due to gravity, ρ_1 is the density of the upper fluid and ρ_2 is the density of the lower fluid, with $\rho_2 > \rho_1$.

- The **dispersion relation** for a linearly stratified fluid in which wave-induced variations occur over much smaller vertical scales than the stratification, the density gradient is linear, and waves are free to propagate in any direction in Cartesian co-ordinates, is given by (4.31) on page 129,

$$\boxed{\omega^2 = \left(\frac{\kappa^2 + \ell^2}{\kappa^2 + \ell^2 + k^2}\right) N^2},$$

where κ, ℓ and k are the wavenumbers in the x, y and z directions, and N is the **buoyancy frequency** (or Brunt-Väisälä frequency) given by (4.21) on page 127,

$$\boxed{N^2 = -\frac{\mathrm{g}}{\rho_0} \frac{\partial \rho_0}{\partial z}},$$

where $\partial \rho_0 / \partial z$ is a constant gradient in the density, ρ_0, that varies in the vertical direction, z.

- Useful **textbooks** include Sutherland (2010), which is specific to internal gravity waves, Turner (1973), which focuses on flows driven by density differences in general, and Gill (1982).

DOI: 10.1201/9780429295263-4

4.2 An example

The island of examples §2.2 and §3.2 has a small marine-research station staffed by a biologist located on the island's sheltered side. An undersea pressure sensor records low-frequency oscillations. Occasional surveys in a small boat measured the temperature and salinity of the ocean. Similarly to many tropical seas, it is stably stratified owing to heat and rainfall, a state re-established after the cyclone and tsunami. It has a density of 1022 kg m^{-3} at a depth of 200 m, increasing linearly to a density of 1032 kg m^{-3} at a depth of 700 m. Just prior to the cyclone, and again over the last week, the biologist had noticed regular, periodic low-frequency signals and asked the chief engineer what they might be.

Solution

The average density, ρ_0, is $\rho_0 = (1022 + 1032)/2 = 1027$ kg m^{-3}. The buoyancy frequency, (4.21),

$$N^2 = -(g/\rho_0)\partial\rho/\partial z,$$

can be simplified since the stratification is linear. Hence $\partial\rho/\partial z = (1022 - 1032)/(-200 - (-700)) = -0.02$ kg m^{-4}, so

$$N^2 = -9.81/1027 \times (-0.02) = 0.00019 \text{ s}^{-2}.$$

Therefore, $N = 0.01382$ rad s^{-1}, and since $f = N/(2\pi)$, natural oscillations of the stratified layer should be at $f = 0.0022$ Hz.

The biologist checks and finds this is indeed the measured frequency. The chief engineer estimates that in order to make these waves, a large object would have to move repeatedly through the depths of 200 and 700 m. The biologist suspects the mining survey ship. The engineer points out that many natural phenomena cause internal waves, and the ship cannot make such disturbances deep underwater.

4.3 Linear internal gravity-wave theory

4.3.1 The influence of gravity within a fluid

When the basics of wave behaviour on the surface of a body of water were derived (chapter 2), the assumptions were that the fluid was inviscid and incompressible, that the height of the waves was small relative to the wavelength, so that the behaviour was linear, that waves were long enough for surface tension to be neglected, and that the density of the air was very much less than the density of the water, so that the momentum of the air could be neglected. Subject to these assumptions, only the magnitude of the acceleration due to gravity, g, affected the relations between wave speed, velocity and surface elevation. The density of the water was irrelevant and therefore did not appear in the formulae of chapter 2.

Very many fluid systems in nature have differences of density within the fluid that are stable, or, if they vary at all, they vary only very slowly with time. Fluids in which such variations in density occur are called *stratified* fluids. Stratification may be due to differences in temperature within the atmosphere and ocean, and in the ocean, to differences in salinity as well.

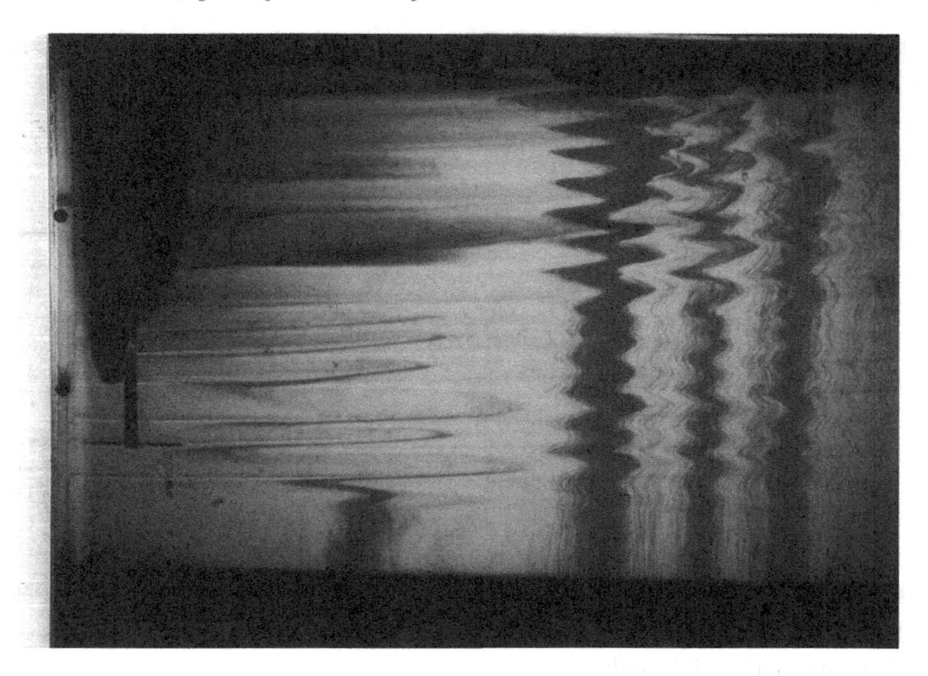

FIGURE 4.1
A block of melting ice (at the top left) releases fresh cold water into a laboratory tank in which there is a gradient of salinity. Vertical lines of dye reveal multiple layers at which intrusions of meltwater penetrate. Similar phenomena may affect the melting of Arctic and Antarctic ice. Photograph by Richard Manasseh, experiment at Dept. of Applied Mathematics and Theoretical Physics, University of Cambridge.

Sometimes, differences in both temperature and salinity occur (figure 4.1) with interesting and complex consequences. These systems feature gravity waves; indeed the familiar gravity waves we may see on the surface of the water could be considered a special case of more general gravity waves. While density variations occur in sound waves, these are small, rapid variations in time that disappear once the wave is over. For sound waves, the compressibility of the fluid provides the restoring-force 'spring' allowing waves to exist, but for stratified fluids, gravity is the restoring force, just as with surface gravity waves.

Civil and environmental engineers are concerned with stratification in water reservoirs and rivers, where heat and fresh rainwater can create a density difference that inhibits mixing, potentially allowing pathogenic bacteria to multiply, or toxic algae to bloom.

There are also many industrial systems where density stratification occurs. For example, in the smelting of metals, a layer of slag, which contains the impurities gradually being removed from the metal ore, floats on top of the layer

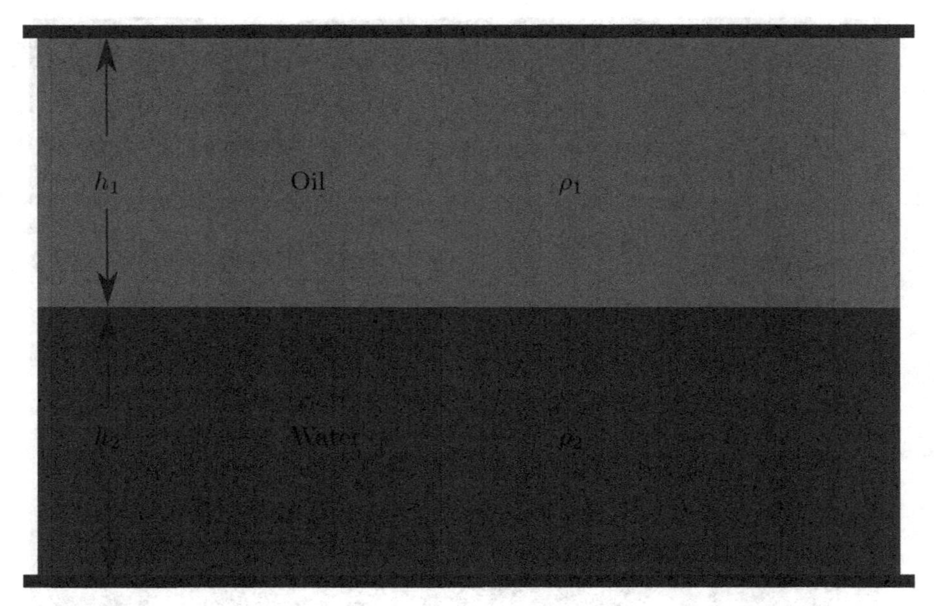

FIGURE 4.2
A two-layer rigid-lid model

of molten metal. Waves and instabilities of this interface affect the efficiency of the process. Oil and water systems also occur in the food and pharmaceuticals industries. Oil may also be split on the water in environmental accidents.

4.3.2 Two-layer rigid-lid interfacial waves

The analysis in §2 is not only valid for water, but for any liquid with a sharp interface, for example, waves in a different world where the value of g is different. The density, of the fluid, ρ, only has a role when pressures and forces need to be calculated from the solution.

Now imagine that over the water surface lies not air, but another liquid of a lower density. The result can be observed in a toy available in many varieties, in which a light, clear oil like kerosine is sealed into a clear container with a similar volume of water. The water can be dyed blue to represent the sea. A gentle shake of the container generates waves on the interface between the two liquids looking similar to waves on the sea, but in miniature. A homemade version consists of a jar half filled with water and half filled with a clear oil. To model this toy system, the subscripts 1 and 2 will be used for the upper and lower layers respectively. The container is sealed at the top, so the oil layer touches the lid without any air gap. This is called a *rigid-lid* system. The container top is made of the same material as the bottom. A schematic is shown in figure 4.2.

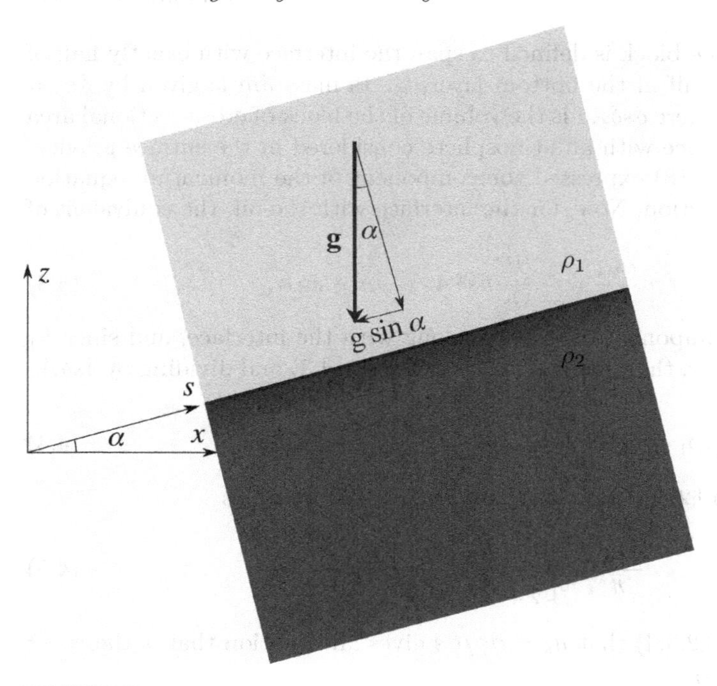

FIGURE 4.3
Interfacial boundary condition for the two-layer rigid-lid model

For the lower, water layer, the analysis is exactly the same as in §2.3.2-2.3.3. The bottom boundary condition will also be the same, given by (2.17). However, the top of the water column is no longer an interface with the atmosphere; instead, it is an interface with the oil layer. Careful consideration will have to be given to the pressure at this interface, which, unlike in §2.3.4, is not zero. The distance from the undisturbed interface to the top boundary (the container lid) is h_1, so that in the absence of any 'background' steady flows (so that the initial pressure equals the ambient pressure, i.e. $P_0 = P_\emptyset$ as discussed in §1.2.2.3), the initial (or undisturbed) pressure at the interface is given by (1.2),

$$P_0 = P_{\text{atm}} + \rho_1 g h_1 + P_s, \tag{4.1}$$

where $\rho_1 g h_1$ is the hydrostatic pressure, P_{atm} is the atmospheric pressure and P_s is some static engineering pressure, which, like P_{atm}, we will consider constant and thus will disappear in the analysis to follow. If the interface is disturbed, so that it has an upwards elevation η, the total pressure at the interface is given by (1.3) i.e. $P = P_0 + p$, where $p = -\rho_1 g\eta$, so that the total pressure P is now

$$P = P_{\text{atm}} + \rho_1 g(h_1 - \eta) + P_s. \tag{4.2}$$

Consider the acceleration of an infinitesimal block of fluid at the interface, in the direction along the interface, which (just as in §2.3.4) will be called the s-direction. The block is defined to span the interface with exactly half of it in the top and half in the bottom layer, so its mass δm is given by $\delta m = \frac{1}{2}(\rho_1+\rho_2)\delta s\delta A_\times$, where $\delta s\delta A_\times$ is the volume of the block of cross-sectional area $\delta A_\times$. For the interface with an atmosphere considered in the surface gravity-wave derivation, (2.18) expressed the component of the momentum equation (2.10) in the s-direction. Now, for the interface with the oil, the equivalent of (2.18) is

$$\delta m \frac{\partial u_s}{\partial t} = -\frac{\partial P}{\partial s}\delta s\delta A_\times - \delta m\, \mathrm{g} \sin\alpha, \tag{4.3}$$

where u_s is the component of velocity along with the interface, and since h_1 does not vary with s, the substitution of (4.2) into (4.3) and dividing by $\delta s\delta A_\times$ gives

$$\frac{1}{2}(\rho_1+\rho_2)\frac{\partial u_s}{\partial t} = \rho_1 \mathrm{g}\frac{\partial \eta}{\partial s} - \frac{1}{2}(\rho_1+\rho_2)\mathrm{g}\sin\alpha, \tag{4.4}$$

and, noting (as in §2.3.4) that $\partial\eta/\partial s = \sin\alpha$, (4.4) becomes

$$\frac{\partial u_s}{\partial t} = \left[\frac{\rho_1}{\frac{1}{2}(\rho_1+\rho_2)} - 1\right]\mathrm{g}\frac{\partial \eta}{\partial s}, \tag{4.5}$$

and noting (as in §2.3.4) that $u_s = \partial\phi/\partial s$ gives an equation that is the exact equivalent of (2.19),

$$\frac{\partial \phi}{\partial t} = -g'\eta \text{ at } z = \eta, \tag{4.6}$$

with the *only* difference between (2.19) and (4.6) being that the acceleration due to gravity, g, has been replaced with the *reduced gravity*, g', where

$$\boxed{g' = \left[\frac{\rho_2-\rho_1}{\rho_1+\rho_2}\right]\mathrm{g}}. \tag{4.7}$$

Note that (2.19) is the only point in all of §2.3 that the acceleration due to gravity is inserted into the analysis. Hence, this simple replacement of g with g' means that *all* of §2.3 after (2.19) applies perfectly well, simply with g′ replacing g. The same phenomena of shallow- and deep-water waves, wakes of vessels, shoaling on a beach and many other phenomena also apply a layer of fluid with less-dense fluid on top rather than the atmosphere. Indeed, this makes perfect sense, since, of course, the atmosphere is a less-dense fluid on top, and if ρ_1 becomes extremely small compared to ρ_2, g' reduces back to g. The ratio of densities in the square brackets in (4.7) represents the extent to which waves are interfacial rather than free-surface waves, and is sometimes called the *Atwood number*.

This analysis may have been of a toy, but the two-layer approximation is a surprisingly good representation of what can sometimes be observed in both the atmosphere and ocean, as well as in a large number of industrial processes, such as the smelting of metals noted in §4.3.1 above.

4.3.3 Waves in continuously stratified fluids

The preceding analysis in §4.3.2 of the two-layer system is correct for any two fluids that are *immiscible*, i.e. cannot mix, such as oil and water (and, of course, air and water, for which the analysis reduces to that of §2.3). It is also a good approximation even where the fluids can mix, such as fresh water and salty water, or cold air and warm air, provided the density difference is fairly large and the interface sharp. In §4.3.2 the density ρ remained constant at ρ_2 in the lower layer and constant at ρ_1 in the upper layer. In more general circumstances, however, the density varies continuously and potential flow cannot be assumed. The relations (2.13)-(2.14), $\boldsymbol{u} = \boldsymbol{\nabla}\phi$ where $\phi = \mathrm{i}/(\rho\omega)p$, are no longer valid, since the density ρ is no longer a constant. Therefore it is necessary to return to the original laws of conservation of mass and momentum and re-derive a new kind of wave solution from there. The Law of Conservation of Mass is

$$\frac{\partial \rho}{\partial t} = -\boldsymbol{\nabla} \cdot (\rho \boldsymbol{u}). \tag{4.8}$$

Although the density may now vary, an important and perhaps surprising point is that we can still deal with a large range of real-world problems by assuming the fluid is incompressible. In other words, increasing the pressure of a *motionless* mass of the fluid does not change its density, and the flow has zero divergence, i.e

$$\boldsymbol{\nabla} \cdot \boldsymbol{u} = 0. \tag{4.9}$$

Since $\boldsymbol{\nabla} \cdot (\rho \boldsymbol{u}) = \boldsymbol{u} \cdot \boldsymbol{\nabla}\rho + \rho \boldsymbol{\nabla} \cdot \boldsymbol{u}$, the Law of Conservation of Mass reduces to

$$\frac{\partial \rho}{\partial t} + \boldsymbol{u} \cdot \boldsymbol{\nabla}\rho = 0. \tag{4.10}$$

Meanwhile, the Law of Conservation of Momentum, or momentum equation, is

$$\frac{\mathrm{D}(\rho \boldsymbol{u})}{\mathrm{D}t} = \boldsymbol{\nabla} \cdot \boldsymbol{\tau} + \rho \mathbf{g}. \tag{4.11}$$

Just as in chapters 2 and 3, the momentum equation (4.11) can be simplified by assuming the fluid is inviscid, so there is no loss due to friction. The momentum equation can also be simplified by linearisation, which, as in chapter 2, is equivalent to saying that the amplitude of any waves is small compared with their length. These assumptions reduce (4.11) to

$$\frac{\partial(\rho \boldsymbol{u})}{\partial t} = -\boldsymbol{\nabla}P + \rho \mathbf{g}, \tag{4.12}$$

where P is the total pressure. For consistency with the linearising assumption on the amplitude, and thus on the velocity of the fluid, the density is supposed to have a small variation from its constant-in-time (or time-averaged) value, ρ_0, so, as with the approximation for sound waves, (3.3), the density can be written

$$\rho = \rho_0 + \varrho \tag{4.13}$$

where ϱ is the small variation of density in space and time. This small variation is due to a new sort of wavelike disturbance in a fluid that we now seek to analyse. It is important to remember that although the time-averaged density ρ_0 does not vary in time, the entire reason for undertaking this analysis is that in a stratified fluid, ρ_0 does vary in space. Typically, ρ_0 will increase with depth so that less-dense fluid floats above more-dense fluid, just as the less-dense oil layer floated above the more-dense water layer in the analysis of §4.3.2. Now, however, the density is a continuous function. The density linearisation (4.13) means that in the term $\boldsymbol{u} \cdot \boldsymbol{\nabla}(\rho_0 + \varrho)$ in the mass conservation equation (4.10), the small quantity $\boldsymbol{u}$ multiplied by the small quantity ϱ must disappear; and because ρ_0 does not vary with time, the derivative with respect to time also simplifies. Therefore, (4.10) reduces to

$$\frac{\partial \varrho}{\partial t} + \boldsymbol{u} \cdot \boldsymbol{\nabla} \rho_0 = 0. \tag{4.14}$$

Meanwhile, the momentum equation (4.12) becomes

$$\frac{\partial}{\partial t}\left[(\rho_0 + \varrho)\boldsymbol{u}\right] = -\boldsymbol{\nabla} P_0 - \boldsymbol{\nabla} p + \rho_0 \mathbf{g} + \varrho \mathbf{g}, \tag{4.15}$$

where P_0 is the constant part of the total pressure and p is the time- and space-varying dynamic pressure. Firstly, the gradient in steady-state pressure, $\boldsymbol{\nabla} P_0$, perfectly balances the part of the gravitational force that is not changing in time, $\rho_0 \mathbf{g}$, just like in §2.3, so they cancel out. Secondly, it can be seen that on the left-hand side of (4.15), the small quantity ϱ multiplies the small quantity $\boldsymbol{u}$. Thus, for consistency with the linearising assumption already applied to the mass conservation equation and to the earlier version of the momentum conservation equation, this small term should be neglected. Because ρ_0 does not vary with time, it disappears from the derivative on the left-hand side, just as in the step leading to (4.14). These steps reduce (4.15) to

$$\rho_0 \frac{\partial \boldsymbol{u}}{\partial t} = -\boldsymbol{\nabla} p + \varrho \mathbf{g}. \tag{4.16}$$

There are three equations, (4.9), (4.14) and (4.16), in the three unknowns, $\boldsymbol{u}$, p and ϱ. We now need to decide which dependent variable is most meaningful, and eliminate the other two. In considering surface or interfacial waves, the surface elevation was the most useful, since it is what one observes. However, for internal waves, it is not so clear.

To decide on a dependent variable, consider (4.14). The time-average density ρ_0 only varies in the vertical; if it did not, there would be a horizontal motion to adjust the ocean back to a state where ρ_0 only varies in the vertical. Hence, the term $\boldsymbol{u} \cdot \boldsymbol{\nabla} \rho_0$ in (4.14) becomes simply $w \partial \rho_0 / \partial z$, so (4.14) becomes

$$\frac{\partial \varrho}{\partial t} + w \frac{\partial \rho_0}{\partial z} = 0. \tag{4.17}$$

This reveals that there is a simple relation between ϱ and the vertical component of velocity, w. Therefore, we can choose to eliminate the pressure, and derive a wave equation in either ϱ or w. Of course, as with all wave-equation derivations, one could choose any other dependent variable; it is simply a matter of what is the least work to derive.

Differentiating (4.17) with respect to time gives

$$\frac{\partial^2 \varrho}{\partial t^2} + \frac{\partial w}{\partial t}\frac{\partial \rho_0}{\partial z} = 0, \tag{4.18}$$

from which w can be eliminated using the vertical component of (4.16), and recalling $\mathbf{g}$ only points in the negative vertical direction, giving

$$\frac{\partial^2 \varrho}{\partial t^2} - \frac{1}{\rho_0}\left(\frac{\partial p}{\partial z} + \varrho g\right)\frac{\partial \rho_0}{\partial z} = 0. \tag{4.19}$$

Now, somewhat anticipating, as with all fluid wave problems, that a natural frequency based on the restoring force will feature in the solution, (4.19) can be written,

$$\frac{\partial^2 \varrho}{\partial t^2} - \frac{1}{\rho_0}\frac{\partial \rho_0}{\partial z}\frac{\partial p}{\partial z} + N^2(z)\varrho = 0, \tag{4.20}$$

where

$$N^2(z) = -\frac{g}{\rho_0}\frac{\partial \rho_0}{\partial z}; \tag{4.21}$$

the quantity N has units of rad s^{-1} and is called the *buoyancy frequency*. It is traditionally called the *Brunt-Väisälä frequency* because it was independently discovered by David Brunt and Vilho Väisälä in the early 20th century. However, the present derivation of linear internal waves was undertaken much earlier, by Rayleigh in 1883. It is important to note that N is, in general, not a constant but is a function of z.

To eliminate the pressure, begin by taking the divergence of (4.16), giving

$$\boldsymbol{\nabla} \cdot \left(\rho_0 \frac{\partial \boldsymbol{u}}{\partial t}\right) = -\nabla^2 p + \boldsymbol{\nabla} \cdot (\varrho \mathbf{g}). \tag{4.22}$$

A new assumption is now required, and it is an assumption that has not been required in earlier chapters of this book. We will assume that the variation in the time-averaged density, ρ_0, (this is the variation that defines the stratification) occurs over a length scale much larger than vertical variations in the time-dependent variables $\boldsymbol{u}$, p or ϱ. This is sometimes called the *Boussinesq approximation* after Joseph Valentin Boussinesq who introduced it in 1903 (Gill, 1982). Thus, spatial derivatives with respect to z of ρ_0 (recalling ρ_0 varies only in z) in the following analysis will be negligibly small relative to derivatives of $\boldsymbol{u}$, p or ϱ with respect to z. The Boussinesq approximation is often described as limiting density variations only to forces driving the flow (to the right-hand side of the momentum equation), not to the inertia of the flow (the left-hand side). However, the preceding definition based on the length scales is sufficient, and maybe less confusing if one loses track of which side of the momentum equation terms came from originally. A rigorous scaling analysis, in which derivatives with respect to space generate a different length scaling when applied to ρ_0 versus $\boldsymbol{u}$, p or ϱ, will demonstrate the validity of the following steps.

The left-hand side of (4.22) is $\dot{\boldsymbol{u}} \cdot \boldsymbol{\nabla}\rho_0 + \rho_0 \boldsymbol{\nabla} \cdot \dot{\boldsymbol{u}}$, in which the first term is negligible *relative to* the terms on the right-hand side, since, following the Boussinesq approximation, its spatial derivative is much smaller; and in which the second term is zero owing to the incompressibility assumption, (4.9). Therefore (4.22) becomes

$$0 = -\nabla^2 p + \boldsymbol{\nabla} \cdot (\varrho \mathbf{g}), \tag{4.23}$$

and since $\mathbf{g}$ is only in the negative vertical direction, (4.23) is more simply written

$$0 = -\nabla^2 p - \frac{\partial \varrho}{\partial z} \mathrm{g}, \tag{4.24}$$

The next step in eliminating the pressure is to take the Laplacian derivative, ∇^2, of (4.20), and to note that the Boussinesq approximation means that derivatives of $\partial \rho_0 / \partial z$ and thus of $N^2(z)$ can be neglected relative to derivatives of ϱ and p. The result is

$$\nabla^2 \frac{\partial^2 \varrho}{\partial t^2} - \frac{1}{\rho_0} \frac{\partial \rho_0}{\partial z} \nabla^2 \frac{\partial p}{\partial z} + N^2(z) \nabla^2 \varrho = 0. \tag{4.25}$$

Differentiating (4.24) with respect to z and substituting that into (4.25) gives

$$\nabla^2 \frac{\partial^2 \varrho}{\partial t^2} - N^2(z) \frac{\partial^2 \varrho}{\partial z^2} + N^2(z) \nabla^2 \varrho = 0, \tag{4.26}$$

and finally

$$\boxed{\nabla^2 \frac{\partial^2 \varrho}{\partial t^2} + N^2(z) {\nabla_\perp}^2 \varrho = 0}, \tag{4.27}$$

where ${\nabla_\perp}^2$ is the Laplacian operator in the horizontal directions only. Incidentally, from (4.17), the vertical component of velocity, w, is related directly to $\partial \varrho / \partial t$ by a function of z only. Thus (4.27) can be differentiated with respect to time, and, on applying the Boussinesq assumption for consistency, $\partial \varrho / \partial t$ ends up simply replaced by w, giving

$$\nabla^2 \frac{\partial^2 w}{\partial t^2} + N^2(z) {\nabla_\perp}^2 w = 0, \tag{4.28}$$

and demonstrating, as mentioned earlier, that the final governing equation with any wave problem can always be expressed in any of the dependent variables.

Now, (4.27) is not a simple wave equation, owing to the Laplacian operator in front of the time derivative but also because the factor N^2 is a function of z. Thus, an important simplification is to assume the density stratification is linear, so that $\partial\rho_0/\partial z$ is a constant and therefore N^2 is a constant, giving

$$\nabla^2 \frac{\partial^2 \varrho}{\partial t^2} + N^2 {\nabla_\perp}^2 \varrho = 0. \tag{4.29}$$

Then, despite the Laplacian operator in front of the time derivative, (4.29) can be solved by standard separation of variables, as detailed in §1.5.3; for example, in Cartesian coordinates (4.29) could be solved by

$$\varrho = A\,\mathcal{X}(x)\mathcal{Y}(y)\mathcal{T}(t). \tag{4.30}$$

Unlike the approach to surface gravity waves in chapter 2, boundary conditions have not yet been considered here, and so a mathematical problem has not yet been posed. In some industrial systems, boundaries might be rather close to the region of interest. However, in the ocean or atmosphere, it is at least possible to consider waves being generated in the mid-depth of the ocean or mid-height of the atmosphere, far from any boundary, so that the sinusoidal functions are the default choice for the spatial directions x, y and z as well as for time. Thus, as usual, $\mathcal{T} = \mathrm{e}^{\mathrm{i}\omega t}$, $\mathcal{X} = \mathrm{e}^{\pm \mathrm{i}\kappa x}$, $\mathcal{Y} = \mathrm{e}^{\pm \mathrm{i}\ell y}$ and $\mathcal{Z} = \mathrm{e}^{\pm \mathrm{i}kz}$, and the separation-of-variables procedure soon yields the dispersion relation,

$$\boxed{\omega^2 = \left(\frac{\kappa^2 + \ell^2}{\kappa^2 + \ell^2 + k^2}\right) N^2}, \tag{4.31}$$

which could also be written

$$\omega^2 = \cos^2\theta N^2, \tag{4.32}$$

where, for plane waves, θ is the angle between the vertical and the planes containing the wave crests and troughs (figure 4.4), given by

$$\cos^2\theta = \left(\frac{\kappa^2 + \ell^2}{\kappa^2 + \ell^2 + k^2}\right). \tag{4.33}$$

This relation (4.31) has some interesting consequences; it is clear that the vertical wavenumber, k, has a different influence on the frequency to the horizontal wavenumbers κ and ℓ, which represents the fact that gravity acts in the vertical direction. If k is extremely large relative to the horizontal wavenumbers, so that the vertical wavelength is extremely small relative to the horizontal wavelength, the frequency ω is extremely low. The other extreme is when $k = 0$, so that the vertical wavelength is infinite. Then $\omega = \pm N$. This corresponds to the fluid oscillating at the buoyancy frequency. The frequency of the waves, ω, depends only on the *direction* of the waves and not on their wavelength. There is an analogy with the effect of background rotation that leads to an equivalent dispersion relation, (5.16); as discussed in §5.3.2, stratification inhibits vertical motion while rotation inhibits horizontal motion.

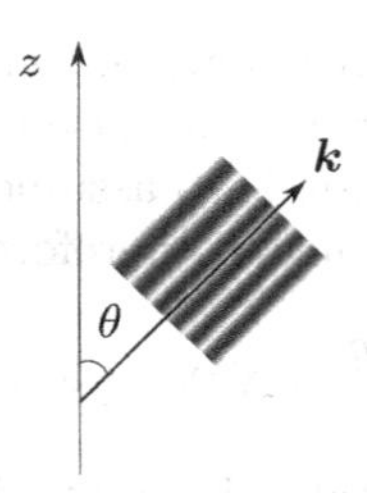

FIGURE 4.4
The orientation of internal waves relative to the vertical; $\boldsymbol{k}$ is the wavenumber vector as defined in §1.5.4. When $\omega \to 0$, $\theta \to 90°$ and variations in the horizontal disappear.

4.4 Problems

1. A mixture of 50% waste cooking oil of density 910 kg m^{-3} and 50% water of density 1000 kg m^{-3} was pumped into a 1 m deep tank 5 m long on a flatbed truck. Overnight, the mixture separated into two layers. At what radian frequency would half-wavelength sloshing waves on the interface occur?
2. A pipeline on the sea bed carries a mixture of crude oil of density 880 kg m^{-3} and seawater of density 1027 kg m^{-3}. Pump noise transmitted along the pipeline has a frequency of 10 Hz. What wavelength waves are excited on the interface?
3. Calculations on internal gravity waves are relevant to

   A Calculating forces in a tanker containing both oil and water

   B Pumping honey in a food-industry plant

   C Estimating stresses created by breaking waves at the sea surface

   D Predicting the formation of tropical cyclones

4. An experimental hydrogen ironmaking converter is 5 m in diameter and contains a layer of molten iron 0.4 m deep under a layer of fluid slag of equal depth. The molten-iron density is 6 800 kg m^{-3} and the slag density is 2 300 kg m^{-3}. Dangerous sloshing of the molten metal could occur if a half-wavelength wave on the metal-slag interface equals the vessel diameter. At what frequency would this be excited?
5. A new design for a low-fat mayonnaise pre-mixing tank in a food factory is 2 m in diameter and 2 m deep. It is 50% filled with an oil of density 920 kg m^{-3} and 50% with a water-based solution of density 1040 kg m^{-3}. In order to prevent a mode-1 wave with a length equal to the tank circumference, what mixer rotation rate in RPM should be avoided?
6. When modelling continuously stratified fluids the most common approach is to

   A divide the fluids into two distinct layers

   B assume the fluid is fully compressible

   C assume differences in density are significant only in the inertia terms

   D assume differences in density are significant only in the gravitational terms

7. A drinking-water reservoir 50 m deep is continuously stratified with a bottom density of 1000 kg m^{-3} decreasing linearly to a surface density of 994 kg m^{-3}. What frequency (in Hz) of internal waves will be generated when its outlet gate is suddenly opened?

8. A container ship undertakes a routine expendable bathythermograph drop to report ocean conditions to weather forecasters. The resulting density profile is given by $\rho_0(z) = B|z|^{-n}$ where z is vertically upwards and B and n are constants. The ocean's buoyancy frequency is given by

   A $N^2 = g/B$

   B $N^2 = -nB|z|^{-n-1}$

   C $N^2 = gn|z|^{-1}$

   D $N^2 = gB|z|^{-n}$

9. Waves in a continuously-stratified atmosphere have a horizontal wavelength of 3.5 km and a vertical wavelength of 700 m. At what angle to the horizontal are they propagating?

   A 0.2°

   B 1.1°

   C 6.1°

   D 11.3°

10. A submarine is operating in a linearly stratified ocean with a buoyancy frequency $N = 0.012$ rad s^{-1}. A signal with a frequency of 0.00175 Hz is recorded. What is the ratio of the horizontal to the vertical wavelengths of internal waves creating this signal?

11. Discuss the relevance of the two-layer assumption for linear inviscid internal gravity waves to

    (a) The flow of mercury under a water layer

    (b) Design of a pharmaceutical reactor containing a mixture of oil- and water-based chemicals

    (c) Movements of the slag layer in a molten-metals refining vessel

    (d) Mixing a water reservoir thermally stratified by hot weather

12. By defining a length scale, L_0, over which the time-averaged density ρ_0 varies in a stratified fluid, and a second length scale, L, over which time-dependent variables such as the velocity vary in the vertical, such that $L \ll L_0$, conduct a scaling analysis of the linear, inviscid equations of motion (4.14) and (4.16), showing that

$$\nabla^2 \frac{\partial^2 w}{\partial t^2} + N^2(z) \nabla_\perp{}^2 w = 0,$$

(the Boussinesq governing equation for the vertical velocity, w).

5

Waves in rotating fluids

5.1 Summary of key points

- The **spatial structure of inertial waves** that may exist in a frame of reference rotating with radian frequency Ω relative to inertial space is given by (5.11) on page 138,

$$\boxed{\frac{\partial^2 \mathcal{P}}{\partial x^2} + \frac{\partial^2 \mathcal{P}}{\partial y^2} - \left[\left(\frac{2\Omega}{\omega}\right)^2 - 1\right]\frac{\partial^2 \mathcal{P}}{\partial z^2} = 0},$$

  where $\mathcal{P}$ is a spatially-varying pressure factor, ω is the radian frequency of an oscillation and x, y and z are Cartesian co-ordinates with the rotation being around the z-axis.

- The **criterion for inertial waves to exist** is given by (5.12) on page 138,

$$\boxed{|\omega| < 2\Omega}.$$

- The **dispersion relation** for unbounded inertial waves is given by (5.16) on page 139,

$$\boxed{\omega^2 = \left(\frac{k^2}{\kappa^2 + \ell^2 + k^2}\right)(2\Omega)^2},$$

  where κ, ℓ and k are the wavenumbers in the x, y and z directions.

- The **Coriolis parameter in the f-plane approximation**, where the local angular rotation may be assumed not to vary over small variations in latitude over the Earth's surface, is given by (5.19) on page 140,

$$f_\oplus = 2\Omega_\oplus \sin\theta,$$

  where $\Omega_\oplus$ is the Earth's angular frequency ($2\pi/[(23\times 60+56)\times 60+4.1) \simeq 7.2921 \times 10^{-5}$ rad s^{-1}), and θ is the latitude on the Earth's surface.

DOI: 10.1201/9780429295263-5

- The **radius of deformation** (Rossby radius), $L_\oplus$, the scale at which the Coriolis force balances the force of gravity, and above which the Coriolis force dominates is given by (5.20) on page 141,

$$\boxed{L_\oplus = \frac{\sqrt{gh}}{|f_\oplus|}},$$

  where h is the depth of the fluid layer.

- Useful **books** include Greenspan (1968) for a detailed applied-mathematical approach, while Pedlosky (1987) and Gill (1982) provide more detail on geophysical applications.

5.2 An example

The backpacker has been watching the mining survey ship (§3.2) from the dockside. It is the best of its kind in the world, with closely-guarded confidential equipment below the waterline. A small drum was seen on the deck inside a spinning frame. He counts approximately $\Omega = 100$ revolutions per minute. The entire spinning frame is driven by a second motor, causing a precessing motion like a spinning top at approximately $\omega = 32$ revolutions per minute. The length of the drum is approximately four times its diameter, so $h = 4/3$. He suspects this is a mixer for chemicals too toxic to be stirred conventionally in an open barrel. Could it be a mixer?

Solution

First, check that using (5.12) that inertial waves could be driven by the rotation rates. Since $|\omega| < 2\Omega$, inertial waves could be the basis of a mixer. Next, an iterative solution of two equations for Λ, (5.33),

$$\left(\frac{2\Omega}{\omega}\right)\frac{m}{R_0}J_m(\Lambda R_0) + \Lambda J_m'(\Lambda R_0) = 0,$$

where R_0 is the drum radius, and (5.34),

$$\Lambda_n^2 = \left(\frac{n\pi}{h}\right)^2\left[\left(\frac{2\Omega}{\omega}\right)^2 - 1\right],$$

for various values of wavenumber indices (n, m), shows that the first inertial-wave mode with $n = m = 1$ would resonate in the drum, likely causing mixing.

He shows this to the island's chief engineer. Although the calculation is detailed, she points out that almost all such devices are mixers anyway, and it is probably just mixing anti-fouling paint for instruments to be left in the sea. The backpacker apparently has an education in the physical sciences but is a troublemaker. He convinces a local fisherman that the survey ship is endangering fishing stocks and needs to be followed.

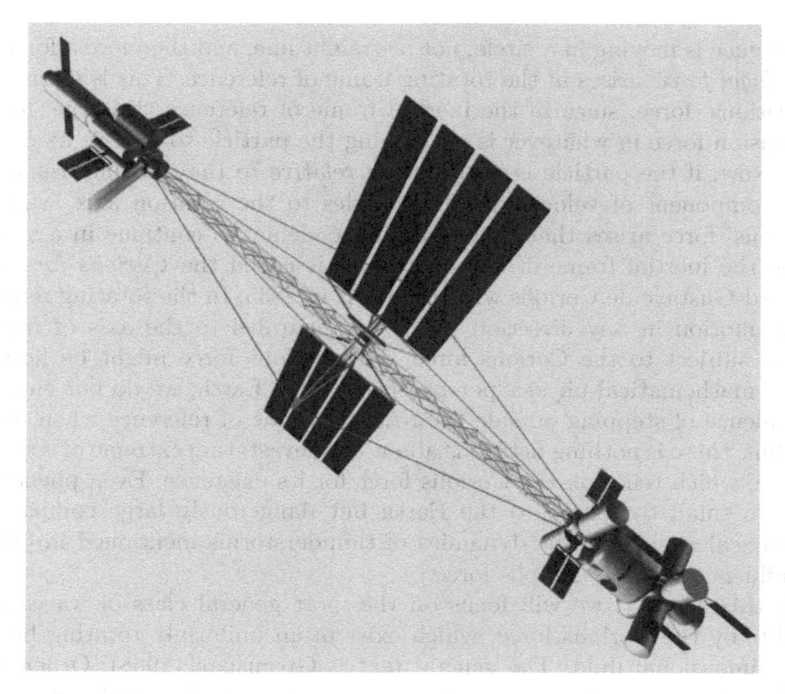

FIGURE 5.1
A rotating spacecraft concept intended to provide artificial gravity on a voyage to Mars (Rousek (2010)). Further details in §11.3.2. Image copyright by and courtesy of Tomas Rousek.

5.3 Linear inertia-wave theory

5.3.1 Coriolis force

Rotating fluids are found in many situations in astrophysics, planetary sciences and geophysics. Rotating fluids are also found in engineering applications that are both exotic, such as spinning spacecraft (figure 5.1), and mundane, such as the humble paint-can mixer. For example, the Earth is rotating on its axis once a day, creating a class of phenomena in its atmosphere and ocean due to rotation. The ubiquitous high- and low-pressure systems we see on the daily weather forecast are marked by winds circulating around their centres because of the Earth's rotation. Some of these applications will be outlined in chapter 11.

Here, by 'rotation' we mean that the entire body of fluid is rotating as if it were a solid body, relative to some inertial reference frame. Recall Newton's First Law: particles continue to move in a straight line at constant velocity unless acted on by an external force. However, a particle in the rotating frame of reference is moving in a circle, not a straight line, and therefore a force, the *centrifugal force*, arises in the rotating frame of reference. This is often called a 'fictitious' force, since in the inertial frame of reference the 'true' force is the tension force in whatever is restraining the particle to stay in its circular path. Now, if the particle is also moving *relative to the rotating frame* with some component of velocity at right angles to the rotation axis, a further 'fictitious' force arises that represents its tendency to continue in a straight line in the inertial frame of reference. This is called the *Coriolis force* after Gaspard-Gustave de Coriolis who derived it in 1835. In the rotating reference frame, motion in any direction that is *not* parallel to the axis of rotation will be subject to the Coriolis force. The Coriolis force might be fictitious from a mathematical-physics perspective, but on Earth, we do not enjoy the convenience of stepping outside the rotating frame of reference when it suits us. Thus, there is nothing fictitious about the devastating extreme of a tropical cyclone, which relies on the Coriolis force for its existence. Even phenomena that are small compared to the Earth but dangerously large compared to human scales, such as the dynamics of thunderstorms mentioned in §12.4.3, are influenced by the Coriolis force.

In this chapter, we will focus on the most general class of waves made possible by the Coriolis force, which exist in an uniformly rotating body of three-dimensional fluid. The general text is Greenspan (1968). Other, more specialised rotating-fluid waves exist, including waves in which gravity and density differences play a role, as well as waves in the thin layer of a sphere that could represent a planet's atmosphere and oceans; and readers are referred to standard texts on geophysical fluid dynamics (e.g. Pedlosky, 1987; Gill, 1982) for more details.

In what follows, the term 'axial' is used to mean 'parallel to the axis of rotation' and the axial direction is denoted as the z-direction. In this elementary derivation, there are no density differences or free surfaces for gravity to act on and otherwise define a vertical direction, for which the symbol z is used in other chapters. In general, however, the rotation axis need not be aligned with gravity, and for rotating celestial bodies like stars and planets, the rotation axis is only parallel to the direction along which gravity acts at the north and south poles.

5.3.2 Inertial oscillations in an unbounded domain

The mass-conservation or continuity equation, (1.15), reduces for an incompressible fluid to

$$\nabla \cdot \boldsymbol{u} = 0, \tag{5.1}$$

which, since it uses vector-calculus notation, is true irrespective of the co-ordinate system. Meanwhile, assuming the flow is inviscid as well as incompressible, the momentum equation, (1.20), can be reduced to a linearised version of Euler's equation in the rotating co-ordinate system, with no density differences or free surfaces, so that gravity is balanced by hydrostatic pressure, is

$$\frac{\partial \boldsymbol{u}}{\partial t} + 2\boldsymbol{\Omega} \times \boldsymbol{u} = -\frac{1}{\rho_0}\boldsymbol{\nabla} p, \tag{5.2}$$

where p is the dynamic pressure relative to the steady pressure fields due to gravity and centrifugal forces. The term $2\boldsymbol{\Omega} \times \boldsymbol{u}$ on the left-hand side is the Coriolis force per unit mass, and it can be derived from first principles by consideration of derivatives in a rotating frame of reference, as shown in Pedlosky (1987). Now, following the first step of the typical derivation of the linear fluid-wave equation, taking the axial component of the curl of (5.2) and using Cartesian coordinates for simplicity, not necessity, gives

$$\frac{\partial}{\partial t}\left(\frac{\partial v}{\partial x} - \frac{\partial u}{\partial y}\right) + 2\Omega\left(\frac{\partial u}{\partial x} + \frac{\partial v}{\partial y}\right) = 0, \tag{5.3}$$

and noticing that u and v components of velocity in the Coriolis term with the 2Ω factor can be replaced using the continuity equation (5.1), so that (5.3) becomes

$$\frac{\partial}{\partial t}\left(\frac{\partial v}{\partial x} - \frac{\partial u}{\partial y}\right) - 2\Omega\frac{\partial w}{\partial z} = 0, \tag{5.4}$$

in which the quantity

$$\frac{\partial v}{\partial x} - \frac{\partial u}{\partial y}$$

is the axial component of the vorticity, which was defined in (1.39). As an aside, it is worth noting an interesting consequence of (5.4): if the flow is steady, so the partial derivative with time is zero, $\partial w/\partial z = 0$, so there is no variation of the axial component of velocity in the axial. In fact, if the flow is steady, there is no axial variation in any component of velocity. Furthermore, even beginning with the nonlinear version of Euler's equation instead of the linear version (5.2), taking the curl and applying incompressibility exactly as was done to get (5.4), the consequence is that if the vorticity is a conserved quantity, as in (1.38), there is also no variation in the axial velocity. Fluid flowing steadily in a rotating frame of reference only flows in planes normal to the axis, creating axial columns of streamlines: tube-like vortices parallel to the axis of rotation. This consequence is traditionally called the *Taylor-Proudman theorem* and the columnar flows are called *Taylor columns*.

In the second step of the typical linear fluid-wave derivation, take the divergence of (5.2) and use the incompressibility condition (5.1) on the local-derivative term, giving

$$2\Omega\left(-\frac{\partial v}{\partial x} + \frac{\partial u}{\partial y}\right) = -\frac{1}{\rho_0}\nabla^2 p. \tag{5.5}$$

Now, eliminate the axial component of vorticity by multiplying (5.4) by 2Ω and by taking the derivative with respect to time of (5.5), allowing the combination of these two forms of the momentum equation into

$$-4\Omega^2\frac{\partial w}{\partial z}=-\frac{1}{\rho_0}\frac{\partial}{\partial t}\nabla^2 p. \tag{5.6}$$

As usual, it is convenient to obtain a wave equation in a scalar like the pressure, so choose to eliminate w from (5.6); taking the derivative with respect to z of the axial component of (5.2) gives

$$\frac{\partial^2 w}{\partial t\partial z}=-\frac{1}{\rho_0}\frac{\partial^2 p}{\partial z^2}, \tag{5.7}$$

and differentiating (5.6) with respect to time and substituting in (5.7) gives

$$4\Omega^2\left(\frac{1}{\rho_0}\frac{\partial^2 p}{\partial z^2}\right)=-\frac{1}{\rho_0}\frac{\partial^2}{\partial t^2}\nabla^2 p, \tag{5.8}$$

which can be written as

$$\frac{\partial^2}{\partial t^2}\left(\frac{\partial^2 p}{\partial x^2}+\frac{\partial^2 p}{\partial y^2}\right)+\left(\frac{\partial^2}{\partial t^2}+4\Omega^2\right)\frac{\partial^2 p}{\partial z^2}=0. \tag{5.9}$$

This does not look at all like the familiar wave equation, $\ddot{p}-c^2\nabla^2 p=0$; rather, it is $\nabla^2\ddot{p}+(2\Omega)^2 p_{zz}=0$. However, despite this complicated form, wavelike behaviour is nevertheless possible. The general separation-of-variables solution to all fluid wave problems, (1.54), applies here too. Thus, adopting the separation-of-variables method, assume the time dependence is sinusoidal, i.e.

$$p=\mathcal{P}\mathrm{e}^{\mathrm{i}\omega t}, \tag{5.10}$$

where $\mathcal{P}$ is a spatially-varying pressure factor, i.e $\mathcal{P}=\mathcal{P}(x,y,z)$, and ω is the radian frequency. Substituting (5.10) into (5.9) yields

$$\boxed{\frac{\partial^2\mathcal{P}}{\partial x^2}+\frac{\partial^2\mathcal{P}}{\partial y^2}-\left[\left(\frac{2\Omega}{\omega}\right)^2-1\right]\frac{\partial^2\mathcal{P}}{\partial z^2}=0}. \tag{5.11}$$

This is traditionally called *Poincaré's equation*, derived by Jules Henri Poincaré in 1910 in his studies of rotating astronomical bodies. Since the time dependence was already assumed, it cannot be called a wave equation, but it can be a hyperbolic partial differential equation like the familiar wave equation: the second derivative of one of the dependent variables can have the opposite sign to the second derivative of the other dependent variables. It is worth emphasising that it *can* be hyperbolic, but does not *have* to be. For (5.11) to be hyperbolic, the term in square brackets must be positive, so that

$$\boxed{|\omega|<2\Omega}, \tag{5.12}$$

so that the radian frequency of the oscillations in time must be less than twice the rotation rate.

If, as in the present section, boundary conditions are not specified, just as (3.21) for sound waves, for example, the general separation-of-variable solution (1.54) can be written as

$$p = A\mathrm{e}^{\mathrm{i}(\pm\kappa x \pm \ell y \pm kz + \omega t)}, \tag{5.13}$$

where, similarly to the notation in §2, κ, ℓ and k are the wavenumbers in the x, y and z dimensions respectively, so that the spatially-varying pressure factor $\mathcal{P}$ defined in (5.10) is

$$\mathcal{P} = A\mathrm{e}^{\mathrm{i}(\pm\kappa x \pm \ell y \pm kz)}. \tag{5.14}$$

Now, substituting (5.14) into (5.11), the relation between the wavenumbers and the temporal frequency becomes clear; it is given by

$$\kappa^2 + \ell^2 - \left[\left(\frac{2\Omega}{\omega}\right)^2 - 1\right] k^2 = 0, \tag{5.15}$$

A re-writing of (5.15) gives

$$\boxed{\omega^2 = \left(\frac{k^2}{\kappa^2 + \ell^2 + k^2}\right)(2\Omega)^2}, \tag{5.16}$$

which can also be written

$$\omega^2 = \sin^2\theta\,(2\Omega)^2, \tag{5.17}$$

where, for plane waves, θ is the angle between the rotation axis and the planes containing the wave crests and troughs, given by

$$\sin^2\theta = \left(\frac{k^2}{\kappa^2 + \ell^2 + k^2}\right). \tag{5.18}$$

In figure 5.2, inertial waves are shown relative to the rotation axis and to a wavenumber vector, $\boldsymbol{k}$, where (as defined in §1.5.4) $\boldsymbol{k} = (\kappa, \ell, k)$.

When the ratio of the axial wavenumber k to axis-normal wavenumbers κ and ℓ is small, the angle θ of the wave crests relative to the axis is small. It is also clear from (5.16) that very small axial-wavenumber waves have very low temporal frequencies ω. As the frequency ω becomes lower and lower, the axial wavelength $2\pi/k$ becomes longer and longer until there is no vertical variation at all. This is no more than a consequence of the Taylor-Proudman theorem mentioned earlier: when the temporal variation disappears, the axial variation in a rotating flow also disappears.

However, as the magnitude of the frequency, $|\omega|$, increases, the right-hand side of (5.18) becomes larger, corresponding to the angle of the waves to the

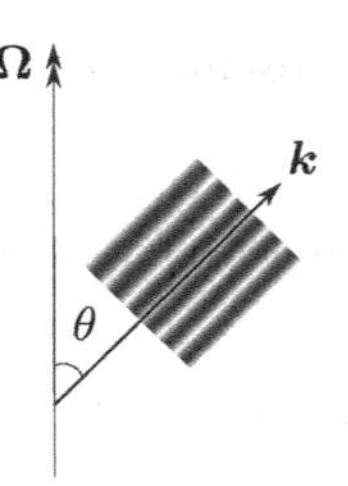

FIGURE 5.2
The orientation of inertial waves relative to the rotation axis, represented by vector $\boldsymbol{\Omega}$. When $\omega \to 0$, $\theta \to 0$ and variations in the vertical disappear.

axis becoming larger, and when $|\omega| = 2\Omega$, wave crests are at right angles to the axis and the axial wavenumber becomes infinite. Since each of κ, ℓ and k are real numbers, Poincaré's equation (5.11) does not have a wavelike solution (5.13) for $|\omega| > 2\Omega$.

These waves are usually called *inertial waves*, because they are made possible by the Coriolis force, a fictitious force that arises from shifting one's frame of reference from the inertial frame of reference to a non-inertial frame of reference: the rotating frame. Provided the condition (5.12) that $|\omega| < 2\Omega$ exists, waves exist and propagate.

The inertial-wave dispersion relation (5.16) is the equivalent of (4.31) for internal waves. There is an analogy between the effect of stratification and that of rotation: stratification inhibits motion in the vertical, causing flows to spread out horizontally, whereas rotation inhibits motion in the horizontal, causing flows to spread out vertically. Furthermore, just like waves in a stratified fluid, the frequency of inertial waves, ω, depends only on the direction of the waves and not on their wavelength.

Pure inertial waves are of interest in some specialised astrophysical, planetary-physical and engineering applications which will be elaborated in chapter 11. However, the effects of rotation are more commonly experienced when combined with the forces due to gravity discussed in chapter 4. The Earth's atmosphere and oceans are very thin layers compared to the Earth's radius, so motion is often considered to be in a two-dimensional plane with the rotation vector normal to this plane. Therefore, an effective variation of the background rotation rate with latitude must be considered; the component of the Earth's rotation normal to the local horizontal has opposite signs in the Northern and Southern Hemisphere and is zero at the Equator. Thus the motion is sometimes considered on an *f-plane*, for which the 2Ω in the foregoing is replaced by $f_\oplus$, so that

$$f_\oplus = 2\Omega_\oplus \sin\theta, \qquad (5.19)$$

where $\Omega_\oplus$ is the Earth's angular frequency ($2\pi/[(23 \times 60 + 56) \times 60 + 4.1] \simeq 7.2921 \times 10^{-5}$ rad s^{-1}), and θ is the latitude on the Earth's surface where the motions are being approximated. There are also waves that occur over larger distances, such that the variation of the background rotation with latitude does matter. The $\sin\theta$ in (5.19), rather than being a constant as in the f-plane approximation, is linearised by making $f_\oplus$ a linear function of North-South distance multiplied by a parameter, $\beta_\oplus$; this is called a *β-plane* approximation. These waves are called *planetary waves* or traditionally *Rossby waves* after Carl-Gustaf Rossby who identified this phenomenon in 1939. Atmospheric planetary waves have a significant influence on the weather; more details are in Pedlosky (1987) and Gill (1982).

5.3.3 Relation to gravity waves

If the fluid is assumed to be a shallow layer of depth h - such as an ocean or atmospheric layer that is very thin compared to the size of the planet - variations in the vertical may be neglected, and the continuity equation takes the same form as for a shallow-water surface-gravity-wave analysis. A scaling analysis of both the continuity and momentum equations shows that forces due to shallow-water gravity waves are similar to the Coriolis force over a length scale $L_\oplus$ of

$$\boxed{L_\oplus = \frac{\sqrt{gh}}{|f_\oplus|}}, \tag{5.20}$$

where as usual g is the acceleration due to gravity, and $L_\oplus$ is traditionally called the *Rossby radius*. Over distances greater than $L_\oplus$, the planet's rotation will begin to dominate the flow. Considering the Earth's ocean basins of typical depth scale 4 km, in the mid-latitudes, $L_\oplus$ is about 2000 km. Only the tides create surface motions on these scales. On a mid-latitude continental shelf of depth 100 m, $L_\oplus \simeq 430$ km. Meanwhile, even in a bathroom sink of depth 0.10 m, $L_\oplus \simeq 14$ km, explaining why the oft-stated principle that water should spin down the plughole in opposite directions in the Northern and Southern Hemispheres is unlikely to be demonstrated: the smallest surface gravity waves would overwhelm the influence of the Earth's rotation as water drains unless the bathroom sink is many kilometres wide!

In the atmosphere, while vertical length scales may be several kilometres, the density differences are much smaller than that between the sea and the air, so that internal gravity waves (detailed in §4.3.3) lose the competition with rotation over a scale of about 1000 km (Gill, 1982), explaining the scale of the rotating mid-latitude high- and low-pressure systems seen on the weather forecast. This is called the *synoptic scale* and will be discussed in §12.4.3 in the context of thunderstorms and aviation safety.

5.3.4 Inertial oscillations with boundary conditions

In many applications, inertial waves occur in some contained system, where the boundary conditions critically define the nature of the waves. Examples are rotating geophysical and astrophysical bodies, such as the liquid outer core of the Earth, and stars and gaseous planets. To solve these problems, it is necessary to re-write Poincaré's equation in appropriate co-ordinates.

As an example, consider a cylindrical container of radius R_0, say, and axial height h. Cylinders are found in engineering systems, rather than geophysical and astrophysical bodies. However, laboratory experiments and numerical simulations intended to improve our understanding of fundamental geophysical and astrophysical flows are often done in cylindrical geometries, owing to the geometrical simplicity, as detailed in §13.4.2, and engineering systems in spacecraft (§11.3) and in terrestrial industry are usually based on simple geometries. The appropriate co-ordinates are cylindrical polars, r, φ, z where z remains the axial dimension, but now the other dimensions are r for the radial and φ for the azimuthal dimensions, respectively. Then, returning to the last equation that could be considered as being in either Cartesian or cylindrical polar co-ordinates, (5.8), express it in cylindrical polars (and cancel the common $1/\rho_0$ factor), giving

$$4\Omega^2 \frac{\partial^2 p}{\partial z^2} = -\frac{\partial^2}{\partial t^2}\left[\frac{1}{r}\frac{\partial}{\partial r}\left(r\frac{\partial p}{\partial r}\right) + \frac{1}{r^2}\frac{\partial^2 p}{\partial \varphi^2} + \frac{\partial^2 p}{\partial z^2}\right]. \tag{5.21}$$

Separation of variables can be applied as before, and since waves are still sought, apply the same assumption of sinusoidal behaviour in time, (5.10), giving Poincaré's equation in cylindrical polars,

$$\frac{1}{r}\frac{\partial}{\partial r}\left(r\frac{\partial \mathcal{P}}{\partial r}\right) + \frac{1}{r^2}\frac{\partial^2 \mathcal{P}}{\partial \varphi^2} - \left[\left(\frac{2\Omega}{\omega}\right)^2 - 1\right]\frac{\partial^2 \mathcal{P}}{\partial z^2} = 0. \tag{5.22}$$

The solution can be written, more generally than at the stage of (5.13), as

$$\mathcal{P} = A\,\mathcal{R}(r)\,\mathfrak{F}(\varphi)\mathcal{Z}(z). \tag{5.23}$$

However, since there are now boundaries, the boundary conditions determine the forms of each of the separated factors. In the azimuthal direction, the boundary conditions must be periodic, because there cannot be any discontinuities in the solution as φ is changed continuously from 0 to 2π. Therefore a sinusoidal form is appropriate for $\mathfrak{F}(\varphi)$. In the axial direction, we can imagine that the flow is occurring within a right circular cylinder, so that at the two ends of the cylinder, there can be no flow through the end. These axial boundary conditions are conveniently satisfied by sinusoidal functions as well. Consideration of the radial boundary condition is not necessary at this point, but will have to be dealt with soon. Thus, (5.23) should be written as

$$p = A\,\mathcal{R}(r)\,\mathrm{e}^{\pm im\varphi}\,\mathrm{e}^{\pm ikz}, \tag{5.24}$$

where, to distinguish the notation in this co-ordinate system from that in §5.3.2, we will use the symbol m for the azimuthal wavenumber. Since the axial direction has special importance in rotating flows irrespective of co-ordinate system, analogous to the importance of the vertical direction in gravity-dominated flows, we will retain the use of k for the axial wavenumber, and use ℓ for the radial wavenumber. Substituting (5.24) into (5.22) and dividing by $\mathcal{P}$ gives

$$\frac{1}{r}\frac{\partial}{\partial r}\left(r\frac{\partial \mathcal{R}}{\partial r}\right) - \frac{m^2}{r^2}\mathcal{R} + k^2\left[\left(\frac{2\Omega}{\omega}\right)^2 - 1\right]\mathcal{R} = 0, \tag{5.25}$$

which can be re-arranged into the standard form of a classical equation,

$$r^2\frac{\partial^2 \mathcal{R}}{\partial r^2} + r\frac{\partial \mathcal{R}}{\partial r} + \left(\Lambda^2 r^2 - m^2\right)\mathcal{R} = 0, \tag{5.26}$$

where

$$\Lambda^2 = k^2\left[\left(\frac{2\Omega}{\omega}\right)^2 - 1\right] \tag{5.27}$$

is the dispersion relation. Now, (5.26) appears more complicated than the familiar equation with a wave solution, $p_{xx} + k^2 p = 0$, but it is just the same a second-order differential equation for a single dependent variable, $\mathcal{R}$, as a function of a single independent variable, r. It is only the cylindrical geometry that makes this equation more complicated, and in fact (5.26) is *Bessel's equation*, to which the solutions are *cylindrical Bessel functions* of the first and second kinds, usually written $J_m(\Lambda r)$ and $Y_m(\Lambda r)$ respectively, where the subscript m is the *order* of the Bessel function. These two functions are analogies in the cylindrical geometry of the two functions that solve $p_{xx} + k^2 p = 0$, $\sin(kx)$ and $\cos(kx)$. However, the Bessel function of the second kind tends to minus infinity as $r \to 0$. Here, therefore, knowledge of the boundary condition at $r = 0$, which is simply that the pressure cannot have an infinite magnitude, allows the Bessel functions of the second kind to be ruled out as valid solutions. In other wave problems cast in a cylindrical co-ordinate system, Bessel's equation crops up again. For example, a drop falling onto a water surface creates a pattern of radially-spreading surface waves, and to correctly model this propagation Bessel functions of the second kind are needed, with the infinite value at $r = 0$ representing the discontinuity in the surface at the instant of drop impact.

Thus, the solution for the pressure, (5.24), becomes

$$p = A\, J_m(\Lambda r)\, \mathrm{e}^{\pm i m\varphi}\, \mathrm{e}^{\pm i k z}. \tag{5.28}$$

The boundary condition at the cylinder side-wall, where $r = R_0$, must now be specified. Since the flow in the radial direction cannot pass through the side-wall, $u(R_0) = 0$, where the velocity components are u, v and w in the radial, azimuthal and axial directions respectively. In order to relate the pressure solution just found, (5.28), to the radial component of velocity, u, return to the momentum equation (5.4), which in component form and cylindrical coordinates is

$$\frac{\partial u}{\partial t} - 2\Omega v = -\frac{1}{\rho_0}\frac{\partial p}{\partial r}, \tag{5.29a}$$

$$\frac{\partial v}{\partial t} + 2\Omega u = -\frac{1}{\rho_0 r}\frac{\partial p}{\partial \varphi}, \tag{5.29b}$$

$$\frac{\partial w}{\partial t} = -\frac{1}{\rho_0}\frac{\partial p}{\partial z}. \tag{5.29c}$$

The azimuthal component of velocity, v, can be eliminated from (5.29a) and (5.29b), giving

$$\frac{\partial^2 u}{\partial t^2} + 4\Omega^2 u + \frac{2\Omega}{\rho_0 r}\frac{\partial p}{\partial \varphi} = -\frac{1}{\rho_0}\frac{\partial^2 p}{\partial r \partial t}, \tag{5.30}$$

and since the velocity must have the same dependence on time and on azimuthal angle as the pressure, and the axial dependence $\mathcal{Z}(z)$ cancels out, on substitution of (5.28) into (5.30), the latter becomes

$$(4\Omega^2 - \omega^2)\mathcal{U} + \frac{\mathrm{i}m2\Omega}{\rho_0 r} J_m(\Lambda r) = -\frac{\mathrm{i}\omega}{\rho_0}\frac{\partial J_m(\Lambda r)}{\partial r}, \tag{5.31}$$

where $\mathcal{U}$ is the radial variation of the radial velocity u (the equivalent of $\mathcal{R}$ for the pressure p). Using the chain rule for the derivative on the right-hand side yields the desired expression for $\mathcal{U}$,

$$\mathcal{U} = \frac{\mathrm{i}}{\rho_0(\omega^2 - 4\Omega^2)}\left[\frac{2\Omega m}{r} J_m(\Lambda r) + \omega\Lambda J_m'(\Lambda r)\right], \tag{5.32}$$

where the notation J_m' denotes the derivative of J_m with respect to its argument, Λr. Finally, applying the boundary condition $u(R_0) = 0$ gives

$$\boxed{\left(\frac{2\Omega}{\omega}\right)\frac{m}{R_0} J_m(\Lambda R_0) + \Lambda J_m'(\Lambda R_0) = 0,} \tag{5.33}$$

and applying the boundary condition $w(\pm h/2) = 0$ requires that the axial wavenumber k takes only the values $n\pi/h$, where n is an integer, so that the dispersion relation (5.27) becomes

$$\boxed{\Lambda_n^2 = \left(\frac{n\pi}{h}\right)^2\left[\left(\frac{2\Omega}{\omega}\right)^2 - 1\right].} \tag{5.34}$$

Although Poincaré's equation was derived in 1910, (5.33) and (5.34) were derived by Kelvin in 1880. They must be solved together to determine the set of

(a)

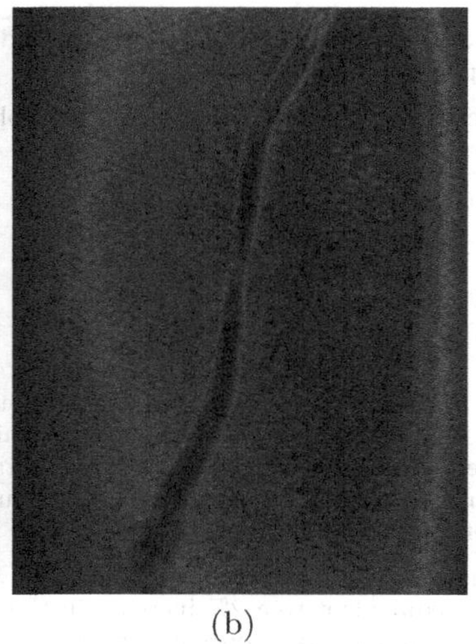

(b)

FIGURE 5.3
Experiment in which inertia waves satisfying (5.33) and (5.34) are forced. (a) Turntable rotating at Ω on which a spinning fluid-filled cylinder is set to rotate at ω at an angle θ to the axis of $\boldsymbol{\Omega}$. Image by Richard Manasseh. (b) Digitally-enhanced image of a dyeline inside the cylinder following the flow when $2(\Omega + \omega)/\omega = 1.28$ and three half-wavelengths satisfy the axial boundary condition (5.34). From Manasseh (1996), reproduced with permission of Cambridge University Press.

eigenmodes of inertial waves that 'fit' within the cylinder, which are sometimes called Kelvin modes (although many types of modes bear Kelvin's name). The need to solve (5.33) and (5.34) simultaneously arises from the same principle as finding the set of modes of sound waves that fit in a duct, for example. However, the cylindrical geometry, causing the Bessel functions to arise, plus the Coriolis term that led to Poincaré's equation makes the determination of the eigenmodes more complicated. In particular, (5.33) is a *transcendental* equation that can only be solved numerically. Nevertheless, when experiments are conducted under appropriate conditions, evidence of inertial-wave modes emerges (figure 5.3). Some applications of contained inertial waves and their nonlinear behaviour will be described in §11.2 and §13.4.

5.4 Problems

1. Observations of the atmosphere of an exoplanet with a rotation period of 10 Earth hours indicate possible inertial waves propagating at an angle of about 30° to the

rotation axis. What would be the period of the waves? What would be the cause of this phenomenon?

2. Which one of the following geophysical phenomena is **not** significantly affected by the Coriolis force?

   A Mid-latitude thunderstorm

   B Typhoon

   C Tsunami

   D Earth's magnetic field

3. Part of a spacecraft on a Mars voyage rotates at 4 RPM to create an artificial gravity field for the passengers. Vibrations at this same frequency will occur throughout the spacecraft. If there are cylindrical water tanks in the rotating section, what is the ratio of tank length to diameter that would cause inertial waves generated at the tank edges to reflect exactly back on their origin?

4. Simulations of future trends in tropical-cyclone formation suggest the effect of warmer seas is equivalent to a 2% increase in the Coriolis parameter per decade. If cyclones in one part of the world presently reach their lifetime-maximum intensity at a latitude of 25°, at what latitude will this maximum occur in 10 years' time?

5. Tidal forcing applied by a moon to its planet, like that applied by Earth's moon to Earth, causes two high tides, one directly under the moon and the other on the opposite side of the planet. Discuss whether this forcing could generate inertial waves in the atmosphere or ocean of a planet.

6. An industrial paint mixer rotates a drum at 240 RPM while revolving it around a second axis at right angles to the first axis. The cylindrical drum is 0.33 m high and 0.25 m in diameter. It is thought that exciting the lowest-order mode of inertial waves in the drum will maximise the mixing efficiency. At what rate should it be revolved around the second axis?

7. Using the identity

$$\left(\frac{\partial \bullet}{\partial t}\right)_I = \left(\frac{\partial \bullet}{\partial t}\right)_\Omega + \boldsymbol{\Omega} \times \bullet,$$

where $\bullet$ denotes any vector, the subscript I indicates inertial space and the subscript Ω denotes a frame of reference rotating with angular velocity $\boldsymbol{\Omega}$, show that the linearised, inviscid momentum equation in the rotating frame is given by (5.2),

$$\frac{\partial \boldsymbol{u}}{\partial t} + 2\boldsymbol{\Omega} \times \boldsymbol{u} = -\frac{1}{\rho_0}\boldsymbol{\nabla} p.$$

8. Considering the inviscid momentum equation in a rotating frame of reference, (5.2), and assuming the fluid moves without any vertical variation in accordance with a shallow-water continuity equation in an ocean or atmospheric layer of depth h, apply scalings $\boldsymbol{u}^* = U\boldsymbol{u}$, $2\boldsymbol{\Omega}^* = f_\oplus \boldsymbol{\Omega}$, $\boldsymbol{x}^* = L_\oplus \boldsymbol{x}$ and $t^* = t/f_\oplus$, and note that the pressure is given by $p^* = \rho_0 g \eta^*$ where $\eta^* = \hat{\eta}\eta$ is the dimensional surface elevation, to show that gravity-wave and Coriolis effects are similar in magnitude over a horizontal scale of

$$L_\oplus = \frac{\sqrt{gh}}{f_\oplus},$$

where L is traditionally called the *Rossby radius*.

6

Introduction to some nonlinear wave theories

6.1 Summary of key points

- The **general Stokes-drift velocity** for one-dimensional plane waves, (6.13) on page 154, is

$$\boxed{u_S = \frac{1}{2}\frac{k}{\omega}\hat{u}^2},$$

where k is the wavenumber, ω is the radian frequency and $\hat{u}$ is the velocity amplitude.

- The **Stokes-drift velocity for one-dimensional plane sound waves**, (6.16) on page 155, is

$$\boxed{u_S = \frac{1}{2\rho_0^2 c^3}\hat{p}^2},$$

where ρ_0 is the undisturbed-fluid density, c is the speed of sound and $\hat{p}$ is the sound-pressure amplitude.

- The **Stokes-drift velocity for plane water-surface waves**, (6.25) on page 157, is

$$\boxed{u_S = \frac{1}{2}\left(\frac{k}{\omega}\right)^3 \mathrm{g}^2 \left[\frac{\cosh 2(kz + kh)}{\cosh^2(kh)}\right]\hat{\eta}^2},$$

where g is the acceleration due to gravity, h is the water depth and $\hat{\eta}$ is the amplitude of the surface elevation. This has the deep-water approximation, (6.27) on page 158, of

$$\boxed{u_S = k\omega \mathrm{e}^{2kz}\hat{\eta}^2},$$

which at the water surface is given by (6.28) on page 158, of

$$\boxed{u_S\big|_{z=0} = k\omega\hat{\eta}^2},$$

and the shallow-water approximation, (6.15) on page 154, of

$$\boxed{u_S = \frac{1}{2}\frac{\mathrm{g}^{1/2}}{h^{3/2}}\hat{\eta}^2}.$$

DOI: 10.1201/9780429295263-6

- The **solitary-wave solution** to the Korteweg-de Vries equation for flows such as tsunamis or large atmospheric disturbances is given by (6.42) on page 162,

$$\boxed{\mathsf{u} = a_{\mathsf{u}}\,\mathrm{sech}^2\left(\frac{\mathsf{x}}{\mathcal{L}}\right)},$$

  where $\mathsf{u} = u/c_0 - 1$, u being the horizontal component of fluid velocity, c_0 is the linear shallow-water speed given by $\sqrt{gh}$ where h is the water depth, and x is related to the horizontal distance x by $x = h\mathsf{x}$; the wavelength $\mathcal{L}$ depends on the wave amplitude a_{u}, according to (6.43) on page 162,

$$\boxed{\mathcal{L}^2 = \frac{B}{a_{\mathsf{u}}}},$$

  where B is a parameter that depends on the initial conditions.

- Useful **textbooks** include Whitham (1999), while Lighthill (1978) also covers several nonlinear wave phenomena. For topics not covered in this book, see Drazin and Reid (1981) for waves due to instabilities, Whitham (1999) for shock waves and Streeter and Wylie (1979) for hydraulic jumps.

6.2 An example

The island's research-station biologist (§4.2) discovers the body of a local fisherman on a steeply-shelving beach on the island's sheltered side. Police forensic experts determine the victim, apparently murdered, had been in the water for 8 hours. Over this time there was no wind and small waves of wavelength 2 m and amplitude 0.1 m were travelling over a seabed of depth 3 m or greater, only breaking right on the beach. Public suspicion of the mining-survey ship is growing. Police need to know the farthest distance away from which the floating evidence could have come. The island's chief engineer is called.

Solution

Since waves were gentle and regular and occurring in a deep-water regime, use the deep-water Stokes-drift relation for surface water, (6.28),

$$u_S = k\omega(\hat{\eta})^2.$$

Using $k = 2\pi/\lambda$ gives $k = 2\pi/2.0 = 3.14\ \mathrm{m}^{-1}$. Thus, $\omega = \sqrt{(}gk)$, so that $\omega = \sqrt{(}9.81 \times 3.14) = 5.55\ \mathrm{rad\ s}^{-1}$. Then $u_S = 3.14 \times 5.55 \times (0.1^2)$, i.e. $u_S = 0.17\ \mathrm{m\ s}^{-1}$.

The distance covered in 8 hours is therefore 5 km. Police determine that the mining-survey ship was at least 10 km away over this time and thus its crew have an alibi. However, the chief engineer remembers the unusual internal waves in §4.2 were on this side of the island. She decides to talk to the backpacker.

6.3 Nonlinearity in fluid waves

As detailed in §1.2.3.2, the nonlinearity in the momentum equation is responsible for most of the complexity of fluid flows, and real waves often stretch beyond the small-amplitude assumption required for their linear derivation. Also, as will be discussed in §6.4 below, the simple sine-wave solution to the linear wave equation is itself a nonlinear function of space and time, which leads to nonlinear phenomena even if the momentum equation can be regarded as linear. In this chapter only two nonlinear wave phenomena will be covered, drifts and solitons, while in chapter 7 further phenomena due to nonlinear wave interactions will be briefly covered. In §9.6 effects of nonlinear *oscillations* caused by sound waves will be covered; wave-breaking, where the linear theory literally breaks down, features in chapter 10, and applications of nonlinear-wave physics are the basis of the applications discussed in chapters 12 and 13.

Nevertheless, a huge area of fluid dynamics is concerned with waves that arise as instabilities in fluid flows, which are not covered in this book. Readers are referred to Drazin and Reid (1981) for coverage of this area. Nonlinear sound waves, including shock waves, are likewise not covered in this book and are treated in Whitham (1999) and Lighthill (1978). Hydraulic jumps, a highly nonlinear flow connected to surface gravity waves, are covered in many engineering texts (e.g. Streeter and Wylie, 1979).

6.4 Stokes drift

6.4.1 Eulerian and Lagrangian displacements

It is commonly observed that seaweed and driftwood are washed up on beaches, even if they originated far away. Sand and sediment can also be slowly and steadily moved considerable distances over very long times by the unrelenting action of waves (figure 6.1). These are only some of the very many situations where waves do not simply cause the fluid to oscillate as predicted by linear theory, returning the fluid and anything suspended in it to its starting point after the passage of each wave. Many of these situations can be explained by *drifts*, which are nonlinear phenomena generated by waves, of which the best-known has the traditional name *Stokes drift* following its analysis by George Gabriel Stokes in 1847.

In some ways, Stokes drift is the simplest of the departures from the mathematically-convenient but physically-unrealistic world of infinitesimally small waves. Unlike the nonlinear waves we will consider later, Stokes drift is not due to the nonlinear term in the momentum equation - the nonlinearity

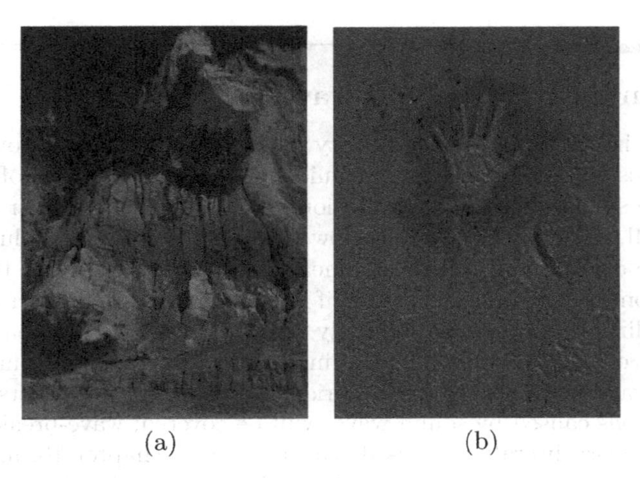

(a) (b)

FIGURE 6.1
Ocean waves create the world's largest longshore-drift system, which has transported sand northwards along the south-eastern Australian coastline for approximately 750 000 years (Boyd et al., 2008). (a) Drift ends north of the continent's easternmost extent, forming dunes at Rainbow Beach, Queensland, where cliffs show sand coloured during different times in Earth's recent history. (b) Particles remain loose and easily removed by hand, demonstrating that the sand has not yet formed into rock. Photographs by Richard Manasseh.

that makes a fluid accelerate differently to a solid when a force is applied. In §7.3, currents created by the nonlinear momentum of waves are studied. Stokes drift is an inevitable consequence of velocity varying sinusoidally in time and also varying in space - and thus an inevitable consequence of the existence of fluid waves.

The fixed-location (Eulerian) one-dimensional plane-wave velocity, u, that might represent the solution to any of the fluid-wave problems considered so far, is

$$u(x,t) = A\,\mathrm{e}^{\mathrm{i}(-kx+\omega t)}, \tag{6.1}$$

where, as usual, A is a complex velocity amplitude, and, as usual, k is the wavenumber of waves propagating in the positive x-direction with radian frequency ω. (Since there is no propagation in any other direction, the symbol reserved for the wavenumber in the x-direction, κ, is equal to the overall wavenumber, k.)

We will now distinguish between the *fluid velocity at a fixed point* (the Eulerian velocity in an Eulerian frame of reference) and the *velocity of fluid particles at a fixed point* (the Lagrangian velocity in an Eulerian frame). This distinction was briefly introduced in §1.3. The 'fluid particles' could be molecules of fluid, or of any particles carried along with the fluid. In a steady flow of fluid where there is no change of velocity with distance, there is no difference between these two velocities. However, in more general flows, including the flows created by waves, this is not true. At any fixed point, the averaged velocity of the fluid particles is *not* the same as the fluid velocity that is given by (6.1) and which we have been dealing with up to now. To emphasise this, denote the fluid-particle location with $\tilde{x}$ and the fluid-particle velocity with $\tilde{u}$, where $\tilde{u}$ is the derivative of $\tilde{x}$ with time.

The difference between the fluid-particle velocity at a fixed location and the fluid velocity at a fixed location arises because there is a nonlinearity in the relation between fluid-particle velocity and fluid-particle location, and that nonlinearity can be understood as arising from the fact that the linear solution to the wave equation is itself a nonlinear function: sine waves. Thus, when the velocity is integrated with respect to time to obtain the fluid-particle location - a simple matter of integrating a sine wave, one would have thought - the equation to be integrated is not trivial.

To understand this non-triviality of integrating the velocity to get location, consider $\tilde{u}(\tilde{x}, t)$, which is the fluid-particle velocity at the fluid-particle location. An expression for $\tilde{u}$ is obtained by replacing x with $\tilde{x}$ in (6.1), giving

$$\tilde{u}(\tilde{x}, t) = A\,\mathrm{e}^{\mathrm{i}(-k\tilde{x}+\omega t)}, \tag{6.2}$$

This velocity will clearly cause particles to move. Define ξ as the particle *displacement* in the x-direction, and where ξ like x and $\tilde{x}$ is a real number (it will shortly be explained why it is best to keep ξ real). Now, the fluid-particle location $\tilde{x}$ is given by

$$\tilde{x} = x + \xi, \tag{6.3}$$

Then, because $\tilde{u}$ is the derivative of $\tilde{x}$ with time (and defining $\tilde{u}_R$ to be the real part of $\tilde{u}$),

$$\Re(\tilde{u}) = \tilde{u}_R = \frac{\partial \tilde{x}}{\partial t} = \frac{\partial}{\partial t}(x + \xi), \tag{6.4}$$

and because the fixed location x (as well as time t) is an independent variable, so x can not vary with time, the substitution of (6.3) and (6.4) into (6.2) shows that the rate of change of ξ with time is given by

$$\frac{\partial \xi}{\partial t} = \Re\left\{A\,\mathrm{e}^{\mathrm{i}[-k(x+\xi)+\omega t]}\right\}, \tag{6.5a}$$

$$\Rightarrow \frac{\partial \xi}{\partial t} = \hat{u}\cos\left[k(x+\xi) - \omega t - \Phi\right], \tag{6.5b}$$

where $\hat{u}$ is the real velocity amplitude and Φ is the phase. Wave-generated drifts are easiest to observe for water-surface waves, and in the notation of §2.3, $\hat{u} = k\mathcal{Z}(z)\hat{\eta}$. However, for now, we will allow $\hat{u}$ to represent the amplitude of any fluid waves that are travelling in one direction only with no variation in the other directions. The addition of ξ to x in the argument of the cosine represents the fact that the location at which the speed of a particle is being calculated *depends on the particle location.* This makes (6.5) non-trivial to solve.

Clearly, if ξ is very small compared with x (or with $(\omega/k)t$), it can be neglected relative to x, so that the argument of the cosine on the right-hand side of (6.5) would become $kx - \omega t$, as in (6.1). Even at $x = 0$ and at $t = 0$, ξ would also be zero, so it could always be neglected. For ξ to remain very small, $\hat{u}$ must be very small so that ξ can not grow significantly with time. The integration of (6.5b) would simply yield another sinusoidal function, showing the fluid particles to be oscillating to and fro forever and always returning to exactly the same location after the passage of each wave. That would be consistent with the predictions of linear wave theory. Indeed, a very small value of $\hat{u}$ means that the speed with which fluid is set into motion is very small, which is the assumption behind the linear theory.

However, if $\hat{u}$ is not negligibly small, the presence of ξ in the argument of the cosine on the right-hand side of (6.5b) means that the solution will not be simple sine waves.

6.4.2 Perturbation approach for the drift velocity of 1D waves

To deal with the difficult-to-solve (6.5b), consider the Taylor-series expansion of (6.5b) about x for small values of ξ, so that ξ is a small perturbation from x. While the Taylor's series used below requires ξ to be small, it is really an expansion about $\xi = 0$, and, in principle, if enough terms were taken, ξ could be arbitrarily large. In practice, a large ξ requires a large wave amplitude, and other nonlinear effects such as wave breaking would probably dominate before more terms in the Taylor's series would need to be considered. First, we should anticipate that the Taylor series will generate terms in which the right-hand side of (6.5b) or derivatives or integrals of it are multiplied together. This is one situation where, unfortunately, we can no longer take advantage of the convenience of the complex exponential representation of sines and cosines. As explained in §1.5.1, the real part of the square of a complex exponential is not the same as the square of the real part. That is why at the point of (6.5), ξ was defined to be real. The Taylor series expansion of $\tilde{u}_R$ is, to first order in ξ,

$$\tilde{u}_R \simeq \tilde{u}_R\big|_{\xi=0} + \left.\frac{\partial \tilde{u}_R}{\partial(x+\xi)}\right|_{\xi=0} (\xi). \tag{6.6}$$

In order to evaluate (6.6), an expression for the small perturbation ξ is required, obtained by noting that since $\xi \ll x$, $k(x+\xi) \simeq kx$, and then integrating (6.5b) with respect to time, giving

$$\xi = \frac{\hat{u}}{\omega}\left[\sin(kx - \Phi) - \sin(kx - \omega t - \Phi)\right], \tag{6.7}$$

in which the first term, $\sin(kx - \Phi)$, arises from a constant of integration and is only included for completeness (it will disappear when the final calculation is done). Now, substituting (6.5b) and (6.7) into (6.6) yields

$$\begin{aligned}\tilde{u}_R &\simeq \hat{u}\cos(kx-\omega t-\Phi)\\ &\quad+\frac{\hat{u}^2}{\omega}\frac{\partial}{\partial(x+\xi)}\cos\left[k(x+\xi)-\omega t-\Phi\right]\Big|_{\xi=0}\left[\sin(kx-\Phi)-\sin(kx-\omega t-\Phi)\right],\\ &=\hat{u}\cos(kx-\omega t-\Phi)-\frac{\hat{u}^2k}{\omega}\sin(kx-\omega t-\Phi)\sin(kx-\Phi)\\ &\quad+\frac{\hat{u}^2k}{\omega}\sin^2(kx-\omega t-\Phi)\,. \end{aligned}\tag{6.8}$$

It can be seen straight away that the $\sin^2$ term in (6.8) will cause different behaviour of the particle-location (Lagrangian) velocity, $\tilde{u}$, to the fixed-location (Eulerian) velocity u in (6.1). The Eulerian velocity oscillates around zero, so its average over time will always be zero; however, because a sine or cosine squared does not oscillate around zero, there will be a non-zero average with time. In other words, the squared trigonometric functions with time 'survive' the time-averaging process. To determine exactly what this non-zero average is, average (6.8) over one period, T, where $T = 2\pi/\omega$. The time average of any quantity, say $\bullet(x,t)$, can be defined just as in (1.47) as

$$\overline{\bullet}=\frac{1}{T}\int_0^T \bullet(x,t)\mathrm{d}t. \tag{6.9}$$

The first and second terms in (6.8) will be time-averaged to zero, while the third, $\sin^2$ term can be evaluated by recalling the standard trigonometric identities,

$$\sin^2\alpha=\tfrac{1}{2}\left[1-\cos(2\alpha)\right], \tag{6.10a}$$

$$\cos^2\alpha=\tfrac{1}{2}\left[1+\cos(2\alpha)\right], \tag{6.10b}$$

(the double-angle formulae). Thus, owing to the 1 inside the square brackets of (6.10), the time-average of (6.8) is not zero. Writing for convenience $\alpha = \omega t + \Phi$, the $\sin^2$ factor in the surviving term of (6.8) can be expanded as $[\sin(kx)\cos\alpha - \cos(kx)\sin\alpha]^2$, giving $\sin^2(kx)\cos^2\alpha - 2\sin(kx)\cos(kx)\cos\alpha\sin\alpha + \cos^2(kx)\sin^2\alpha$. The time-average of the $\cos\alpha\sin\alpha$ is zero, while the rest time-averages to $\frac{1}{2}[\sin^2(kx)+\cos^2(kx)] = \frac{1}{2}$. Therefore the x-dependence disappears and (6.8) reduces to

$$\overline{\tilde{u}_R}=\frac{1}{2}\frac{k}{\omega}\hat{u}^2. \tag{6.11}$$

This is the nonzero average velocity of the fluid (and thus of any particles travelling with the fluid) that was first derived by Stokes in 1847. Since the analysis in this §6.4.2 is not specific to any particular type of fluid wave, it is clear that there will *always* be an averaged motion (sometimes called a 'net' or 'mean' motion) of fluid where there are travelling waves. Of course, in some cases, for example, water waves travelling along a river, or sound waves in a wind, the fixed-location velocity would not time-average to zero either. Thus, in order to make the distinction between the averaged drift generated by waves and any other net flow, the *Stokes drift velocity*, u_S, is formally defined as the difference between the time-averaged particle-location (Lagrangian) and the time-averaged fixed-location (Eulerian) velocities,

$$u_S = \overline{\Re(\tilde{u})} - \overline{\Re(u)}; \tag{6.12}$$

in the derivation above, $\Re(u)$ is simply the sine or cosine from (6.1), so $\overline{\Re(u)} = 0$ (but in a river or wind, $\overline{\Re(u)}$ would not be zero). Thus, (6.11) and (6.12) give

$$\boxed{u_S = \frac{1}{2}\frac{k}{\omega}\hat{u}^2.} \tag{6.13}$$

In all fluid-wave problems, (for example in §2.3.6) the choice of a positive value of k corresponded to waves propagating in the positive x-direction. Since the drift velocity in (6.13) is also positive, the drift is in the same direction as the wave propagation; reversing the sign of k and hence the direction of wave propagation would also reverse u_S.

Stokes's drift theory explained why seaweed and driftwood get washed up on beaches, for example. However, there was some confusion between the class of averaged *drifts* derived above, and the averaged *wave-generated currents* that are often also observed. In an observation or an experiment, there is no obvious way to tell them apart, and the words 'drift' and 'current' have been used interchangeably to refer to the net or mean movement of fluid and any particles travelling with them. The difference was clarified by Andrews and McIntyre (1978), who derived a 'Generalised Lagrangian Mean' theory. The essential difference between wave-generated drifts and wave-generated currents (as will be outlined in §7.3), is that wave-generated drifts occur whenever there are waves, but wave-generated currents require a gradient in the wave amplitude with distance.

If the waves are shallow-water plane water-surface waves, (6.13) can be applied directly, noting that $\hat{u} = k\mathcal{Z}(z)\hat{\eta}$ and since, for shallow-water waves there is no variation in the vertical, $\mathcal{Z} = \mathrm{g}/\omega$, so (6.13) becomes

$$u_S = \frac{1}{2}\left(\frac{k}{\omega}\right)^3 \mathrm{g}^2\hat{\eta}^2, \tag{6.14}$$

and noting that for shallow-water waves the wave speed $c = \omega/k = \sqrt{\mathrm{g}h}$, where as usual g is the acceleration due to gravity and h is the water depth,

$$\boxed{u_S = \frac{1}{2}\frac{\mathrm{g}^{1/2}}{h^{3/2}}\hat{\eta}^2.} \tag{6.15}$$

The drift velocity for shallow-water plane water-surface waves depends only on the wave amplitude and the water depth. For example, if one-metre-high shallow-water waves (therefore waves of amplitude $\hat{\eta} = 0.5$ m) are travelling in water 10 m deep, (6.15) predicts a drift velocity of about 0.012 m s^{-1}, enough to move seaweed, driftwood and fine sand several tens of metres over an hour or so. It is also clear from (6.15) that as the water becomes shallower and the wave height stays the same, the drift velocity becomes higher. If they have not already broken, waves must break when the amplitude equals the depth, as noted in §2.3.11, so the maximum possible shallow-water plane-wave drift velocity is $\frac{1}{2}\sqrt{gh}$. In practice, other nonlinear processes would come into play well before the waves begin to break.

If the waves are one-dimensional plane sound waves, (6.13) can also be applied directly, just as for shallow-water water-surface waves. It is convenient to convert the velocity amplitude to pressure amplitude, since pressure is more easily measured, just like surface elevation is more easily measured for water-surface waves. This relation is given by (3.28), $\hat{u} = \hat{p}/(\rho_0 c)$, where ρ_0 is the undisturbed-fluid density and c is the speed of sound that for all waves is given by $c = \omega/k$; substituting this into (6.13) gives

$$\boxed{u_S = \frac{1}{2\rho_0^2 c^3}\hat{p}^2.} \tag{6.16}$$

6.4.3 Perturbation approach for the drift velocity in 2D waves

For plane sound waves, or for plane shallow-water water-surface waves, a one-dimensional approximation could represent reality quite well. However, even if the waves are propagating in one direction only, there could be a variation in their structure in another dimension, and the most common such circumstance is water-surface waves in water that is not shallow. As detailed in §2.3, the shallow-water approximation begins to be a good approximation only when the water depth h is less than about 1/20 of the wavelength. To deal with the situation typical of water-surface waves, the displacement is now a vector, $\boldsymbol{\xi}$, in the vertical plane, so that $\boldsymbol{\xi} = (\xi_x, \xi_z)$, and the Taylor's series expansion to the first order of the particle-location (Lagrangian) horizontal component of velocity, $\tilde{u}$, needs to be extended to include the vertical dimension, z, so that the equivalent of (6.6) is

$$\tilde{u}_R \simeq \tilde{u}_R|_{\boldsymbol{\xi}=\mathbf{0}} + \left.\frac{\partial \tilde{u}_R}{\partial(x+\xi_x)}\right|_{\boldsymbol{\xi}=\mathbf{0}}(\xi_x) + \left.\frac{\partial \tilde{u}_R}{\partial(z+\xi_z)}\right|_{\boldsymbol{\xi}=\mathbf{0}}(\xi_z). \tag{6.17}$$

The particle-location (Lagrangian) velocity, the equivalent of (6.5) specific to water-surface waves so that $\hat{u} = k\mathcal{Z}(\tilde{z})\hat{\eta}$, is the vector $\tilde{\boldsymbol{u}} = (\tilde{u}, \tilde{w})$ where $\tilde{u}$ and $\tilde{w}$ are the particle-location speeds in the horizontal and vertical directions, respectively, whose real parts $\tilde{\boldsymbol{u}}_R = (\tilde{u}_R, \tilde{w}_R)$ are given from §2.3.5.6 by

$$\tilde{u}_R = k\mathcal{Z}(\tilde{z})\hat{\eta}\cos(k\tilde{x} - \omega t - \Phi),$$
$$\tilde{w}_R = \frac{\partial \mathcal{Z}(\tilde{z})}{\partial \tilde{z}}\hat{\eta}\sin(k\tilde{x} - \omega t - \Phi), \tag{6.18}$$

where $\mathcal{Z}(\tilde{z})$ for water-surface waves in water of any depth is given by the general Airy-solution vertical structure, (2.51),

$$\mathcal{Z}(\tilde{z}) = \frac{\mathrm{g}}{\omega}\frac{\cosh(k\tilde{z} + kh)}{\cosh(kh)}. \tag{6.19}$$

Like the one-dimensional particle-velocity relation (6.5), the two-dimensional particle-velocity relation (6.18) is non-trivial to solve analytically for the particle location. However it is easy to solve numerically, and an example of a solution to (6.18) is given in figure 6.2.

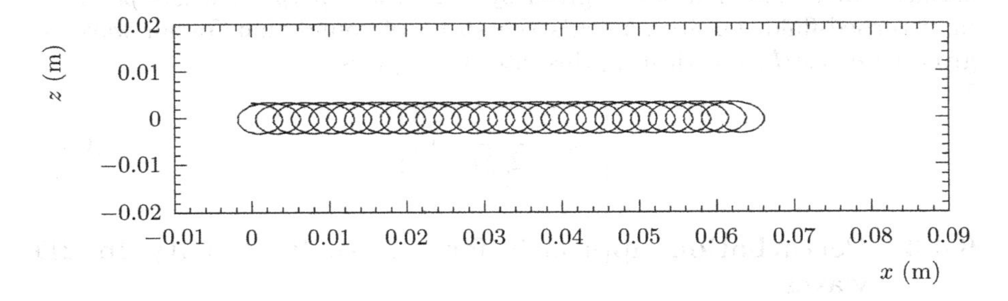

FIGURE 6.2
Solution of (6.18) over 10 s, giving the trajectory of a fluid particle (e.g. an individual water molecule or neutrally-buoyant particle much smaller than the wavelength) at the surface of a laboratory wave tank 0.25 m deep, with waves of wavelength 0.2 m and amplitude 0.0034 m. The particle drifts in the positive x-direction at about 0.0061 m s^{-1}.

Just as with the 1D waves, a Taylor's series approximation can be used to obtain a formula for wave-generated drift in the 2D case. The first term in the series given by (6.17) requires a version of (6.18) when the particle displacement is zero, i.e. when $\boldsymbol{\xi} = \mathbf{0}$, the particle-location velocity is $\tilde{\boldsymbol{u}}_R|_{\boldsymbol{\xi}=\mathbf{0}}$, and using (6.19) for $\mathcal{Z}(z)$, (6.18) gives

$$\tilde{u}_R|_{\boldsymbol{\xi}=\mathbf{0}} = \hat{u}_h\cosh(kz + kh)\cos(kx - \omega t - \Phi),$$
$$\tilde{w}_R|_{\boldsymbol{\xi}=\mathbf{0}} = \hat{u}_h\sinh(kz + kh)\sin(kx - \omega t - \Phi), \tag{6.20}$$

where

$$\hat{u}_h = k\frac{\mathrm{g}}{\omega}\frac{1}{\cosh(kh)}\hat{\eta}. \tag{6.21}$$

The two partial derivatives in (6.17) can now be evaluated as

$$\left.\frac{\partial \tilde{u}_R}{\partial(x+\xi_x)}\right|_{\boldsymbol{\xi}=\mathbf{0}} = -k\hat{u}_h \cosh(kz+kh)\sin(kx-\omega t-\Phi),$$
$$\left.\frac{\partial \tilde{u}_R}{\partial(z+\xi_z)}\right|_{\boldsymbol{\xi}=\mathbf{0}} = k\hat{u}_h \sinh(kz+kh)\cos(kx-\omega t-\Phi). \quad (6.22)$$

Next, the components ξ_x and ξ_z of the particle-displacement vector $\boldsymbol{\xi}$ are needed in (6.17). These can be obtained by integrating the particle-location velocity vector $\tilde{\boldsymbol{u}}_R$ in (6.18) with respect to time to give the particle-location vector $\tilde{\boldsymbol{x}}$, giving

$$\tilde{x}|_{\boldsymbol{\xi}=\mathbf{0}} = x + \frac{\hat{u}_h}{\omega}\cosh(kz+kh)\left[\sin(kx-\Phi)-\sin(kx-\omega t-\Phi)\right],$$
$$\tilde{z}|_{\boldsymbol{\xi}=\mathbf{0}} = z + \frac{\hat{u}_h}{\omega}\sinh(kz+kh)\left[\cos(kx-\omega t-\Phi)-\cos(kx-\Phi)\right], \quad (6.23)$$

from which $\boldsymbol{\xi}$ is immediately obtained since $\boldsymbol{\xi} = (\xi_x, \xi_z) = (\tilde{x}, \tilde{z}) - (x, z)$. Substituting (6.22) and (6.23) into (6.17) would involve some algebra with trigonometric functions. However, this can be side-stepped by recalling that applying the time-averaging operation, (6.9), causes any term not containing a factor with a $\sin^2$ or $\cos^2$ with time to disappear. Meanwhile, the squared factors in the surviving terms reduce to 1/2. Therefore, the time average of the particle-location velocity in (6.17) becomes

$$\overline{\tilde{u}_R} = \frac{1}{2}\frac{k}{\omega}\hat{u}_h^2\left[\cosh^2(kz+kh)+\sinh^2(kz+kh)\right]. \quad (6.24)$$

Now, using a standard identity for the hyperbolic functions, $\cosh^2\alpha + \sinh^2\alpha = \cosh(2\alpha)$, replacing $\hat{u}_h$ with its constituent variables from (6.21), and recalling the definition (6.12) of the wave-generated drift velocity, $u_S = \overline{\Re(\tilde{u})} - \overline{\Re(u)}$, yields

$$\boxed{u_S = \frac{1}{2}\left(\frac{k}{\omega}\right)^3 g^2\left[\frac{\cosh 2(kz+kh)}{\cosh^2(kh)}\right]\hat{\eta}^2}. \quad (6.25)$$

This formula (6.25), applied to the case of figure 6.2, predicts a drift velocity of 0.0063 m s^{-1}, very close to the value of 0.0061 m s^{-1} obtained by numerically integrating the particle-velocity relations, (6.18). As in the general one-dimensional drift velocities derived earlier, this water-wave drift velocity is proportional to the amplitude of the linear wave, squared. This is an inevitable consequence of the squared trigonometric terms of time being the 'survivors' of the time-averaging process. As a check, note that for shallow-water waves ($k \to 0$) the term in square brackets in (6.25) becomes unity, and (6.25) reduces to the 1D shallow-water drift velocity (6.14). Recalling the dispersion relation, (2.35), $\omega^2 = gk\tanh kh$, it is possible to re-arrange

(6.25) into a form that emphasises the dimensionless small parameter $k\hat{\eta}$ that represents the wave steepness (which in oceanography is generally given the notation ka),

$$u_S = \frac{1}{2}c\left[\frac{\cosh 2(kz + kh)}{\sinh^2(kh)}\right](k\hat{\eta})^2. \tag{6.26}$$

Now, from §2.3.9, deep water means that $kh \to \infty$, and noting from (2.74) that for deep water, $\omega^2 = gk$ causes (6.25) to simplify to

$$\boxed{u_S = k\omega \mathrm{e}^{2kz}\hat{\eta}^2}. \tag{6.27}$$

It can be seen that the drift velocity falls with depth much faster than the first-order orbital velocity, (e^{2kz} is the square of e^{kz}), and hence deep-water drifts are concentrated near the surface. At the surface itself, (6.27) becomes

$$\boxed{u_S\big|_{z=0} = k\omega\hat{\eta}^2}; \tag{6.28}$$

This deep-water approximation (6.28) to the formula (6.25), applied to the case of figure 6.2, predicts a drift velocity of 0.0064 m s^{-1}. Not surprisingly, this is very close to the result from the 2D drift formula (6.25) of 0.0063 m s^{-1}, since the case of figure 6.2 was a situation in which $h > \lambda/2$ where the deep-water approximation should hold; and, as noted earlier, the result from the 2D drift formula (6.25) is already very close to the full numerical solution.

In an ocean example, an object floating in a deep-ocean swell with a wavelength of 100 m, a frequency of 0.1 s and an amplitude $\hat{\eta} = 0.5$ m would drift in the direction of wave propagation at about 0.01 m s^{-1}. Although a centimetre per second may appear to be a small speed, after a month, even this very modest-height swell would have transported a floating object over 25 km. If the object gets close enough to shore, the wave regime becomes that of shallow water, and now depends on depth as shown by (6.15), increasing the drift velocity. Furthermore, as noted in §2.3.12 and detailed in §3.3.5, shallow-water waves are turned by refraction so that they head towards the shore. Finally, objects will get washed up on the beach, and left there on the high tide.

Of course, there are currents in the sea transporting water in addition to the wave-generated drift, and these currents may have much larger velocities than the wave-generated drift. Currents could transport a floating object in some direction other than the wave-propagation direction. The most powerful ocean currents form 'gyres' thousands of kilometres across, so that surface water could travel in a circuit, never entering shallow water. Thus, it is perfectly possible that a floating object far from land could take a very long time to reach any shore. Eventually, however, a storm would cross the path of the current, generating waves and thus a wave-generated drift at some angle to the current that takes the object out of the current and providing the opportunity for a drift to carry it towards land. The chaotic nature of the weather and, consequently, of the wave systems generated by the weather, could result in widely differing fates for floating objects released at approximately the same location. For example, of six hydrographic survey bottles released off the coast of North America in 1959, four were discovered washed up within two months. The fifth bottle was discovered in 2013, 54 years later. The sixth bottle has not been found yet.

While the movement of dense particles like sand is more complicated, if waves are in intermediate-depth water, there will be some drift just above the seabed. At depths offshore where refraction has not completely turned the waves towards the coast, the drift velocity will therefore have some component parallel to the coast, a phenomenon called *longshore drift*. Over decades, centuries, and millennia, enormous amounts of sand can be steadily shifted along a coast by waves, resulting in deposits such as that of figure 6.1.

6.5 Solitary waves

6.5.1 Balancing nonlinear momentum and dispersion

If a disturbance is made in a fluid system that allows waves to exist, unless the disturbance happens to be exactly sinusoidal in space and time, the result will not be a single sinusoidal wave. The disturbance will result in the creation of as many waves as necessary to synthesise the original disturbance when summed up. Mathematically, the result would be the Fourier transform of the disturbance; if the disturbance has a sharp edge to it; an infinity of wavelengths each with a specific amplitude and phase is required to perfectly synthesise the original. Considering dispersive waves such as water-surface waves, the longest wavelengths would travel fastest, leaving behind the shorter wavelengths and destroying the initially-perfect synthesis of the original disturbance. Soon, the original shape of the disturbance would be lost. For example, dropping a rectilinear object with sharp edges (such as a brick) into water will result in a disturbance consisting of very many wavelengths. The original rectangular 'footprint' formed at the instant the brick touched the water will soon disappear owing to wave dispersion.

However, in many fluid-wave systems, a single large disturbance, rather than a series of waves, is observed to propagate through a fluid apparently unchanged in shape. The earliest and best-known report was made by Scott Russell in 1834 when observing disturbances in the canals that had been constructed for transportation of heavy goods. Careful consideration of these observations in the late 19th century led to the hypothesis that phenomena of wave dispersion and wave steepening could reach a balance, allowing a disturbance to propagate without losing its shape.

Consider waves propagating in one direction only, and neglecting viscosity, so that the incompressible momentum equation along the sea surface is

$$\frac{\partial u}{\partial t} + u\frac{\partial u}{\partial x} = -g\frac{\partial \eta}{\partial x}. \tag{6.29}$$

This could be obtained in the same way as (2.4), but without neglecting the nonlinear term. Now, if the nonlinear term were zero, the solution would simply represent shallow-water waves for which (as noted in §6.4.2) there is a direct relation between the surface elevation and velocity given by $u = k\mathcal{Z}(z)\eta$, and, for shallow-water waves, $\mathcal{Z}(z) = g/\omega$. Substituting results from the linear solution back into the nonlinear equation can sometimes give a useful if informal representation of the physics; this technique falls into the class of mathematical approximations loosely called the *weakly nonlinear* approach. (A more detailed weakly nonlinear technique will be used in §7.4 when wave interactions are considered.) Hence, (6.29) can be written

$$\frac{\partial u}{\partial t} + u\frac{\partial u}{\partial x} = -\frac{\omega}{k}\frac{\partial u}{\partial x}, \tag{6.30}$$

and since for any waves, the phase speed is given by $c_0 = \omega/k$, where the subscript 0 has been added to emphasise this is the phase speed from the linear solution, (6.30) becomes

$$\frac{\partial u}{\partial t} + u\frac{\partial u}{\partial x} = -c_0\frac{\partial u}{\partial x}. \tag{6.31}$$

The full dispersion relation for water-surface waves in water of any depth is given by (2.35), reproduced here as

$$\omega^2 = gk\tanh kh, \tag{6.32}$$

as already noted in §2.3.11, it can be approximated by noting the Taylor's series expansion of the tanh function, $\tanh(\epsilon) \simeq \epsilon - \frac{1}{3}\epsilon^3 + \ldots$ for some argument ϵ that is small; in (6.32), $kh = 2\pi h/\lambda$ is small because the depth h is small compared with the wavelength λ. Considering only the leading-order term in the expansion, which implies the water is extremely shallow compared to the wavelength, allowed the approximation of (6.32) to $\omega^2 = gk^2h$ and thus the phase speed c_0 was given by $c_0 = \omega/k = \sqrt{gh}$; this is the shallow-water limit. Since this shallow-water phase speed does not depend on k, as noted in §2.3.11, these waves are not dispersive.

Now, however, consider waves that are in fairly shallow water, but not so shallow that they are pure shallow-water waves. Hence the waves are slightly dispersive, so that ω^2 must be represented by more than just the first term in the Taylor's series expansion of $\tanh(kh)$. With only the first term in the expansion, simply taking the square root of the leading-order term, gk^2h, immediately gave ω and hence c_0. However, including more terms requires a Taylor's series expansion for small kh of

$$\omega = \sqrt{gk\tanh kh}, \tag{6.33}$$

rather than (6.32). Writing for convenience $\epsilon = kh$, the expansion is

$$\omega \simeq \omega|_{\epsilon=0} + \left.\frac{\partial \omega}{\partial \epsilon}\right|_{\epsilon=0} \epsilon + \frac{1}{2!}\left.\frac{\partial^2 \omega}{\partial \epsilon^2}\right|_{\epsilon=0} \epsilon^2 + \frac{1}{3!}\left.\frac{\partial^3 \omega}{\partial \epsilon^3}\right|_{\epsilon=0} \epsilon^3 + \ldots, \tag{6.34}$$

and on evaluating ω and its first, second and third derivatives at $\epsilon = 0$, their values (after a considerable amount of algebra) turn out to be $\sqrt{\mathrm{g}/h}$ multiplied by 0, 1, 0 and -1 respectively, reducing (6.34), up to third order in the expansion, to

$$\omega \simeq \left(\frac{\mathrm{g}}{h}\right)^{1/2}(kh) - \frac{1}{3!}\left(\frac{\mathrm{g}}{h}\right)^{1/2}(kh)^3, \tag{6.35}$$

which simplifies to

$$\omega \simeq c_0 k - \frac{1}{6} c_0 h^2 k^3, \tag{6.36}$$

where $c_0 = \sqrt{\mathrm{g}h}$ as before.

A formal derivation would non-dimensionalise the momentum equation including both vertical variation and nonlinearity, and then expand the momentum equation in small parameters representing *both* the depth (kh) and the nonlinearity (ratio of amplitude to depth). Here, only an informal 'sketch' derivation will be undertaken. Firstly note that if the nonlinear term were neglected, the shallow-water momentum equation (6.31) would be solved by $u = \hat{u}\mathrm{e}^{\mathrm{i}(kx-\omega t-\Phi)}$ as usual, giving the trivial relation

$$-\mathrm{i}\omega = -c_0\,\mathrm{i}k. \tag{6.37}$$

While (6.37) might be trivial, comparing it to a version of (6.36) trivially modified to be $-\mathrm{i}\omega = -c_0\,\mathrm{i}k - (1/6)c_0 h^2 \mathrm{i}^3 k^3$ reveals that the effect of slight dispersion (up to third order), but still with the linear solution $u = \hat{u}\mathrm{e}^{\mathrm{i}(kx-\omega t-\Phi)}$, is represented by the equation

$$\begin{aligned} -\frac{\partial u}{\partial t} &= c_0\frac{\partial u}{\partial x} + \frac{1}{6}c_0 h^2\frac{\partial^3 u}{\partial x^3}, \\ \Rightarrow -\frac{\partial u}{\partial t} &- c_0\frac{\partial u}{\partial x} - \frac{1}{6}c_0 h^2\frac{\partial^3 u}{\partial x^3} = 0. \end{aligned} \tag{6.38}$$

Effectively, (6.38) is no more than a modification of the linear inviscid momentum equation such that it admits solutions that are almost, but not quite, pure shallow-water waves. Now assume the nonlinear term is not small enough to be neglected and is the same order as the term representing the slight dispersion. We can infer that the resulting equation is

$$-\frac{\partial u}{\partial t} + u\frac{\partial u}{\partial x} - c_0\frac{\partial u}{\partial x} - \frac{1}{6}c_0 h^2\frac{\partial^3 u}{\partial x^3} = 0. \tag{6.39}$$

Next, shift into a frame of reference moving at the original linear phase speed, c_0. In this new frame of reference, the velocity in the x-direction, say u, is given by $\mathrm{u} = u - c_0$, changing (6.39) to

$$\frac{\partial \mathrm{u}}{\partial t} - \mathrm{u}\frac{\partial \mathrm{u}}{\partial x} + \frac{1}{6}c_0 h^2 \frac{\partial^3 \mathrm{u}}{\partial x^3} = 0. \tag{6.40}$$

Finally, a scaling operation, $\mathrm{u} = c_0\mathsf{u}$, $t = (6h/c_0)\mathsf{t}$ and $x = h\mathsf{x}$ gives

$$\boxed{\frac{\partial \mathsf{u}}{\partial \mathsf{t}} - 6\mathsf{u}\frac{\partial \mathsf{u}}{\partial \mathsf{x}} + \frac{\partial^3 \mathsf{u}}{\partial \mathsf{x}^3} = 0.} \tag{6.41}$$

The nonlinear partial differential equation (6.41) was derived by Diederik Johannes Korteweg and Gustav de Vries in 1895, so it is traditionally called the *Korteweg-de Vries equation* or more simply the 'KdV equation'. Although it is nonlinear, it is remarkable in that it has an analytic solution. This solution represents *solitary waves*, often called *solitons*, which occur in very many branches of physics, including optical systems, and the soliton solution to (6.41) is given by

$$\boxed{\mathsf{u} = a_{\mathsf{u}} \operatorname{sech}^2\left(\frac{\mathsf{x}}{\mathcal{L}}\right),} \tag{6.42}$$

where a_{u} is a dimensionless nonlinear 'amplitude', and $\mathcal{L}$ is a dimensionless nonlinear 'wavelength' given by

$$\boxed{\mathcal{L}^2 = \frac{B}{a_{\mathsf{u}}},} \tag{6.43}$$

where B is a constant. This solution is shown in figure 6.3. Versions of (6.41) also occur in stratified fluids, for which B can be calculated from the buoyancy frequency, N, that was derived for stratified fluids in §4.3.3.

While this analytic solution represents the sort of waves seen by Scott Russel in 1834 in a canal, by far the most significant solitary-wave phenomenon is caused by a geological event that suddenly creates a discontinuity in the seabed so large the entire depth of the ocean is set into motion. This is the most feared of waves and is known by the combination of the Japanese words for 'harbour' and 'wave': *tsunami.* Tsunamis will be discussed in more detail in §12.3.

6.6 Problems

1. An oil slick that arrived on a beach between 06:30 and 07:00 is suspected to have come from a passing cargo ship. Two ships passed in the last four hours in the shipping lane 2.3 km offshore, ship 'A' at 01:30 and ship 'B' at 04:30. Over the last 7 hours, waves

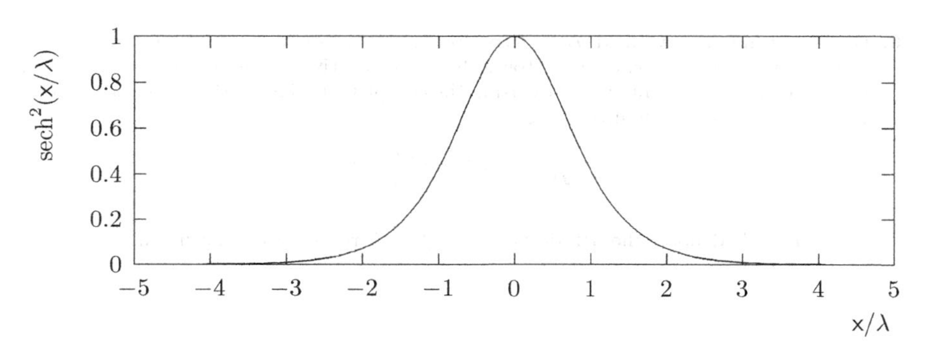

FIGURE 6.3
The solitary-wave or 'soliton' solution (6.42), with $a_u = 1$, to the kDV equation, (6.41).

of wavelength 2.5 m and amplitude 0.10 m were travelling over a sea of depth 4 m, becoming shallow only right at the beach. There was no significant wind. Which ship was most likely to have caused the oil slick?

2. Solitary waves travel unchanged in shape and form because

   A There is a high ratio of nonlinearity to dispersion

   B There is no wave dispersion

   C Nonlinearity is negligible

   D There is a balance between dispersion and nonlinearity

3. In the Korteweg-deVries equation,

$$\frac{\partial \upsilon}{\partial \tau} - 6\upsilon\frac{\partial \upsilon}{\partial x} + \frac{\partial^3 \upsilon}{\partial x^3} = 0,$$

   what represents the effect of nonlinearity?

   A The difference between the first and third terms

   B The third term

   C The second term

   D The difference between the first and second terms

4. Stokes drift occurs because of

   A The nonlinearity in the sinusoidal solution to the wave equation

   B The nonlinearity in the advective term in the momentum equation

   C A balance between nonlinearity and wave dispersion

   D A balance between the nonlinear and viscous terms in the momentum equation

5. An unusually large number of dead shellfish washes up on a steeply-shelving beach approximately 6 hours after a ship passes offshore that is suspected of illegally dumping toxic waste. There was no wind and waves of amplitude 0.08 m and wavelength 1.7 m were occurring over the sea of depth 7 m. At what distance (in m) from shore should divers be sent to take samples of the seabed?

6. Given a standing-wave field $u_R = \hat{u}\cos(kx - \omega t) + \hat{u}\cos(kx + \omega t)$, prove that there is zero particle-velocity drift (zero Stokes drift) irrespective of whether the amplitude $\hat{u}$ may be considered small or not, by using the complete Taylor's series for some function $f(x+\xi)$ for a small perturbation ξ,

$$f(x) = \sum_{n=0}^{\infty} \frac{f(x)^{(n)}(\xi)}{n!}\xi^n,$$

where $f(x)^{(n)}$ denotes the nth derivative of f with respect to its argument.

7

Nonlinear wave interactions

7.1 Summary of key points

- In the absence of external pressure gradients, the magnitude of a wave-driven **streaming flow** is proportional to the square of the wave amplitude, and proportional to the gradient in wave amplitude with distance.
- Waves nonlinearly coupled by the momentum equation can interact in groups of three, called **resonant triads**, given by (7.17) on page 176,

$$\boxed{\begin{aligned} \frac{\partial A_i}{\partial t} + \sum_{j=0}^{N}\sum_{l=0}^{N} A_j A_l &= 0, \\ \pm k_j \pm k_l \pm k_i &= 0, \\ \pm \omega_j \pm \omega_l \pm \omega_i &= 0 \end{aligned}}$$

  where the A_i are time-varying amplitudes of the waves with wavenumbers k_i and frequencies ω_i.
- Surface gravity waves couple nonlinearly owing in part to the nonlinear boundary condition at the surface, even if the momentum balance remains linear. This leads to **four-wave resonance** as waves become steeper, transferring energy to longer and longer wavelengths and thus creating the ocean swell from short waves generated by local winds.
- A useful **reference** for streaming flows is the review by Riley (2001). For nonlinearly coupled waves, a useful **book** is Craik (1988).

7.2 An example

The island mentioned in §2.2, §3.2, §4.2, §5.2 and §6.2 is exposed to ocean swell on one side, creating a surf beach. The backpacker tourist goes for a swim, picking a narrow zone where wave breaking appears absent. The chief engineer has come to the beach to find him. In the breaking zone, the wave height is known to reduce by about 0.1 m for every 10 m distance

DOI: 10.1201/9780429295263-7

towards the shore, whereas the seabed slope is very gentle. The backpacker is swept out to sea in a rip current. At what distance from the beach will the radiation stress causing the rip reach zero?

Solution

The radiation stress depends on $u\partial u/\partial x$, and from (2.84),

$$\hat{u} = |A|\sqrt{\frac{g}{h}},$$

where $|A| \simeq A_0(1 - \alpha x)$, where $\alpha = 0.1/10$. Ignoring the variation of h with x, the stress reaches zero at $x = 1/\alpha = 100$ m, and he is nearly about that far out. She motions him to swim parallel to shore, and eventually, he is returned to the beach. Fortunately, he has found a large floating object beyond the breaking zone and clung to it. On examination, it appears to be a composite-foam panel with periodically-spaced voids engineered into it. The chief engineer determines the voids are bubbles serving an acoustic-scattering application (§9.6.2), and may have come from a damaged submarine. They agree to work together.

7.3 Mean flows driven by waves

The nonlinearity in the acceleration terms in the momentum equation (the Navier-Stokes equation and also its inviscid version, the Euler equation) is a velocity multiplied by a velocity gradient, $\boldsymbol{u} \cdot \boldsymbol{\nabla u}$, so it is proportional to the velocity magnitude squared. Hence it is a quadratic nonlinearity. Following the reasoning of perturbation theory, the quadratic nature of the nonlinearity means that it has a particular tendency to create second-order flows that are two first-order flows multiplied together. This allows two waves to interact, rather than simply superposing (as was illustrated in figure 2.1), but it also allows a wave to interact with itself.

The simplest of these second-order flows, but in many ways, the most widespread, is a flow that does not vary with time at all: a *mean flow*, where the words 'mean' or 'net' refer to the concept that the flow is what one observes as an average over time. There will always be a nonzero average motion with time generated by all fluid waves, usually called the Stokes drift, as detailed in §6.4, due to the nonlinearity in the *velocity*. However, mean flows driven by fluid waves also exist due to the quadratic term - the nonlinearity in the *acceleration*. The elementary trigonometric identities behind this phenomenon (encountered in §6.4, and repeated here for convenience) is

$$\sin^2\alpha = \tfrac{1}{2}\left[1 - \cos(2\alpha)\right], \tag{7.1a}$$

$$\cos^2\alpha = \tfrac{1}{2}\left[1 + \cos(2\alpha)\right], \tag{7.1b}$$

(the double-angle formulae). If any term with a time-dependent factor $\cos\omega t$ is integrated with respect to time over a period, the result is of course zero, representing the fact that wave motion according to linear wave theory is an oscillation where there is zero average motion. However, if any term containing

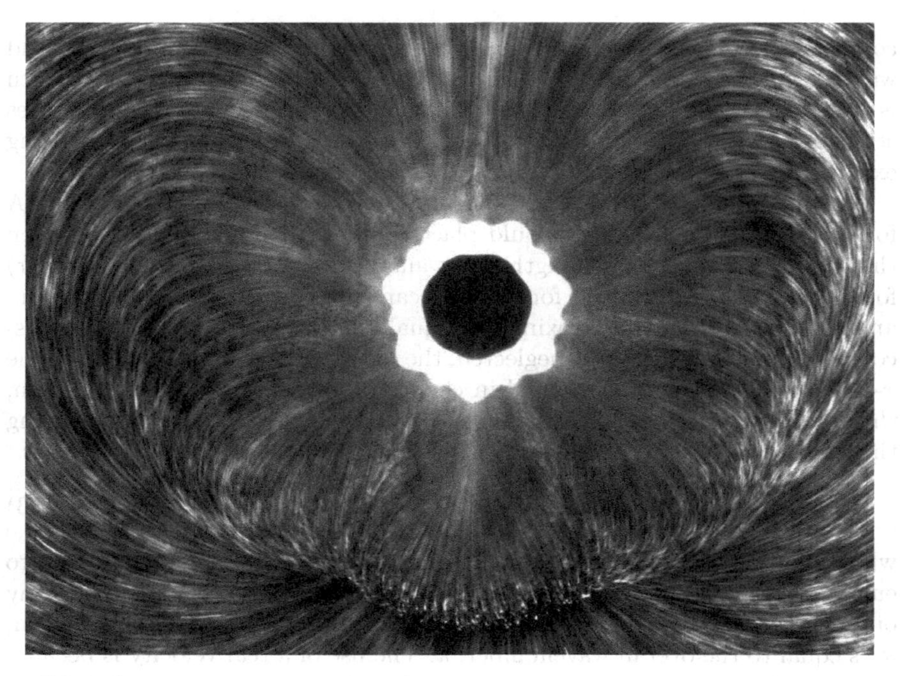

FIGURE 7.1
Acoustic microstreaming field created around a bubble of approximate radius 200 μm driven at 12.94 kHz. Similar phenomena have been utilised in biomedical micromixing technologies (e.g. Boon et al., 2011). From Tho et al. (2007), reprinted under a CSIRO Licence to Publish.

the factor $\cos^2 \omega t$ is integrated over a period, the $+1$ in the square brackets in (7.1) means the result is not zero, but a constant, $\frac{1}{2}$. Thus, the time average of the quadratic term is not zero. In §6.4 we saw how (7.1) meant that the time average of the particle-displacement velocity was non zero, but now we will see that it means the time average of the acceleration and thus the momentum is also nonzero.

To see the effects of this simple but profound result, consider the incompressible momentum equation in one dimension (the x-direction) only,

$$\frac{\partial u}{\partial t} + u\frac{\partial u}{\partial x} = -\frac{1}{\rho_0}\frac{\partial p}{\partial x} + \nu\nabla^2 u. \tag{7.2}$$

The viscous term contains second derivatives in all three spatial dimensions which can be simplified later, depending on the system to be modelled. For now, imagine the system is the coastal ocean, and waves are propagating towards the shore. Equally, one could imagine planar sound waves propagating in the x-direction, since a very similar argument to the following would apply, although for sound waves, as in chapter 3, we would begin with the compressible momentum and continuity equations. Furthermore, for sound waves, as mentioned in 1.2.2.4, there is a stress due to volume viscosity which is not included in (7.2). For purely planar sound waves, volume-viscous stresses still exist even though shear stresses are zero, because the fluid is still being expanded and compressed.

Assume that the viscous term is of the same order as the nonlinear term. A formal perturbation analysis would place conditions on the relation between the velocity-amplitude and length scales and viscosity (the Reynolds number) for this to be true. However, for now, we can simply assume the viscous term and nonlinear term are approximately equally negligible. Thus, when the viscous and nonlinear terms are neglected, the momentum equation is exactly the x-component of (2.18) considered in chapter 2 or (3.26) in chapter 3. Then, the analysis would lead to the standard solution of the wave equation, giving the real part of the linear inviscid velocity field, denoted here u_{0R}, which is

$$u_{0R} = \hat{u}\cos(kx - \omega t), \tag{7.3}$$

where $\hat{u}$ is the real velocity amplitude and the subscript 0 has been added to emphasise that this is the linear solution. Since there is no propagation in any other direction, the symbol reserved for the wavenumber in the x-direction, κ, is equal to the overall wavenumber, k. The use of a real velocity is because we are only interested in waves propagating in one direction without any reference point so that the phase is not relevant; furthermore, just like in 6.4, squared quantities will occur in the derivation to follow, so we can no longer exploit the convenience of the complex exponential, as explained in §1.5.1. The amplitude $\hat{u}$ includes all the other factors from (2.53). Although $\hat{u}$ does not vary with x under the linear inviscid theory, when we use (7.3) below in a mean-flow derivation, we will relax this invariance with x, so that $\partial\hat{u}/\partial x$ is small but nonzero. For example, there may be a gentle decrease of $\hat{u}$ with x due to the small effects of viscosity that were neglected to arrive at (7.3).

The goal of the present analysis is an illustration of how a current that is an average over time can arise from waves. Hence, first note (as defined in (1.47)) that the time average of any quantity, say $\bullet(x,t)$, is

$$\overline{\bullet(x)} = \frac{1}{T}\int_0^T \bullet(x,t)dt, \tag{7.4}$$

where T is some timescale that here is the period of the waves. Following perturbation-theory reasoning, the real parts of the velocity u and pressure p in (7.2) are written as

$$u_R = u_{0R} + \bar{u}, \tag{7.5a}$$

$$p_R = p_{0R} + \bar{p}, \tag{7.5b}$$

where u_{0R} and p_{0R} satisfy the linear inviscid momentum equation and $\bar{u}$ and $\bar{p}$ are the small time-averaged quantities that we wish to study. Then, (7.5) can be substituted into (7.2). The terms $\partial u_{0R}/\partial t$ and $(1/\rho_0)\partial p_{0R}/\partial x$ cancel since by definition they are satisfied by the linear inviscid solution. The term $\bar{u}\partial\bar{u}/\partial x$ is at a higher order (it is the product of two small quantities, so is very small) so it is neglected, leaving behind the terms

$$u_{0R}\frac{\partial u_{0R}}{\partial x} + u_{0R}\frac{\partial \bar{u}}{\partial x} + \bar{u}\frac{\partial u_{0R}}{\partial x} = -\frac{1}{\rho_0}\frac{\partial \bar{p}}{\partial x} + \nu\nabla^2\bar{u}, \tag{7.6}$$

or, in an equivalent form that will shortly be useful,

$$\frac{\partial}{\partial x}\left(\frac{1}{2}{u_{0R}}^2\right) + u_{0R}\frac{\partial \bar{u}}{\partial x} + \bar{u}\frac{\partial u_{0R}}{\partial x} = -\frac{1}{\rho_0}\frac{\partial \bar{p}}{\partial x} + \nu\nabla^2\bar{u}, \tag{7.7}$$

The linear inviscid solution (7.3) for the velocity, $\hat{u}\cos(kx - \omega t)$, the simple sine waves in x and t, should be expanded, using the standard trigonometric identity (1.62), giving $\hat{u}\cos kx \cos\omega t + \hat{u}\sin kx \sin\omega t$, and substituted into (7.7), generating three terms from the binomial expansion of ${u_{0R}}^2$. Now apply the time-average operation (7.4), affecting only the left-hand side since the right-hand side is already time-averaged. The second and third terms on the left-hand side time-average to zero. In the first term on the left-hand side, the time-dependent factor in the middle term of the binomial expansion, $2\hat{u}^2\cos\omega t\sin\omega t$, time-averages to zero and the first and last terms combine since $\cos^2 kx + \sin^2 kx = 1$, so that (7.7) becomes

$$\frac{1}{4}\frac{\partial}{\partial x}\hat{u}^2 = -\frac{1}{\rho_0}\frac{\partial \bar{p}}{\partial x} + \nu\nabla^2\bar{u}. \tag{7.8}$$

To solve this new equation for the mean flow, it must be combined with a version of the law of conservation of mass, and suitable boundary conditions must be specified. Furthermore, the time-averaged flow, if incompressible, must in reality be two-dimensional. However, we can understand the implications of (7.8) without undertaking a full solution, by considering two simple cases and making some rough assumptions. Thus, it must be emphasised that the rest of this section is descriptive and not rigorous.

In the original linear solution, $\hat{u}$ was not a function of x, so the left-hand side of (7.8) was zero. Hence, a mean flow of any strength could still exist, provided its pressure $\bar{p}$ as well its speed $\bar{u}$ vary so as to satisfy (7.8) with the left-hand side zero. In this trivial situation, there would be no connection at all to the linear solution; in other words, waves satisfying the linear inviscid theory can be superimposed on a steady flow (in the sea, a steady current), but the waves would not *create* the steady flow. However, as noted above, we will relax this invariance of $\hat{u}$ with x, so the left-hand side of (7.8) is not zero. Then, a mean pressure difference $\bar{p}$ or a mean flow $\bar{u}$ (or both) *must* exist, and they would be *nonlinearly driven* by the variation in $\hat{u}$ with distance x. This applies to all fluid waves. They all have the potential to drive a mean flow via the nonlinear term $\boldsymbol{u}\cdot\boldsymbol{\nabla}\boldsymbol{u}$, provided there is a variation in the wave amplitude with distance. Often, it is viscous or turbulent losses that provide this variation, but any variation will do.

The nonlinear forcing due to waves - the term on the left-hand side of (7.8) - is usually called the *radiation stress*. The same process of time-averaging the nonlinear term composed of fluctuating leading-order velocities is also undertaken in turbulence modelling; in this case, the same term is called the *Reynolds stress*.

Now consider the two cases, re-iterating that these will be simplifications for the purposes of illustration that can not properly represent any real case; rigorous analyses have been undertaken in the literature, and depend on the specific circumstances. Firstly, imagine there is a zero pressure gradient. If sound waves are considered, the pressure is atmospheric pressure everywhere, for example. If the system is the coastal ocean, that means there is no change in average sea-surface elevation with distance x; $\partial\bar{p}/\partial x$ is zero since pressure and sea-level elevation are related. In this case, (7.8) reduces to

$$\frac{1}{4}\frac{\partial}{\partial x}\hat{u}^2 = \nu\nabla^2\bar{u} \tag{7.9}$$

indicating a balance between the nonlinear forcing due to waves - the radiation stresses - and viscous stresses. In this first case, there is a mean flow $\bar{u}$ - sometimes called a *streaming flow*. If incompressible, mass conservation implies that this mean flow must be at least two dimensional, implying, as just-noted, that the assumption of one-dimensional motion made earlier is not valid. Furthermore, (7.9) needs careful consideration of the boundary conditions for a solution, and that will depend on the particular case studied. Irrespective of the specific case, however, it is clear that in the absence of external pressure gradients, the streaming flow magnitude is proportional to the *square of the wave amplitude*, and proportional to the *gradient in wave amplitude* with distance.

For laminar systems, it is simply molecular viscosity that provides the factor ν in (7.9). If the waves were sound waves, (7.9) predicts that sound waves would drive a mean flow $\bar{u}$ owing to the presence of fluid viscosity (including volume viscosity). Gradients in $\hat{u}$ could be created by the sound waves spreading out, or could also be created by viscous friction or thermal losses. Applications of such *acoustic streaming* will be discussed in §13.2. There are situations in which the gradients in velocity are large even if the magnitudes of velocity are not so large. With bubble-acoustic oscillations, mean-streaming flows seem to be observed around acoustically-driven bubbles whenever appropriate visualization techniques are employed in experiments, leading to patterns such as that in figure 7.1, which will be studied in §13.2.

If the waves are ocean waves, the factor ν is less likely to be molecular viscosity, and more likely to be an 'eddy viscosity' due to turbulence. Nevertheless, the outcome is the same: provided there is a gradient in wave amplitude $\hat{u}$, waves drive a mean flow $\bar{u}$: a current. Such currents become important near the shore where gradients in $\hat{u}$ are likely to be large, and bring nutrients from

FIGURE 7.2
Illustrations of rip currents on surf beaches; note the rips occur where there is less breaking owing to lower wave amplitudes and hence less white water, and may also involve 'feeder' rips at an angle to the beach. Images taken from a public-education video, copyright by and courtesy of Surf Life Saving Australia.

the deep ocean into the littoral zone, and thus sustaining nearshore marine life. Gradients in $\hat{u}$ could be created by the waves spreading out, friction with the bottom, or by waves breaking.

While many variations of $\hat{u}^2$ with x and many solutions to (7.9) are possible in general, imagine a simple case where the left-hand side of (7.9) is a negative constant, representing a linear decrease in wave energy with distance in which the wave is travelling; this could be a crude model of the situation where waves are breaking when approaching the shore. Further, suppose the second derivative of $\bar{u}$ in the y direction is negligible compared to that in the x-direction. The consequence is that $\partial^2\bar{u}/\partial x^2$ is negative, and applying the boundary conditions $\bar{u} = 0$ at the shoreline, and at some point in the sea beyond the breaking zone, immediately requires that the streaming flow must be positive: a flow towards the shore.

Conservation of mass requires that if the flow towards shore is steady when time-averaged over several cycles, there must be a return current. This return current usually forms in some zone along the shore where the wave radiation stress is weaker owing to the wave amplitudes and gradients being lower in that zone, often because there is less breaking in that zone. Since waves are always directed shorewards, the mean flow maybe towards the shore over much of the shoreline, with the exception of the zones where return currents form. If the zone of the return current is narrower than the zone where flows are shorewards, conservation of mass therefore, dictates that the return-current speed is much faster than the shorewards flow. This return current is called the *rip current*. Rip currents may attain mean speeds of 0.5-1.0 m s^{-1} (Moulton et al., 2017). Even the fastest Olympic swimmers have only reached about 2 m s^{-1} over short distances in heated swimming pools, explaining why rip currents in highly turbulent, cold water are a danger to any swimmer. Since it is an inherently recirculating mean flow, anyone carried out to sea should eventually be returned, but many have become exhausted and drowned before then.

For the second case, imagine there is no significant viscous or turbulent friction. In this case, the wave radiation stress must be balanced by a pressure gradient, so (7.8) reduces to

$$\frac{1}{4}\frac{\partial}{\partial x}\hat{u}^2 = -\frac{1}{\rho_0}\frac{\partial \bar{p}}{\partial x}, \tag{7.10}$$

indicating that the pressure must rise in the direction that the wave height is falling. Rising pressure means that the sea level must rise. Most often, the wave height falls as the waves head towards shore, as in the discussion above that led to the prediction of a rip current, creating as before a negative gradient of $\hat{u}$ with x. Now, however, it can be seen that if no significant current is created, the sea level must rise toward the shore. Effectively, this corresponds to the seawater flooding over the shore. This circumstance is most marked when $\hat{u}$ is large (in other words, large waves occur) and the gradient in $\hat{u}$ is large (the large gradient is usually due to these large waves breaking). This phenomenon is called *wave set-up* and is one contributor to coastal inundation during storms.

The temporary rise in sea level at the shoreline during storms has several components:

1. the additional sea-level elevation due to the low atmospheric pressure in the storm;
2. the stress of the wind, which is balanced by an additional sea-level gradient, causing sea-level rise if the winds blow onshore, but sea-level fall if winds blow offshore;
3. wave set-up noted above;
4. wave *run-up*, the cyclic increase in water level as individual waves break on a shore;
5. wave-generated drifts towards shore (§6.4);
6. astronomical tides, which may add to or subtract from the effects listed above;
7. rainfall, increasing water levels where rivers meet the sea.

Sometimes, the components 1 and 2 together are called the 'storm surge' but sometimes 1-4 together are called the 'storm surge', and, confusingly, sometimes astronomical tides are added to descriptions in the media as 'storm tides'. For accurate forecasting of emergencies, the different physical phenomena must be carefully separated. The total rise in sea level during a storm is most severe during tropical cyclones. Nonetheless, even temperate-climate storms can generate conditions causing catastrophic loss of life and property, compounding the damage from high winds.

In most physical circumstances, the reality would be somewhere in between the simplified cases represented by (7.9) and (7.10); both currents and pressure gradients could arise to balance the wave radiation stress. Furthermore, the current must be two-dimensional as noted earlier (and could also be three-dimensional). For ocean waves approaching a shore, by far the most common circumstance is the wave amplitude $\hat{u}$ decreasing towards shore (due to friction and breaking), with the largest frictional gradients also in the shorewards direction. The consequence is a general movement of water towards the shore. We should also recall that the *drift* due to waves derived in §6.4 also causes water to move towards shore, even in the absence of a gradient in $\hat{u}$.

7.4 Nonlinearly coupled waves

The time-averaged mean flows outlined in §7.3 could be considered a special case where a wave with a single wavenumber k and frequency ω interacts with itself via the nonlinear term in the momentum equation. In the process, owing to the cosine and sine double-angle formulae (7.1), it creates another wave with twice the frequency: the $\cos(2\omega t)$ in the square brackets of the double-angle formula $\cos^2(\omega t) = \frac{1}{2}\left[\cos(2\omega t) + 1\right]$. Previously, in §7.3, the focus was on the $+1$ in the square brackets; this survived a time-averaging procedure, whereas the $\cos(2\omega t)$ did not. Here, however, we will consider what happens when we do not just consider the average with time.

The statement (7.5) was rooted in perturbation theory, the mathematical technique that allows nonlinear equations to be dealt with by separating them into a succession of linear problems at higher and higher orders in which each problem is 'driven' by the solution to the previous, lower-order problem. The lowest-order problem is simply the problem giving the linear solution. For the approach of section §7.3 to be valid, the mean flow velocity $\bar{u}$ had to be at the order above the linear-solution velocity u_{0R}. The mean flow was thus 'driven' by the wave radiation stress of the linear solution. Importantly, the mean flow could not affect the linear solution.

Instead of the perturbation-based approach (7.5), say that the nonlinear equations (7.2) are satisfied by

$$u = A_0 u_0 + A_1 u_1 + A_2 u_2 + \ldots, \tag{7.11a}$$

$$p = A_0 p_0 + A_1 p_1 + A_2 p_2 + \ldots, \tag{7.11b}$$

where u_0, u_1, and so on are all solutions to the linear wave equation for the system being considered. Unlike the derivation in the previous section §7.3, here we will return to the use of complex values, since it will permit a less troublesome deviation to follow. Thus, considering a one-dimensional problem for simplicity, $u_i = \mathrm{e}^{\mathrm{i}(kx-\omega t)}$ is the 'structure' of the velocity solution of the ith wave that has a specific value of k and of ω. The complex amplitudes of these solutions are A_0, A_1 and so on. The solutions u_i differ only in that they have a different combination of wavenumber k and frequency ω. Since any of them satisfies the linear wave equation, any sum of them satisfies the linear wave equation just as well, provided each solution is multiplied by a constant (A_0, A_1, etc). For example, if the waves were water-surface waves, each $|A_i|$ would equal the velocity amplitude $\hat{u}_i$ where $\hat{u}_i = k\mathcal{Z}(z)\hat{\eta}_i$ and η_i is the surface elevation amplitude of the ith wave that has a specific value of k and of ω. This is nothing more than the statement of linear superposition that is fundamental to the assumption of linear behaviour and illustrated (as in figure 2.1) by a superposition of waves.

However, in (7.11), the 'amplitudes' will become functions of time rather than constants, and the supposition is that allowing the amplitudes to vary with time will permit the nonlinear equations to be satisfied by the sum in (7.11), whereas none of the $A_i u_i$ by itself can satisfy the nonlinear equations. So (7.11) can be written

$$u = \sum_{i=0}^{N} A_i(t) u_i, \tag{7.12a}$$

$$p = \sum_{i=0}^{N} A_i(t) p_i, \tag{7.12b}$$

where N is some number of solutions sufficient to represent most of the physics. This technique of substituting linear solutions back into the nonlinear equations falls under a class of techniques loosely called the *weakly nonlinear* approach. A weakly nonlinear approach was already used in the derivation of solitary waves in §6.5.1. This specific weakly-nonlinear technique of substituting a sum of wave-like solutions into a nonlinear equation is sometimes called a *Galerkin* approximation after Boris Galerkin, who published this approach in 1915 based on earlier work by Walther Ritz. Substituting (7.12) into (7.2) will clearly create a rather long expression, particularly because the nonlinear term will contain the product of two sums which could produce very many terms if N is large. Some of the most relevant applications of this approach are when the effects of viscosity are small enough to neglect. Thus, substituting (7.12) into the one-dimensional inviscid momentum equation gives

$$\frac{\partial}{\partial t}\left[\sum_{i=0}^{N} A_i(t) u_i\right] + \sum_{i=0}^{N} A_i(t) u_i \frac{\partial}{\partial x}\left[\sum_{j=0}^{N} A_j(t) u_j\right] = -\frac{1}{\rho_0}\frac{\partial}{\partial x}\left[\sum_{i=0}^{N} A_i(t) p_i\right]. \tag{7.13}$$

The weakly nonlinear approach drastically simplifies this long expression by exploiting the fact that the solutions u_i possess the mathematical property of *orthogonality*. From now on in this section, we will refer to the solutions as *eigenfunctions* to emphasise that this property is being used. The orthogonality property is easily illustrated by an integral with an outcome similar to the time-averaging operation already discussed. Now, however, the integral is over space and not time, and is called the *inner product* of the eigenfunction u_i, and is written with the notation $\langle , \rangle$. It is defined in three dimensions (for a vector $\boldsymbol{u}_i$) by

$$\langle \boldsymbol{u}_i, \boldsymbol{u}_i{}^{\dagger} \rangle = \int_V \boldsymbol{u}_i \boldsymbol{u}_i{}^{\dagger} \mathrm{d}V, \tag{7.14}$$

where V is some volume (in a three-dimensional fluid-wave application, a volume over which the nonlinear wave interactions are to be considered) and $\boldsymbol{u}_i{}^{\dagger}$ is the *complex conjugate* of $\boldsymbol{u}_i$ (the conjugate of a complex number $a + ib$ is $a - ib$). In the present one-dimensional example (7.14) becomes

$$\langle u_i, u_i{}^{\dagger} \rangle = \int_{x=0}^{n\lambda} u_i u_i{}^{\dagger} \mathrm{d}x, \tag{7.15}$$

where n is some number of wavelengths. (Shortly, it will become clear that the value of n does not matter, since it will be cancelled out.) Since the eigenfunctions are $u_i = \mathrm{e}^{\mathrm{i}(kx-\omega t)}$, multiplying each eigenfunction by its own conjugate $u_i = \mathrm{e}^{-i(kx-\omega t)}$ gives $\mathrm{e}^0 = 1$. The integral on the right-hand side of (7.15) reduces to the integral of unity over $x = 0$ to $x = n\lambda$, with the result being simply a constant, $n\lambda$. However, multiplying an eigenfunction, say u_i, by the conjugate of a *different* eigenfunction, say u_j, gives $\exp\{\mathrm{i}[(k_i - k_j)x - (\omega_i - \omega_j)t]\}$. When $u_i u_j$ is integrated in (7.15), the result is zero, since the average of the sine or cosine of some constant times x (here, $k_i - k_j$ times x) is always zero. Only if $i = j$, and thus $k_i = k_j$, does the argument reduce to $\mathrm{e}^0 = 1$, allowing the right-hand side of (7.15) to survive.

If the inner-product operation based on an eigenfunction u_i eliminates all eigenfunctions apart from u_i itself, that eigenfunction is called an orthogonal eigenfunction. The sine and cosine functions are therefore orthogonal eigenfunctions, but many other functions also possess this important and useful property.

Now, apply the inner-product operation based on one eigenfunction u_i to (7.13). This will cause the sums to disappear in the first term on the left-hand side leaving only the time-derivative of the amplitude of the chosen eigenfunction u_i. The pressure term on the right-hand side is always in quadrature to the velocity (it is a sine when the velocity is a cosine, for example), so it disappears entirely. In the next step, divide by $\langle u_i, u_i{}^{\dagger} \rangle$ to cancel out the constant $n\lambda$ (which is called *normalising* the inner product). The result is

$$\frac{\partial A_i}{\partial t} + \left\langle \sum_{j=0}^{N} A_j(t) u_j \frac{\partial}{\partial x} \left[\sum_{l=0}^{N} A_l(t) u_{0R} \right] \right\rangle = 0, \tag{7.16}$$

where a different index, l, is being used in one of the sums in the nonlinear term, to make the reasoning to follow clearer. The inner-product of the nonlinear term will consist of the integral of a large number of terms, but it is not necessary to write them all out. Every term will contain a factor like $\exp\{\mathrm{i}[(k_j + k_l - k_i)x - (\omega_i + \omega_l - \omega_i)t]\}$. When integrated over x, each of these terms will be zero, unless $k_j + k_l - k_i = 0$. This simple result reduces

(7.16) to

$$\boxed{\begin{aligned} \frac{\partial A_i}{\partial t} + \sum_{j=0}^{N}\sum_{l=0}^{N} A_j A_l &= 0, \\ \pm k_j \pm k_l \pm k_i &= 0, \\ \pm \omega_j \pm \omega_l \pm \omega_i &= 0, \end{aligned}} \tag{7.17}$$

where the condition $\pm k_j \pm k_l \pm k_i = 0$ is written with plus-or-minus symbols to emphasise that the waves could be propagating in either positive or negative directions, giving a different combination. A matching condition is also required on the frequencies ω, because a dispersion relation relates ω to k (for non-dispersive waves, it is simply $\omega = ck$ where c is the wave speed).

The fact that the nonlinearity in the momentum equation is a quadratic nonlinearity means that waves must interact nonlinearly in groups of three. This is called *three-wave resonance*, and sets of three waves with their wavenumbers (and therefore also their frequencies) adding up are called *resonant triads*.

The wavenumber indices begin from zero, because it is possible to have a combination where waves travelling in the positive direction and in the negative direction combine nonlinearly to force waves with zero wavenumber. In some wave systems, this zero-wavenumber flow would be a mean flow. Unlike the mean flow derived in §7.3, which is a higher-order phenomenon and thus must be much lower in velocity than the original waves that created it, the mean flow participating in three-wave resonance could be the same strength as the waves going in one direction and force a weaker wave in the opposite direction. In §13.4.2 this possibility will be discussed in the context of explanations of the magnetic fields of the Earth and other planets, which help to make planets habitable.

Finally, the procedure applied in this section led to a condition where, owing to the quadratic nonlinearity in the momentum equation, three waves of different frequencies and wavenumbers nonlinearly interact, transferring energy between themselves. However, for surface gravity waves in particular, even if the momentum equation is regarded as linear, there is nonlinearity in the dynamic boundary condition at the surface. There is also, as detailed in §6.4.2, nonlinearity in the sinusoidal linear solution itself that is increasingly significant as the waves become steeper. A weakly-nonlinear analysis considering these aspects was undertaken by Phillips (1960), which led to the prediction that surface gravity waves with different frequencies and wavenumbers would nonlinearly interact in groups of four, called *four-wave resonance*. Furthermore, the nature of the interaction tends to transfer energy from shorter wavelengths to longer wavelengths. This discovery by Phillips (1960) and others in the 1950s and 1960s explained the creation of the long-wavelength ocean swell from the short, choppy waves generated where the wind is blowing. More details are in §8.2.1.

7.5 Problems

1. A new design of a hand-held instant virus-test device uses ultrasonic streaming to mix sample fluids from a patient with test chemicals. Ultrasound propagates along a fluid-filled microchannel of constant depth that widens with a gradient of 200 μm per mm into the test chamber. If the design is altered to make the gradient 400 μm per mm and the power is increased by a factor of 1.3, by what factor will the streaming flow causing mixing change?

2. Ultrasound-generated steady fluid flows occur

   A where there is a gradient in velocity

   B where the fluid is incompressible

   C when the fluid is a liquid

   D where the nonlinear terms are small

3. Ocean waves passing over a reef double in height because of a distant storm. If the percentage gradient of wave height with distance is the same, the wave-generated currents change by a factor of

   A 1

   B $\sqrt{2}$

   C 2

   D 4

4. The radiation stress due to wave set-up during a storm can

   A only cause the sea level to slope upwards onto land

   B only cause a steady current to flow onto land

   C cause either a slope of the sea level onto land or a current to flow onto land but not both

   D cause both a slope of the sea level onto land and current to flow onto land

5. Ultrasound is used in a soft-drink bottling plant to collapse the foam created as bottles are filled, enabling more rapid filling. However, wave-generated airflow can blow the foam away instead of collapsing it. The pressure amplitude decreases by 10% per cm from the transducer, mostly due to geometric-spreading effects. If the power is increased by 20%, by what factor would the radiation stress causing a steady airflow to change?

6. Waves in an industrial algae-cultivation pond 10 m long and 0.4 m deep are forced by periodic pumping at two simultaneous frequencies above 0.5 Hz. A sloshing problem emerges with a half-wavelength wave fitting the pond length. Using calculations, discuss what the cause may be.

Part II

Further applications

8

Ocean wave energy conversion

8.1 Summary of key points

- The **wave-energy resource** in terms of cycle-averaged **power per unit length of wave crest**, $\bar{\mathbb{P}}_{\hat{\eta}}$, for a hypothetical swell of a single frequency in deep water, in terms of the available power per metre of wave crest, can be estimated from the relation (8.3) on page 185,

$$\boxed{\bar{\mathbb{P}}_{\hat{\eta}} = \frac{1}{8\pi}\rho_0\frac{g^2}{f}\hat{\eta}^2},$$

  where ρ_0 is the density of seawater, g is the acceleration due to gravity, f is the frequency of the waves in Hz and $\hat{\eta}$ is the wave amplitude (half the wave height). In a real ocean with a spectrum of waves, the cycle-averaged **power per unit length of wave crest**, $\bar{\mathbb{P}}_H$, is given by (8.4) on page 185,

$$\boxed{\bar{\mathbb{P}}_H = \frac{1}{64\pi}\rho_0\frac{g^2}{f_H}{H_{m0}}^2},$$

  where quantities normally reported in ocean weather forecasts and measurements are H_{m0}, the significant wave height, and the corresponding frequency f_H.

- The **forcing amplitude**, $\mathcal{F}$, i.e. the force per unit mass applied by the waves to a Wave-Energy Converter (WEC), may be roughly estimated from (8.30) on page 199,

$$\boxed{\mathcal{F} = \frac{\cosh(kh - kd)}{\cosh(kh)}\frac{g}{L_v}\hat{\eta}},$$

  where d is the depth of the WEC's actuator (for example, a float), k is the wavenumber given by $k = 2\pi/\lambda$ where λ is the wavelength, h is the depth of the sea where the WEC is located, and L_v is the vertical length scale of the actuator (given below for three device types) such that $L_v = V/A_\times$ where V is the volume of the WEC actuator and $A_\times$ is the cross-sectional area of the actuator to which the pressure of ocean waves is applied.

DOI: 10.1201/9780429295263-8

- The **displacement amplitude**, a, of a generic resonating WEC when optimally designed (for parameters appropriate to a hypothetical swell of a single frequency) and operating at resonance, is given by (8.31) on page 199,

$$\boxed{a = \frac{\mathcal{F}}{4\zeta_\mu {\omega_0}^2}},$$

  where ζ_μ is the damping ratio owing to 'parasitic' fluid-dynamical losses incurred in transferring energy from the ocean waves into the motion of the WEC and ω_0 is the natural frequency of the machine (which, since the device is resonating, must equal the radian frequency of the waves, i.e. $\omega_0 = \omega = 2\pi f$). In some cases, ζ_μ may be of order 10^{-2}, and values between 0.03 and 0.07 may be appropriate.

- The **practical limit on displacement amplitude** is a_{max}, hence for specific device designs a is replaced by a_{clip} given by given by (8.34) on page 200,

$$a_{\text{clip}} = \min\left(a, a_{\text{max}}\right).$$

- The **maximum cycle-averaged power** extracted by any optimally-designed, resonating WEC operating in the linear regime, $\bar{\mathbb{P}}_{P\text{max}}$, is given by (8.35) on page 200,

$$\boxed{\bar{\mathbb{P}}_{P\text{max}} = m\zeta_\mu {a_{\text{clip}}}^2 \,{\omega_0}^3},$$

  where m is the mass of water displaced by the device. This can be used to roughly estimate the cycle-averaged power obtained for highly-simplified different designs of WEC, the Rigid Pendulum (RP), the liquid pendulum or Oscillating Water Column (OWC), and the Heaving Buoy (HB), using parameters and formulae summarised from §8.4 and shown below.

	RP	OWC	HB
ω_0	$\sqrt{g/L}$	$\sqrt{g/L}$	$\sqrt{K/m}$
m	$\rho_0(1/6)\pi D^3$	$\rho_0(1/4)\pi D^2 L$	$\rho_0(1/6)\pi D^3$
d	$\min(h, D/2 + L/\sqrt{2})$	$\min(L, h, D/2 + \hat{\eta})$	$\geq D/2$
$a_{\max}$	$L\pi/2 - D/2$	L	$d - D/2$
Geometric constraint	$L = h - d$	$D + \hat{\eta} < h$	$D/2 + d < h$
L_v	$(2/3)D$	L	$(2/3)D$

- A useful **book** is McCormick (2013) while there have been several useful **reviews** on technologies and resources (Falcão, 2010; Manasseh et al., 2017a,b).

8.2 Introduction to wave-energy conversion

8.2.1 The wave-energy resource

Illustrations of the beginning of the wave-generation process, and of an advanced stage of wave generation, are in figure 8.1. The process by which the wind generates waves is still an area of active research. Nonetheless, the basic principles are understood. A wind with speed U initially creates ripples: waves with wavelengths significantly less than a decimetre that are dominated by surface tension. Specifically, the wind creates shear stress on the water surface, causing instability in the surface manifested by ripples. The ripples roughen the water surface, allowing wind to more efficiently transfer momentum to water-surface gravity waves. Once waves appear, the water surface has a sinusoidally-varying slope. Conservation of momentum means that as the airflow is forced to divert over the slope, a force must arise to balance this change in momentum, and this force exerts a pressure on the water that drives water motion. Turbulent shear stresses also transfer momentum to the water.

As mentioned at the end of §7.4, as waves travel, they transfer their energy into longer and longer waves via a nonlinear wave-interaction process,

(a) (b)

FIGURE 8.1
The beginning of wind-wave generation, and an advanced stage. (a) Capillary-gravity waves created by a breeze blowing over the sea surface in Port Phillip Bay, Australia. Image by Richard Manasseh. (b) Large swell in the North Sea. Image courtesy of Aneta Nikolovska.

elucidated by Phillips (1960) and others in the 1950s and 1960s. As briefly noted in §2.3.10.1, the maximum wavelength attained may be limited by the fetch: the size of the body of water in the direction the wind is blowing and hence in which the waves are travelling. In the open ocean, however, other factors limit the wavelength. Longer waves travel with faster speed c because they are deep-water waves and hence dispersive. Once $c = U$, the wind can no longer transfer energy to the waves by a simple pressure mechanism, because the sloping sea surface is at rest relative to the air. Energy can continue to be transferred from winds to shorter waves and thence to longer waves, but wave breaking tends to limit further wave growth.

Further details on wave breaking will be given in §10.2.1 and §10.2.2. While wave breaking, particularly in the open ocean, is also an area of active research, it is generally assumed that waves break when some criterion based on $k\hat{\eta}$ is exceeded, where as usual $k = 2\pi/\lambda$ is the wavenumber, λ is the wavelength and $\hat{\eta}$ is the amplitude. In other words, once the wave gets too steep, it breaks. Once the wave breaks, some energy is lost, preventing further growth in wave amplitude. Therefore, the longer the wave, the higher it can be before it breaks and thus the more energy it transports. Thus, the highest energy is found in the longest waves, and the longest waves emanate from regions of the highest wind speed U.

First, define the cycle average (or time average) of any quantity that is a function of time, say $\bullet(t)$, just as in (1.47), reproduced here as

$$\bar{\bullet} = \frac{1}{T}\int_0^T \bullet(t)\mathrm{d}t, \tag{8.1}$$

where T is the period of the waves. The cycle-averaged energy per unit surface area, $\bar{e}_{\hat{\eta}}$, of a surface gravity wave of surface-elevation amplitude $\hat{\eta}$, can be calculated from the linear solution, (2.61) and (2.64), by multiplying distance (surface elevation) by force per unit area (pressure) to get energy per unit area, then cycle-averaging and integrating over the depth, while using the deep-water approximation for simplicity. (Recall that the real parts of these quantities must be used when multiplying them together). This yields

$$\bar{e}_{\hat{\eta}} = \frac{1}{2}\rho_0 \mathrm{g}\hat{\eta}^2, \tag{8.2}$$

where as usual g is the acceleration due to gravity and ρ_0 is the assumed-constant density of water. Now, the rate with which this energy is being transported by waves in deep water is given by the deep-water group velocity,

$$c_g = \frac{1}{2}\sqrt{\frac{\mathrm{g}}{k}}.$$

Multiplying the energy per unit surface area by the group velocity of a hypothetical wave with wavenumber k and frequency f gives the cycle-averaged power of this wave per unit length of wave crest, i.e. $\bar{\mathbb{P}}_{\hat{\eta}} = \bar{e}_{\hat{\eta}} c_g$, giving

$$\boxed{\bar{\mathbb{P}}_{\hat{\eta}} = \frac{1}{8\pi}\rho_0 \frac{\mathrm{g}^2}{f}\hat{\eta}^2}. \tag{8.3}$$

For example, a 0.1 Hz wave with an amplitude of 1 m has a power of 39 kW m^{-1}.

However, the relation (8.3) is not the one used in practice to estimate ocean-wave energy. That is because ocean waves do not in practice have a single frequency, but rather have a spectrum of frequencies, each frequency having a different amplitude, so that $\hat{\eta}$ is not a constant but a statistical variable, say $\hat{H}/2$. In order to represent this, the amplitude of the hypothetical pure wave, $\hat{\eta}$, is replaced by $\hat{H}/2$ and (8.3) is re-written in terms of the *significant wave height*, H_{m0}. Originally defined as the average of the highest one-third of waves, it is now defined as four times the standard deviation in the surface elevation, η. The standard deviation is the square root of the variance, i.e. the average of the squares of the differences from the mean, which for a sinusoidal quantity varying about zero is $(1/2)\hat{\eta}^2$. Therefore $H_{m0} = 4\sqrt{((1/2)\hat{\eta}^2)}$ so that $\hat{\eta}^2 = H_{m0}^2/8$. The frequency used, f_H, is that corresponding to the significant wave height, and (8.3) becomes

$$\boxed{\bar{\mathbb{P}}_H = \frac{1}{64\pi}\rho_0 \frac{\mathrm{g}^2}{f_H}{H_{m0}}^2}. \tag{8.4}$$

The significant wave height is what is typically reported in ocean weather forecasts and measurements of ocean waves, and the corresponding frequency is sometimes reported as its reciprocal, sometimes called the *wave energy period.*

Strong steady winds occur in the Earth's mid-latitudes, which are defined as being between 30° and 60° in both Northern and Southern Hemispheres. This is because of two factors. Firstly, the Earth, like all planets, is spheroidal. Thus, a ray of energy from the Sun is spread over a much larger area at the Poles than at the Equator, heating the surface much more at the Equator relative to the Poles. Hot air rises from the heated surface at the Equator to high in the *troposphere*, the part of the atmosphere where most weather occurs, and flows both north and south to the mid-latitudes. Secondly, the planet is rotating, and (like most but not all planets) the Earth rotates about an axis not far from right-angles to the line from its centre from the Sun. Therefore, the air travelling to the mid-latitudes is also travelling towards the axis of rotation. Conservation of angular momentum demands that this part of the atmosphere should rotate faster than the planet, creating winds from the west.

Thus, the very strongest winds are found in the centre of the mid-latitudes, which is at 45° latitude, earning these winds the label 'The Roaring Forties'. At the surface, the air turns back towards the Equator to complete the circuit. Eventually, the air turns to blow from the east as it nears the Equator to form what is called the Trade Winds. These fundamental circulation cells of the atmosphere are called the *Hadley cells* after George Hadley who explained this process in 1735.

The consequence of the strongest winds being in the mid-latitudes is that the largest waves are in the mid-latitude oceans. Furthermore, these large waves come from the west. Thus, it is the coasts on the eastern boundaries of mid-latitude oceans that are impacted by large waves, for example, the coasts of Western Europe from Portugal to Scotland and southern Norway, the north-west of North America, the south-western coast of Africa, the southwestern and southern coasts of Australia, the west coast of New Zealand, and the coast of Chile. The global distribution of wave energy is shown in figure 8.2.

8.3 Issues with wave-energy conversion

8.3.1 A plethora of inventions

Wave-Energy Converters (WECs) are machines that convert the energy of ocean waves to electricity or to other useful forms of energy. The first WEC patent was filed in 1799, and at the time of writing it is estimated that there are over 250 companies each with its own patented technology for wave-energy

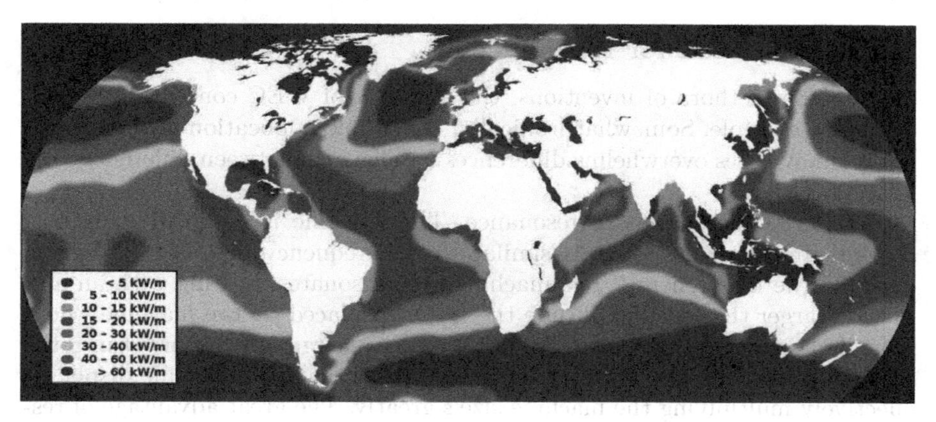

FIGURE 8.2
Global distribution of wave energy. Image by Ingvald Straume in the public domain from `https://upload.wikimedia.org/wikipedia/commons/2/21/World_wave_energy_resource_map.png`

conversion. However, two issues have retarded WEC developments to date. Firstly, very many WEC concepts have been invented; and secondly, most of these concepts must be very large, for the reason outlined in §8.3.2 below.

The large number of inventions is due to the fundamentals of water motion created by surface gravity waves. It is a reciprocating motion; the momentum in the water is constantly reversing. For most practical uses, the motion must be rotary, not reciprocating, so that it can turn a shaft. Contrast this reciprocating motion with the motion created as water exits a dam; that is in a single direction, downhill. For such unidirectional flow, extracting useful power is simple, requiring only a single moving part: the turbine. However, for reciprocating flow, a mechanism of some sort is required to efficiently extract power from the flow. Once there is a mechanism, there are inevitably joints, raising the risk of fouling from the marine life that is particularly abundant in the mid-latitude seas, corrosion from salt, damage from sand, and most concerning, destruction from storm waves. There is a concomitant need to keep the mechanism simple to minimise such risks, but a simple mechanism may entail lower energy-transformation efficiency.

The absence of an obvious solution to this conundrum, plus the fact that devising mechanisms that can convert reciprocating to rotary motion requires only an inventive mind and not an engineering degree, has lead to an enormous number of inventions. Moreover, the present-day imperative to find renewable energy sources has enhanced the availability of venture capital and other financial incentives for inventions to be patented and made the focus of small start-up companies. Thus, there is a large number of small companies globally, each promoting its own concept for wave-energy conversion, and backed by the small amount of capital typically made available to start-up companies.

8.3.2 The need for resonance

Despite the plethora of inventions, the majority of WEC concepts rely on a common principle. Somewhat ironically, the simple application of this principle in many cases overwhelms differences in efficiency between different WEC concepts.

The principle is that of resonance. The machine is designed to have a natural frequency, ω_0, that is similar to the frequency of the ocean waves carrying the most energy. The machine then resonates, moving with an amplitude *larger* than the amplitude the water displaced by the machine would have if the machine were not there. Therefore, energy is transferred into the machine's motion from a larger volume of water than the machine displaces, effectively multiplying the machine size's greatly. The great advantage of resonance is that the amplitude of the machine's motion may be many times higher that of a non-resonating machine, largely compensating for inefficient mechanisms. Thus, whether by calculated intent or trial-and-error discovery, most WEC inventions end up being resonators

Imagine for a moment that the machine is in a completely calm sea, then its mechanism is displaced by a disturbance. It would radiate waves around it in a pattern defined by the degrees of freedom of its mechanism. Essentially, there are only two such patterns: a monopole pattern, in which the radiated waves are perfectly circular and centred on the machine, and a dipole pattern in which positive and negative waves radiate in opposite directions along an axis. In practice, all resonating WECs exhibit aspects of superposed monopolar and dipolar motions, the latter possibly along multiple axes. Energy extraction from ocean waves can be thought of as the superposition of these machine-generated patterns with the ocean-wave field, resulting in a net reduction of the amplitude of ocean waves passing the machine; the averaged energy corresponding to this loss has been transferred into the kinetic and potential energy of the machine's motion. The resonance effect means that the machine may exert this influence on the ocean-wave field over an area much larger than the machine's physical size.

Ocean waves carrying high energy, following the explanation in §8.2.1, are very long waves with correspondingly low frequencies. As with any mechanical resonator, achieving a low natural frequency ω_0 inevitably requires that the machine size should be very large. For example, for the pendulum-type devices outlined shortly, $\omega_0 = \sqrt{(g/L)}$, where L is the pendulum length. For high-energy ocean swell, the frequency may be less than 0.1 Hz, and therefore L must be larger than 25 m. This is an enormous pendulum, the height of an eight-storey building. It would be a challenge to construct on land, let alone to deploy at sea in an ocean that has, of necessity, large and at times dangerous waves.

Thus, there is the unfortunate coincidence of two factors, both originating in fundamentals of physics and engineering. Small companies, backed by correspondingly small amounts of capital, are forced to develop gigantic machines

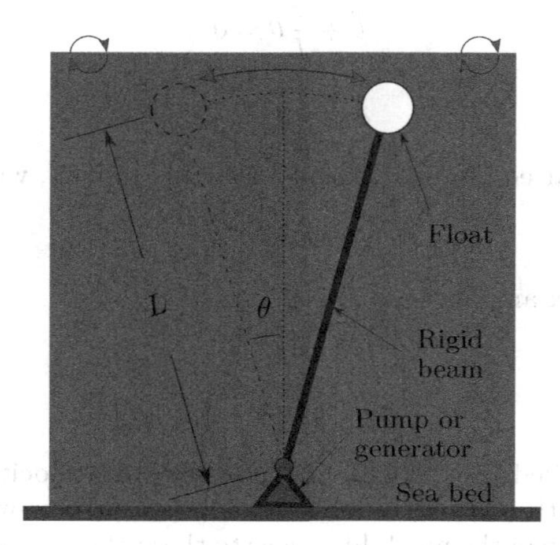

FIGURE 8.3
Schematic and definition diagram for the derivation of the linear operation of a rigid pendulum. The linearising assumption is $\theta \ll 1$

that by the definition demanded by patent-focused investors, have not been developed before. The consequence has been engineering or financial disasters. Nevertheless, in recent years engineering disasters have been rare, and the problem remains the adequate capital resourcing of developments.

8.4 Wave-energy converter technologies

8.4.1 Rigid pendulum

Considering figure 8.3, the elementary linear analysis of a Rigid-Pendulum (RP) type WEC is exactly the same as that for the pendulum of a clock. Newton's Second Law is applied to the tangential motion of the mass of the float: the motion when the pendulum is at an angle θ to the vertical, in the direction tangential to the pendulum. (The mass of the float is effectively the buoyancy of the float, $m = \rho_0 V$, where V is the volume of the float, but m does not need to be determined in order to calculate the natural frequency, since it will cancel out from the calculation below.) This gives

$$mg \sin\theta = m\frac{\partial^2 (L\theta)}{\partial t^2}, \tag{8.5}$$

since $L\theta$ is the displacement of the float along its arc. For a small angle, $\theta \ll 1$, $\sin\theta \simeq \theta$, so (8.5) becomes

$$\ddot{\theta} + \frac{g}{L}\theta = 0. \tag{8.6}$$

or, setting $\xi = L\theta$,

$$\ddot{\xi} + {\omega_0}^2\xi = 0. \tag{8.7}$$

This is the familiar equation of simple harmonic motion, with solution

$$\xi = A\mathrm{e}^{\mathrm{i}\omega_0 t}, \tag{8.8}$$

where the complex amplitude is $A = a\mathrm{e}^{\mathrm{i}\Phi}$ and

$$\boxed{\omega_0 = \sqrt{\frac{g}{L}}} \tag{8.9}$$

Since (as detailed in §2.3.5.6), the pressure and velocity of ocean waves decrease with depth (unless the waves are in the shallow-water regime) it is best to have the top of the pendulum close to the surface. However, a geometric constraint exists: the pendulum length L can be no greater than the depth of the sea, and since the float (the *actuator*) has a finite size, say a diameter D for a spherical float, the length L is further constrained to be $L = h - d$, where h is the depth of the sea and $d = D/2$ is the actuator depth.

The absolute limit on the motion of the rigid pendulum is where the float at the end of the pendulum hits the sea bed at either end of its stroke, and thus the maximum value of a is given by $a_{\mathrm{max}} = L\pi/2 - D/2$. In reality, the pendulum pivot would be some height above the sea bed, but this extreme limit is in any case unrealistically large, because the assumptions of linear inviscid behaviour would have broken down well before this extreme is reached, drastically increasing the effective damping and thus reducing the stroke amplitude. Furthermore, unless the machine is installed in a shallow-water wave regime, the forcing would fall off at the limits of the stroke, further reducing the stroke.

8.4.2 Liquid pendulum (oscillating water column)

Although it is a fluid that is moving, not a solid mechanism, the linear analysis of the Oscillating Water Column (OWC), like the rigid pendulum, can also be made simply using Newton's Second Law, without even considering the complications of fluid dynamics. Considering figure 8.4, the mass set into motion is that contained in the submerged part of the tube, $m = \rho_0 A_\times L$ where $A_\times$ is the cross-sectional area, so Newton's Second Law becomes

$$F = m\ddot{\xi}, \tag{8.10}$$

where the force applied to the mass m is that due to the weight of a very small mass of water temporarily displaced upwards a very small distance ξ,

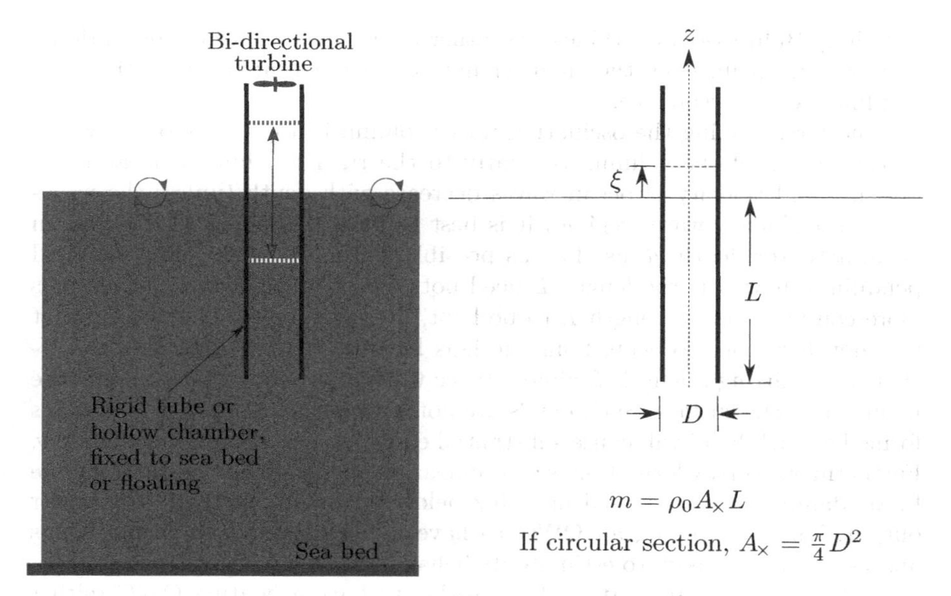

FIGURE 8.4
Schematic and definition diagrams for the derivation of the linear operation of the liquid pendulum, usually called the Oscillating Water Column. The linearising assumption is $\xi \ll L$

given by

$$F = -\rho_0 g A_\times \xi, \tag{8.11}$$

Thus, equating (8.10) and (8.11) gives

$$-\rho_0 g A_\times \xi = \rho_0 A_\times L \ddot{\xi}, \tag{8.12}$$

once again giving the equation of simple harmonic motion

$$\ddot{\xi} + \frac{g}{L}\xi = 0, \tag{8.13}$$

or

$$\ddot{\xi} + {\omega_0}^2 \xi = 0, \tag{8.14}$$

with $\xi = A\exp(i\omega_0 t)$ as with the rigid pendulum, and the natural frequency of a pendulum again,

$$\boxed{\omega_0 = \sqrt{\frac{g}{L}}}. \tag{8.15}$$

Just like the rigid pendulum (and the heaving buoy analysed below), the liquid pendulum is designed to resonate with the frequency of ocean swell. The water surface of the column pushes and pulls air through a bidirectional turbine: a device that turns in the same direction irrespective of whether the air is rising or falling. Bidirectional turbines are rather inefficient relative to conventional turbines, an inefficiency the engineer hopes is compensated for by the large amplification at resonance.

The force driving the oscillating water column is applied by ocean waves at the mouth of the column. Similarly to the rigid pendulum, because the pressure and velocity of ocean waves decrease with depth (unless the waves are in the shallow-water regime) it is best to have the mouth of the column at an actuator depth, d, as close as possible to the surface. Unlike the rigid pendulum, however, the length L need not be in a straight line; indeed it is more common for the length L to be bent, literally into an L-shape, so that the mouth is not too deep. Some designs have the length L bent into a U-shape, permitting a length L almost twice the water depth while keeping the mouth near the surface. Such bends are not without a cost, however: vortices formed at each bend will cause substantial energy losses, decreasing efficiency. Furthermore, vortex formation is a nonlinear phenomenon not captured by the linear damping ratio ζ_μ used in §8.5.2 below to roughly estimate the power output. Thus, arranging an OWC to have L achieved over multiple bends and using linear theory to estimate its behaviour may give a misleading over-estimate of performance. It is also possible to have a floating OWC (either straight, L- or U-shaped) in water much deeper than L, and indeed the first OWCs deployed were floating devices. If the OWC is floating, it may be part of a large floating platform that is effectively motionless as waves pass, so its analysis would be the same as that above. However, if the OWC's structure can also move with the waves, its behaviour could be much more complicated.

Despite the apparent flexibility in geometry, there are geometric constraints on L. Firstly, the actuator depth, d, which here is the depth of the mouth of the OWC, can not be less than the wave amplitude, or the forcing waves would cease to drive the water column at their trough. If the column is indeed in an L-shape, this means that the depth of the centreline of the column mouth can not be less than the wave amplitude plus the column half-width, $D/2$ say, so $d \geq \hat{\eta} + D/2$. Furthermore, the greater the column cross-sectional area, the greater the power obtained, as will be evident in §8.5.2. However, apart from a large D making the device more expensive to construct and install, there is a limit on D in the case of an L-shaped OWC: the column width plus the wave amplitude can be no greater than the depth of the sea, so $D + \hat{\eta} \leq h$.

The absolute limit to the motion of the water in the column is evidently $a_{\max} = L$, since an amplitude any larger than this would result in air being exhausted from the mouth into the sea at one extreme of the motion, while at the other extreme, depending on the height of the turbine above sea level, seawater could flood the turbo-generator. In reality, the assumptions of linear inviscid motion would have broken down well before these extremes are reached.

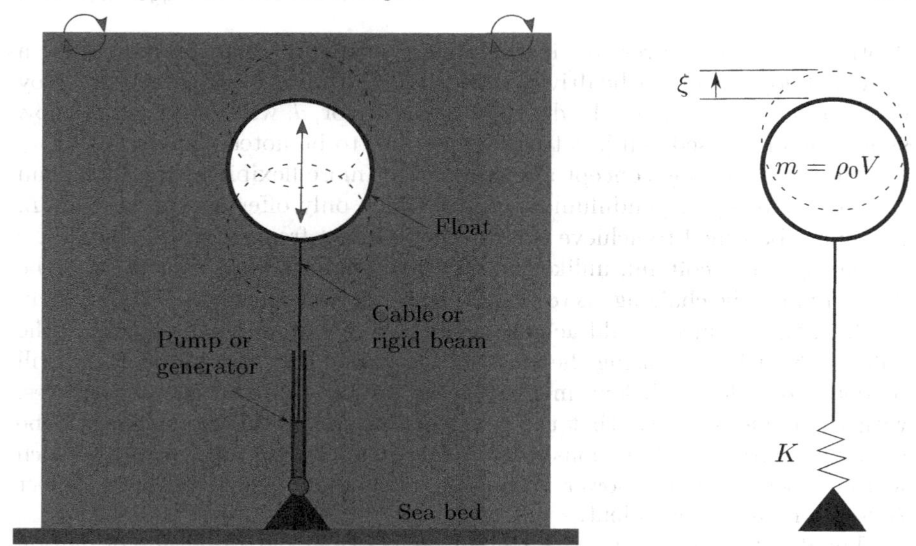

FIGURE 8.5
Schematic and definition diagrams for the derivation of the linear operation of a heaving buoy. The linearising assumption is $\xi \ll R$ where, for a spherical buoy, its volume V is $\frac{4}{3}\pi R^3$

8.4.3 Heaving Buoy

The linear analysis of a Heaving Buoy (HB) is exactly the same as for a simple mass on a spring, i.e. $F = K\xi$ where K is the stiffness of whatever system restores the float to equilibrium when it is displaced. For simplicity, assume the buoy is a sphere. Newton's Second Law is used again, giving

$$F = m\ddot{\xi}, \tag{8.16}$$

and substituting for the force in the spring,

$$-K\xi = m\ddot{\xi}, \tag{8.17}$$

or, once again, the simple-harmonic-motion equation

$$\ddot{\xi} + {\omega_0}^2\xi = 0, \tag{8.18}$$

with $\xi = A\exp(\mathrm{i}\omega_0 t)$ where the natural frequency is given by

$$\boxed{\omega_0 = \sqrt{\frac{K}{m}}.} \tag{8.19}$$

As with the rigid-pendulum float and the mouth of the oscillating-water-column tube, because the pressure and velocity of ocean waves decrease with depth (unless the waves are in the shallow-water regime, in which case a heaving buoy would not be driven vertically at all), it is best to have the buoy close to the surface. Thus the depth of the actuator, d, which here is the buoy, should be minimised, subject to the constraint to be noted shortly.

The heaving-buoy concept appears to offer more flexibility in design than the rigid and liquid pendulum concepts, which only offer one parameter, L, that must be tuned to achieve the correct resonant frequency. (Although the oscillating water column, unlike the rigid pendulum, does permit the L to be 'bent'.) Since the challenge is to match a typically very-low frequency of ocean swell, ω, the engineer could achieve a suitably low ω_0 either by reducing the stiffness K or by increasing the mass m. Increasing the mass m means, as will be detailed in §8.5.2 below, increasing the power extracted from the waves, with the obvious caveat that the construction and installation costs will be increased. However, alterations to K are also not without consequences, which will be discussed when power is considered in §8.5.2 below, so that in effect K is not really independent.

The depth of the actuator, d cannot be less than the amplitude of motion of the buoy, a, minus the buoy's radius $R = D/2$, or the buoy would break the surface at its upper limit of travel, reducing contact with the water and limiting its power extraction from waves. Some care should be taken in reaching this conclusion, because as the buoy is moving up and down the waves are also causing the water to move up and down. At resonance, like any resonator, mechanical or electrical, (as will be shown in §8.5.1), there is a phase difference of 90° between the forcing waves and the buoy motion. Therefore, at the instant when the buoy is at its upper limit of travel, the water level is at its equilibrium level. Thus, $a_{\max}$ does not need to involve the wave amplitude $\hat{\eta}$ as for the OWC, and it is possible to say that $a_{\max} = d - D/2$. As for the value of d, as just-noted, it should be no less than $D/2$, or the buoy would regularly break the surface. A suitable value to consider initially for d may be D, but in practice, some iteration may be required to find the best value of d. That is because increasing d may *permit* a larger a to occur, but as already noted, a larger d also *decreases* a since there is a lower force from waves at greater depth.

Of course, the buoy need not be spherical; a flatter shape would permit the desired m to be achieved with a smaller d than that of a sphere. As with other WEC designs, such geometric alterations are not without consequences. The more disk-like the body becomes, the greater the risk of large vortices being shed, losing more energy to nonlinear processes.

It is also possible to have a floating buoy. In this case, the buoyant force, as well as the spring force, depends on the displacement ξ and the mass of the hull of the buoy must be included in the analysis. Floating heaving-buoy WECs have been proposed and built; they have the advantage of having access to the maximum energy of waves which is at the surface, but that also entails a higher risk of damage from waves.

Having the generator (or a hydraulic cylinder) mounted on the seabed might be impractical if the machine is to operate in very deep water. Furthermore, the use of a flexible tether to the seabed means the heaving buoy can only deliver power on the upstroke. One might imagine that this asymmetry would halve the power that can be extracted from waves, but in reality, it makes the damping of the WEC highly nonlinear, raising the prospect that the amplitude could be even higher at resonance than predicted by linear theory, and compensating for the asymmetry of the power extraction. As an alternative, *two-body systems* have been developed. Here, the heaving buoy floats on the surface, and a second body or buoy is located several metres below, where the amplitude of velocity and pressure oscillations is much less owing to the exponential decrease of wave motion with depth in deep-water waves. The second body may be shaped like a horizontal flat plate, which in addition to the lower wave motion at depth, provides drag resisting motion. Hence the second body can be relatively motionless. The surface buoy slides on a rigid shaft connected to the second body, and a generator is driven by the difference in motion between the two bodies.

8.5 Analysis of a generic WEC

8.5.1 Response of a generic WEC

The simplest model for a wave energy converter capable of resonance is the classical, linearly damped, one-dimensional forced oscillator. This is given by (1.73) which is reproduced here for convenience as

$$\ddot{\xi} + 2\zeta\omega_0\dot{\xi} + \omega_0^2\xi = \mathcal{F}\mathrm{e}^{\mathrm{i}\omega t}, \tag{8.20}$$

where for a mechanical system, ξ is its displacement, ζ is the damping ratio, ω_0 is the natural frequency in radians per second, and the force per unit mass has an amplitude $\mathcal{F}$ and frequency ω. For a WEC, $\mathcal{F}$ is the force per unit mass applied by ocean waves to the machine.

The solution to the classical oscillator equation (8.20) is

$$\xi = A\mathrm{e}^{\mathrm{i}\omega t}, \tag{8.21}$$

where the complex amplitude A has magnitude (real amplitude) a and phase Φ given by (1.83), reproduced here as

$$\begin{aligned} a &= \frac{\mathcal{F}}{\sqrt{\left(\omega_0^2 - \omega^2\right)^2 + \left(2\zeta\omega\omega_0\right)^2}}, \\ \Phi &= -\tan^{-1}\left(\frac{2\zeta\omega\omega_0}{\omega_0^2 - \omega^2}\right). \end{aligned} \tag{8.22}$$

Since $\mathcal{F}$ is force per unit mass whereas a is displacement, dividing $\mathcal{F}$ by ω_0^2 gives a dimensionless ratio, $a\omega_0^2/\mathcal{F}$. The result was shown in figure 1.6.

For a rigid pendulum, $\xi = L\theta$, and ω_0 is given by (8.9). For an oscillating water column, neglecting motion of water outside the device, ξ is simply the height that water inside the column is temporarily raised above its equilibrium position. In this case, ω_0 is given by (8.15); the same as the rigid pendulum, albeit without the constraint that L should be the length of a straight line. For the heaving buoy, ξ is simply the distance the buoy moves during its oscillation and ω_0 is given by (8.19).

8.5.2 Useful power extracted from ocean waves

Now consider the useful power actually extracted from the WEC, assuming for the moment that the power extracted by the WEC from ocean waves has already been calculated and is therefore embodied in the value of $\mathcal{F}$ in §8.5.1 above; we will return to the issue of the value of $\mathcal{F}$ at the end of the present section. Recalling the three classes of WEC in §8.4, for the rigid pendulum, the Power Take-Off (PTO) may be a dual-acting hydraulic cylinder and piston that pressurises hydraulic fluid; the pressurised fluid may be accumulated in pressure vessels and exhausted through a conventional hydraulic turbine connected to an electric generator. For a heaving buoy tethered by a flexible cable, a hydraulic cylinder would only deliver power on the upstroke, as noted earlier, but the rest of the equipment could be the same. For an OWC, as mentioned, the PTO is a bidirectional turbine in the air. For the time being, assume that the effect of the PTO is felt by the oscillator as another form of linear damping. Then the damping ratio in (8.20) is composed of two parts,

$$\zeta = \zeta_\mu + \zeta_P, \tag{8.23}$$

where ζ_μ is the damping ratio due to all forms of 'parasitic' mechanical loss from the system, such as all kinds of thermal losses, bearing friction, fluid turbulence, water-wave radiation, etc, all for the time being assumed linear; and ζ_P is the assumed-linear damping due to the PTO load, which represents the 'useful' loss from the system. In general, ζ_μ represents sources of parasitic power loss that cannot be controlled once the device is built, though of course engineers can minimise these losses by good design. Meanwhile, ζ_P represents the useful load put on the device, and one might naïvely imagine increasing this up to 100 percent of the power extracted from the waves, by, for example, trying to power an increasing number of homes or factories with the WEC. However, we will shortly see that there is a limit to the load we can put on any such device. This separation of ζ_μ and ζ_P allows (8.20) to be re-written as

$$\boxed{\ddot{\xi} + 2\zeta_\mu\omega_0\dot{\xi} + 2\zeta_P\omega_0\dot{\xi} + \omega_0^2\xi = \mathcal{F}e^{i\omega t}}. \tag{8.24}$$

It should be noted in passing that the design of PTO systems is as diverse as the design of the WECs themselves. In reality, a PTO system would have some mass and stiffness as well as causing the purely dissipative load on the WEC that is represented in (8.24) by ζ_P, and, of course, these need not be linear. Therefore, a still more general, if less definitive version of (8.20) is

$$\ddot{\xi} + 2\zeta_\mu \omega_0 \dot{\xi} + \mathcal{F}_P(t) + \omega_0^2 \xi = \mathcal{F} e^{i\omega t}, \tag{8.25}$$

where $\mathcal{F}_P(t)$ is some general, time-dependent force per unit mass that could vary with ξ and $\ddot{\xi}$, variations due to the electrical load on the PTO and the dynamics of the electrical machine, as well as varying with $\dot{\xi}$ as in (8.24).

Since (8.24) has the dimensions of force per unit mass, and all forms of energy input and output are accounted for, multiplication of the real part of (8.24) by the real part of the oscillator velocity yields the power budget of the mechanical system. Here, in effect, the third of the conservation laws of fluid dynamics introduced in §1.2.3, the Law of Conservation of Energy, is invoked. The relation Power = Force × Velocity may be used. Thus, the useful cycle-averaged power per unit mass, $\bar{\mathfrak{P}}_P$, is given by

$$\bar{\mathfrak{P}}_P = \frac{1}{2} \frac{2\zeta_P \omega_0 \mathcal{F}^2 \omega^2}{\left(\omega_0^2 - \omega^2\right)^2 + \left(2\zeta\omega_0\omega\right)^2}. \tag{8.26}$$

(In (8.26), the factor of 1/2 comes from the integral of the $\sin^2(\omega t + \Phi)$ factor due to the multiplication of the $\dot{\xi}$ in (8.20) representing the damping force, by another $\dot{\xi}$ representing velocity.) The parasitic loss, $\bar{\mathfrak{P}}_\mu$, is given by (8.26) with the ζ_P in the numerator replaced by ζ_μ.

Equation (8.26) raises an interesting problem in the operation of WECs. Indeed, this problem affects any system in which useful power is to be extracted from a natural oscillator that is in turn driven by waves or any kind of vibration. Of course, the useful power should be as high as possible, which one might think implies ζ_P should be made as high as possible by connecting as much useful load to the device as possible. However, ζ_P is on the numerator of (8.26), but also in the denominator (recalling $\zeta = \zeta_P + \zeta_\mu$). If a very heavy load is put on the device, the damping will kill the resonance effect, reducing the amplitude a and hence reducing the useful power. On the other hand, if a very light load is put on the device, the device will resonate with a large amplitude a, but owing to the light loading not much of this power is extracted. Clearly, therefore, there is an optimum value of ζ_P that leads to maximum useful power.

Now, by differentiating $\bar{\mathfrak{P}}_P$ with respect to ζ_P, the optimal-loading damping ratio, $\zeta_{P\max}$, is easily found to be

$$\zeta_{P\max} = \frac{1}{2}\sqrt{4\zeta_\mu^2 - \frac{\left(\omega_0^2 - \omega^2\right)^2}{\omega_0^2\omega^2}}. \tag{8.27}$$

At resonance, $(\omega = \omega_0)$, it is clear from (8.27) that

$$\boxed{\zeta_{P\max} = \zeta_\mu.} \tag{8.28}$$

Equation (8.28) is the easily-derived relation that has already been mentioned in both the microsystems energy harvesting literature (Beeby et al., 2006) and the WEC literature (Falcão, 2010). In electrical engineering, the relation (8.28) is better known as an *impedance matching* condition. The ratio ζ between a pressure (or a force) and a velocity is an impedance, just like the acoustic impedance derived in §3.3.6 is a ratio between pressure and velocity. Considering electricity, a relation between voltage and current is an impedance. Impedance matching is perfectly general in physics and engineering; it permits sound waves to pass from one medium to another without refraction or reflection, and it permits alternating electrical current to pass from one circuit to another without reflection or power loss. Having derived this condition, however, it is now clear that (8.28) is only true at resonance and that more general power take-off strategies given by (8.27) are appropriate if the system can not, for whatever reason, operate at resonance.

The immediate consequence of (8.28) is that at resonance, for the optimal loading, the useful power must equal the power dissipated by parasitic losses of all kinds. Since all the power entering and leaving the device is accounted for, this means that the maximum efficiency of the mechanical step of wave energy conversion is 50 percent.

At resonance, with the optimal loading (8.28), the total damping becomes $\zeta = 2\zeta_\mu$, so that the maximum useful power per unit mass is obtained from (8.26) as

$$\bar{\mathfrak{P}}_{P\max} = \frac{1}{2}\frac{2\zeta_\mu \mathcal{F}^2\omega^3}{\left(4\zeta_\mu\omega^2\right)^2}. \tag{8.29}$$

Now, provided $\mathcal{F}$ is known, (8.29) gives the true estimate of useful power per unit mass.

In general, $\mathcal{F}$ is difficult to calculate theoretically, even under the linear inviscid approximation. The calculation is possible and has been done in a number of research contexts. Effectively, the mathematical projection of the ocean-wave field onto the field representing the device motion has to be undertaken, and depending on the size of the machine relative to the wavelength, this may need to include diffraction effects that depend on the geometry of the machine. The forcing $\mathcal{F}$ would also not be a simple scalar but a matrix that depends on its degrees of freedom of motion.

However, for the purposes of making a rough estimate, noting $\mathcal{F}$ is the amplitude of the force per unit mass applied by ocean waves to the device, it may be possible to approximate it, using the amplitude of the pressure at the depth where the force is applied to the device. For a buoy floating at the surface, or for the rigid pendulum with a float just touching the surface, this pressure amplitude is equal to the wave height multiplied by $\rho_0 g$. However, for any device with its part that drives the PTO (its actuator) at some depth d significantly below the surface, $\mathcal{F}$ will need to be reduced from its value at the surface by the factor in the solution for surface gravity waves. This is given by combining (2.53) and (2.13), $p = -\mathrm{i}\rho_0\omega\phi$, where ϕ is the velocity potential solution. Hence the pressure amplitude, $\hat{p}$ at depth d is given by

$$\hat{p} = \rho_0 g \frac{\cosh(kh - kd)}{\cosh(kh)} \hat{\eta}$$

where $\hat{\eta}$ is the surface-wave elevation amplitude and d is the depth at which the force from waves is presumed to apply to the device actuator. This might be the case for a heaving buoy that is designed to be entirely submerged, or for an oscillating water column with a mouth significantly below the surface. Multiplying $\hat{p}$ by an appropriate cross-sectional area, $A_\times$, gives force, then dividing by the mass of the device, m, gives the force per unit mass applied to the device, $\mathcal{F}$. For thin-walled floats or for the water in an OWC, $m = \rho_0 A_\times L_v$ where $L_v = V/A_\times$ and V is the device volume, Thus, $\mathcal{F}$ could be approximated by

$$\boxed{\mathcal{F} = \frac{\cosh(kh - kd)}{\cosh(kh)} \frac{g}{L_v} \hat{\eta}}, \tag{8.30}$$

Clearly, as d gets close to the depth of the sea, h, the force driving the device becomes very small.

Alternatively, $\mathcal{F}$ can be immediately related to a, since a could be measured in an operating device or possibly a laboratory model, using (8.22), which at resonance gives $\mathcal{F}^2 = (2\zeta{\omega_0}^2)^2 a^2$, and recalling that optimal loading at resonance given by (8.28) means that under optimal conditions $\zeta = 2\zeta_\mu$,

$$\boxed{a = \frac{\mathcal{F}}{4\zeta_\mu {\omega_0}^2}}, \tag{8.31}$$

giving an alternative to (8.29),

$$\boxed{\bar{\mathfrak{P}}_{P\max} = \zeta_\mu a^2 {\omega_0}^3} \tag{8.32}$$

Now (8.32) is perfectly general and applies to any natural oscillator device extracting power from waves or vibrations, but it is not immediately useful since it only gives the power per unit mass. In order to obtain the actual power extracted from ocean waves, it is necessary to multiply by the mass m of the water set into motion, giving

$$\bar{\mathbb{P}}_{P\max} = m\zeta_\mu a^2 {\omega_0}^3. \tag{8.33}$$

Clearly, the mass of water displaced by the device is not the only mass set into motion; significantly more mass may be set into motion. Nonetheless, for the purposes of making an engineering design estimate, conventional engineering principles suggest that it may be appropriate to make a conservative 'lower bound' estimate. In this way, the engineer errs on the side of an underestimate, rather than exaggerating the performance of the design. In this context, it may be acceptable to simply use the mass of water displaced by the device for m. Some examples of m for the three types of WEC considered in §8.4 will be given shortly.

The question now arises as to the value of ζ_μ. Like $\mathcal{F}$, this is difficult but in some cases possible to estimate theoretically. Indeed, if one assumes this linear parasitic loss of energy is dominated by the radiation of wave energy from the device, the calculation of ζ_μ is tractable using the principles of surface-gravity-wave theory and some matching of the ocean-wave velocity field to that generated by the motion of the machine. Such calculations can be organised to give a value of the natural frequency of the machine that is a complex number. This can be manipulated to give both a value of ζ_μ and an additional mass loading representing the mass of water radiating from the machine, which results in lower values of ω_0 than those in §8.4. For those cases where the analysis is possible, as well as from data from laboratory experiments, the value of ζ_μ appears to be of order 10^{-2}. Since even a rough engineering estimate requires a number, values of ζ_μ between 0.03 and 0.07 may be appropriate.

The amplitude of oscillation, a, already includes the resonance effect. Examining (8.22), it is theoretically possible for a to be very large at resonance. For example, the classical oscillator with $\zeta = 0.1$ predicts that a will be a factor of five times the equivalent forcing amplitude $\mathcal{F}/\omega^2$. A wave height of several metres could, in some designs, result in a value of a so high that, for example, a heaving buoy should be leaping out of the water at one end of its cycle, or that the underwater duct of an oscillating water column would end up completely filled with air. Apart from violating the assumption of linear behaviour, such extremes would never occur, since some engineering constraint would preclude them well before the displacement became that large. For example, a heaving buoy cannot rise any higher than the maximum travel of its piston. Even if there were no engineering constraints, nonlinearities would render the linear model invalid, drastically increasing the equivalent of ζ when the amplitude is high and thus diverting energy from the oscillation of the device into other motions such as vortex generation in the water. In order to take an engineering constraint into account, it is necessary to replace a with a_{clip} where

$$a_{\text{clip}} = \min(a, a_{\max}), \tag{8.34}$$

where $a_{\max}$ is some practical constraint on a, so that (8.33) has a more practical version,

$$\boxed{\bar{\mathbb{P}}_{P\max} = m\zeta_\mu a_{\text{clip}}^{\;2}\, \omega_0^{\;3}}\,. \tag{8.35}$$

For a rigid pendulum, m might be estimated from the volume of the float at the top of the pendulum, assuming the beam connecting the float to the pivot is negligibly small in effective volume; for example for a spherical float of diameter D, $m = \rho_0(4/3)\pi(D/2)^3 = \rho_0(1/6)\pi D^3$, and the maximum limit of its travel is clearly an arc of 90°, so $a_{\max} = L\pi/2$. Thus a rough estimate of the RP power might be given by

$$\boxed{\bar{\mathbb{P}}_{P\text{max,RP}} = \frac{1}{6}\rho_0 \pi D^3 \zeta_\mu a_{\text{clip}}^{\;2}\, \omega_0^{\;3},} \tag{8.36}$$

For an oscillating water column, $m = \rho_0 AL$ (where $A_\times$ is the cross-sectional area of the 'tube', for example $(1/4)\pi D^2$ for a circular cross-section), and $a_{\text{max}} = L$ so that (8.32) gives the rough estimate of the OWC power as

$$\boxed{\bar{\mathbb{P}}_{P\text{max,OWC}} = \rho_0 A_\times L \zeta_\mu a_{\text{clip}}^{\;2}\, \omega_0^{\;3}}, \tag{8.37}$$

or, using (8.15),

$$\bar{\mathbb{P}}_{P\text{max,OWC}} = \zeta_\mu \rho_0 A_\times g \omega_0 a_{\text{clip}}^{\;2}. \tag{8.38}$$

For a heaving buoy, m might be estimated from the volume of the buoy; for a spherical buoy of diameter D, $m = \rho_0(1/6)\pi D^3$ as for the rigid-pendulum float. The value of a_{max} is governed by the depth at which the buoy is designed to rest when at equilibrium. However, as noted in §8.4.3, it is best to have the buoy close to the surface, or the power extracted will be less. Hence the rough estimate of the HB power appears functionally identical to that of the RP (although of course, ω_0 for the HB is quite different to that of the RP), giving

$$\boxed{\bar{\mathbb{P}}_{P\text{max,HB}} = \frac{1}{6}\rho_0 \pi D^3 \zeta_\mu a_{\text{clip}}^{\;2} \omega_0^{\;3}}, \tag{8.39}$$

It is now possible to consider the constraint on the stiffness, K, of the heaving-buoy system. If a hydraulic system is used to transform the energy of the buoy's motion to energy stored in the form of pressurised hydraulic fluid in an accumulator, since force equals pressure multiplied by area, at the maximum displacement (say at $\xi = a$) of the piston, $Ka = \hat{p_h} A_h$, where $\hat{p_h}$ is the peak hydraulic-cylinder pressure (which might be assumed to be the pressure maintained in the hydraulic accumulator), and A_h is the cross-sectional area of the hydraulic cylinder. The cycle-averaged useful power of the heaving buoy, $\bar{\mathbb{P}}_{P\text{max,HB}}$, should on average be equal to the power steadily withdrawn from the hydraulic system by the turbo-generator, which is the pressure in the accumulator, $\hat{p_h}$, multiplied by the volumetric flow rate of hydraulic fluid, $\dot{V}$. Hence $K = (\bar{\mathbb{P}}_{P\text{max,HB}}/\dot{V})(A_h/a)$ and it is now clear that the stiffness of the system is not really independent, but is related to the size of the system. Lowering the natural frequency towards the ocean-swell frequency by lowering K, rather than the expensive resort of increasing the size of the machine (increasing m), may appear superficially attractive, but would require a very large volumetric flow rate of hydraulic fluid, again requiring very large equipment.

9

Bubble acoustics

9.1 Summary of key points

- The **undamped natural frequency** in rad/s of a bubble vibrating volumetrically with small-amplitude oscillations is given by (9.19) on page 210,

$$\boxed{\omega_0 = \sqrt{\frac{3\kappa_p P_0}{\rho_0}}\frac{1}{R_0}},$$

where $\omega_0 = 2\pi f_0$ and f_0 is the frequency in Hz, κ_p is the polytropic index for the gas compression (which for millimetre-sized air bubbles is approximately equal to the ratio of specific heats, i.e. $\kappa_p \simeq \gamma = 1.4$), P_0 is the constant part of the total pressure, for example, the atmospheric pressure plus hydrostatic pressure due to depth (a few centimetres below a water surface under a standard atmosphere, P_0 is about 101 kPa), ρ_0 is the assumed-constant ambient density of the liquid (for water at 20°C, $\rho_0 = 998$ kg m^{-3}), and R_0 is the equilibrium radius of the bubble. For the air-bubble-in-water conditions noted above, the frequency can be approximated conveniently by (9.21) on page 210,

$$\boxed{f_0 \simeq \frac{(3.3 \text{ m s}^{-1})}{R_0} \quad \text{Hz}}.$$

- The **bubble radius** as a function of time, $R(t) = R_0 + \delta(t)$ where $\delta(t) \ll R_0$, in the absence of interactions with other bubbles, surfaces or boundaries nearby is given by a standard damped-oscillator equation, (9.36) on page 216,

$$\boxed{\ddot{\delta} + 2\zeta\omega_0\dot{\delta} + \omega_0^2\delta = -\frac{p_\infty(t)}{\rho_0 R_0}},$$

where $p_\infty(t)$ is some driving pressure far from the bubble (which would be zero for a bubble freely 'ringing' from a disturbance), and the overall damping ratio, ζ, shown in figure 9.3 on page 217, is given by (9.38) on page 216,

$$\boxed{\zeta = \zeta_\mu + \zeta_T + \zeta_r},$$

DOI: 10.1201/9780429295263-9

in which the **viscous damping ratio**, ζ_μ from (9.33) on page 215, is

$$2\zeta_\mu\omega_0 = \frac{4\mu}{\rho_0 R_0^2},$$

the **thermal damping ratio**, ζ_T from (9.34) on page 215, is

$$2\zeta_T = \frac{\left[\dfrac{\sinh(R_0/l_T) + \sin(R_0/l_T)}{\cosh(R_0/l_T) - \cos(R_0/l_T)} - \dfrac{1}{R_0/(2l_T)}\right]}{\left[\dfrac{R_0/l_T}{3(\gamma-1)} + \dfrac{\sinh(R_0/l_T) - \sin(R_0/l_T)}{\cosh(R_0/l_T) - \cos(R_0/l_T)}\right]},$$

where, as noted above, γ is the adiabatic index that is very close to 1.4 for air, l_T is the thermal boundary-layer thickness, given by $l_T = \sqrt{\alpha_T/(2\omega)}$, where ω is the actual frequency at which the bubble is oscillating, and α_T is the thermal diffusivity of the gas (for air, $\alpha_T = 2.239 \times 10^{-5}\mathrm{m^2s^{-1}}$ at 27°C and atmospheric pressure), and the **radiation damping ratio**, ζ_r from (9.35) on page 216, is

$$2\zeta_r = \frac{R_0\omega_0}{c},$$

where c is the speed of sound.

- The **pressure perturbation** in the liquid at the equilibrium radius of the bubble, $p(R_0)$, is related to the bubble displacement δ by (9.42) on page 218,

$$p(t, R_0) = -\left(P_0 + \frac{2\sigma}{R_0} - p_v\right)\frac{3\kappa_p}{R_0}\,\delta(t),$$

where σ is the gas-liquid surface tension force per unit length ($\sigma = 0.07275 \pm 0.00036$ N m^{-1} at 20°C (Vargaftik et al., 1983)) and p_v is the vapour pressure (2340 Pa at 20°C).

- The **pressure at any radial distance** r in the liquid around the bubble is given by (9.44) on page 219,

$$\boxed{p(r) = -\frac{A}{r}\left(P_0 + \frac{2\sigma}{R_0} - p_v\right)\frac{3\kappa_p}{R_0}\,\mathrm{e}^{\mathrm{i}(\omega_{0\zeta}t - \kappa r) - \zeta\omega_0 t}},$$

where A is the amplitude of the radial displacement of the bubble wall, κ is the wavelength of sound in the liquid given by $\omega_{0\zeta}/c$ and the damped natural frequency of the bubble, $\omega_{0\zeta}$, is given by (9.40) on page 217,

$$\omega_{0\zeta} = \left(\sqrt{1-\zeta^2}\right)\omega_0.$$

- The **Mechanical Index** (MI), an empirical measure used by medical clinicians that is thought to represent the risk of cavitation damage, is given by 9.45 on page 226,

$$\mathrm{MI} = \frac{\hat{p}_-}{\sqrt{f}},$$

where $\hat{p}_-$ is the peak negative pressure measured in MPa, f is the frequency in MHz and diagnostic ultrasound scans are generally recommended to be for $\mathrm{MI} \leq 0.3$.

- Useful **books** include Leighton (1994) and Brennen (1995).

9.2 Volumetric oscillations of bubbles

9.2.1 The collapse of a spherical cavity

By the late 19th century, mariners had noticed pitting on the metal ships' propeller blades (figure 9.1) that had been recently developed. The phenomenon was termed *cavitation.* Rayleigh (1917) analysed the collapse of a spherical cavity of liquid, mentioning both cavitation and the sound emitted by water in a kettle as it comes to the boil. In the case of the propeller, the pressure behind the blade drops below the vapour pressure of water owing to the large change in flow speed as the blade passes. In the case of the kettle, the temperature on the hot surface rises above the boiling point. In both cases, the water undergoes a phase change to its gaseous form (steam), creating bubbles on a surface. However, if the bubbles travel a short distance away from the surface where they were created, they find themselves in a zone where the pressure and temperature are no longer consistent with the water existing in gaseous form. The consequence is that the bubble becomes unstable and collapses in on itself. On the propeller blade, the collapse is so rapid that an intense liquid jet is formed that can damage the metal surface. Cavitation that is caused by a large increase in flow speed or a large increase in elevation, (and hence, in the steady incompressible inviscid approximation of Bernoulli's equation, (1.42), a large drop in pressure), is called *hydrodynamic cavitation.* It occurs not just on ships' propellers but inside pumps, the fuel injectors of cars, in pipelines, and many other engineering systems.

While this may not seem like a fluid wave problem, the study of the volumetric oscillations of bubbles is intimately connected to sound waves in the liquid - including the shock waves that can damage engineering surfaces or biological cells. In general, this topic is called *bubble acoustics* and useful texts on bubble acoustics are by Leighton (1994) and Brennen (1995).

Cavitation underwater not only creates physical damage to the propellers of ships and submarines; it also sends sound waves through the ocean, possibly

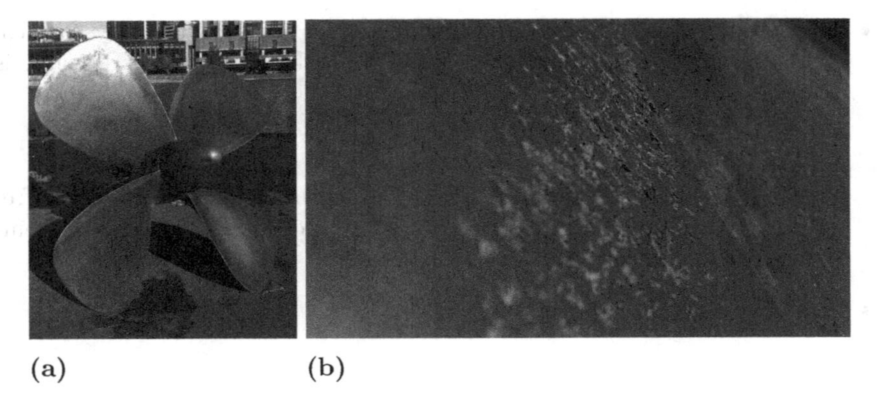

(a) (b)

FIGURE 9.1
Cavitation damage. (a) Ship's propeller. (b) Close-up of low-pressure surface of blade, showing pitting. Photographs by Richard Manasseh.

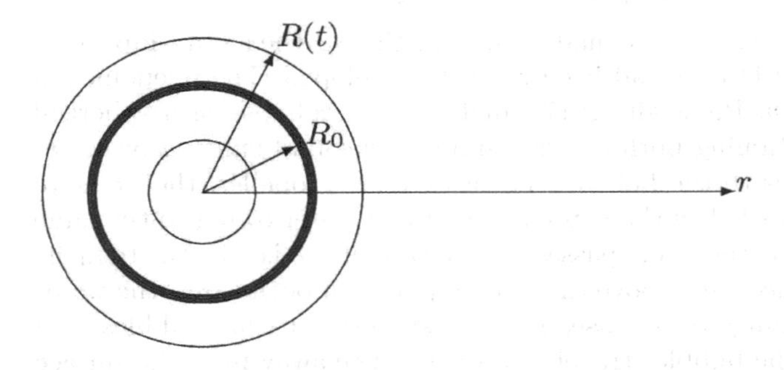

FIGURE 9.2
A bubble undergoing radially-symmetric volumetric oscillations

creating unwelcome attention in a naval-operations environment. Cavitation is also utilised in several medical applications mentioned in §9.6 below. Further to cavitation, and despite its specialised nature, the parent topic of bubble acoustics has a surprisingly large number of applications that will be outlined in this chapter, ranging from the prediction of volcanic eruptions to the targeted treatment of diseases.

Although highly nonlinear processes such as cavitation originally motivated the study of bubble acoustics (and we will return to applications of nonlinear bubble acoustics in §9.6), the first aspect to study is the small-amplitude vibrations that can be analysed by linear theory. To get to this point, we begin as always with the full equations. The bubble is assumed to be at the centre of a spherically-symmetric co-ordinate system, which means that the only motions possible are radial.

The Law of Conservation of Mass (1.15) in spherically-symmetric coordinates is given by

$$\frac{\partial(r^2\rho u)}{\partial r} = -r^2\frac{\partial \rho}{\partial t}. \tag{9.1}$$

where $u = u(t, r)$ is the velocity of the liquid and is positive in the radially-outwards direction. The Law of Conservation of Momentum (Newton's Second Law, in the form of Euler's equation) assuming spherical symmetry, and applied to the liquid only, where the liquid is assumed to be inviscid, is

$$\frac{\partial(\rho u)}{\partial t} + u\frac{\partial(\rho u)}{\partial r} = -\frac{\partial P}{\partial r}, \tag{9.2}$$

where, as usual, P is the total pressure in the liquid.

The aim is to analyse an *acoustic* phenomenon, thus there is definitely propagation of sound through the fluid. For the existence of sound, as discussed in chapter 3, the compressibility of the fluid is essential. However, a remarkable feature of bubble acoustics is that a basic understanding of the phenomenon, and indeed an understanding sufficient for many applications, can be achieved by assuming the liquid is incompressible. The compressibility of the gas in the bubble is far greater than the compressibility of the surrounding liquid. Meanwhile, the density of the surrounding liquid is far greater than the density of the gas. Effectively, the bubble is a spherical 'spring' connected to the surrounding mass of liquid and it is possible to derive the natural frequency with which the bubble oscillates volumetrically as the 'mass bounces on the spring', while ignoring liquid compressibility. Once the natural frequency (usually called f_0) is known, admitting that the liquid really is compressible and applying the relation for all waves, $\lambda = c/f_0$, to the liquid allows the calculation of the wavelength λ of the resulting sound. We will soon see why this procedure of ignoring compressibility in the liquid surrounding the bubble and then re-introducing it is a very good approximation. Thus, begin by considering the bubble only, and how its radius varies with time.

Under the assumption of liquid incompressibility, the right-hand-side of (9.1) is zero, and ρ in (9.2) can be replaced by a constant, ρ_0, reducing these two conservation laws to

$$\frac{\partial(r^2 u)}{\partial r} = 0, \tag{9.3}$$

and

$$\frac{\partial u}{\partial t} + u\frac{\partial u}{\partial r} = -\frac{1}{\rho_0}\frac{\partial P}{\partial r}. \tag{9.4}$$

Now, since $R(t)$, the time-varying radius of the bubble once there is some oscillation, does not depend on r, and defining R_0 as the equilibrium radius of the bubble (its radius 'at rest' before there is any oscillation), integrate (9.3) with respect to r, giving

$$r^2 u(t, r) = R(t)^2 u(t, R_0), \tag{9.5}$$

in which the right-hand side arises from the constant of integration. Noting that $u(t, R_0) = \dot{R}$ reduces (9.5) to

$$u = \frac{R(t)^2}{r^2} \dot{R}. \tag{9.6}$$

This mass-conservation relation is now used, in a procedure familiar from earlier chapters, to eliminate the velocity from the momentum equation (9.4). The derivatives of (9.6) in time and radius are

$$\frac{\partial u}{\partial t} = \frac{2R\dot{R}^2}{r^2} + \frac{R^2}{r^2}\ddot{R}, \quad \text{and} \quad \frac{\partial u}{\partial r} = -2\frac{R^2}{r^3}\dot{R}, \tag{9.7}$$

and substituting these into (9.4) gives

$$\frac{2R}{r^2}\dot{R}^2 + \frac{R^2}{r^2}\ddot{R} - 2\frac{R^4}{r^5}\dot{R}^2 = -\frac{1}{\rho_0}\frac{\partial P}{\partial r}. \tag{9.8}$$

Since the objective is to consider the bubble's radius only, the r-dependence of (9.8) should be removed. This can be achieved by integrating (9.8) from $r = R(t)$ to some arbitrary radius $r = D$, giving,

$$\left[-\frac{2R}{r}\dot{R}^2 - \frac{R^2}{r}\ddot{R} + \frac{1}{2}\frac{R^4}{r^4}\dot{R}^2\right]_R^D = -\frac{1}{\rho_0}\left[P(D) - P(R)\right]. \tag{9.9}$$

Now send $D \to \infty$ and define the value of P at infinity to be the constant part of the total pressure, P_0, plus some time-varying pressure $p_\infty(t)$ due to fluid-dynamical phenomena such as waves. For now, p_∞ will be assumed to be zero, but when bubbles forced by ultrasound are considered, p_∞ will represent this forcing. This gives

$$2\dot{R}^2 + R\ddot{R} - \frac{1}{2}\dot{R}^2 = -\frac{1}{\rho_0}\left[P_0 - P(R)\right]. \tag{9.10}$$

Now $P(R)$ the only unknown that needs to be eliminated to get an equation in just one variable, $R(t)$. This time-varying pressure $P(R(t))$ is the *total pressure* at the bubble radius, and since the momentum equation was written for the liquid only, for consistency it must be considered to be the *pressure in the liquid just outside the bubble.* One more relation must be used to eliminate $P(R)$, and that is a Constitutive Law relating pressure to density for the gas. The liquid has been assumed incompressible, but the gas is definitely compressible. The appropriate constitutive relation is the Ideal Gas Law, introduced in chapter 1 as (1.5),

$$P_g(R)[V(R)]^{\kappa_P} = P_0[V(R_0)]^{\kappa_P}, \tag{9.11}$$

where $P_g(R)$ is the time-varying pressure of the gas inside the bubble and $V(R)$ is the volume of the bubble; since $V(R) = (4/3)\pi R^3$, (9.11) can be re-arranged to give $P_g(R)$ as a function of R only,

$$P_g(R) = P_0 \left(\frac{R_0}{R} \right)^{3\kappa_p} . \tag{9.12}$$

In (9.11), care was taken to introduce a separate symbol, P_g, to distinguish the pressure of the gas inside the bubble from the pressure in the liquid just outside the bubble, because the two pressures are not exactly the same, as will be detailed in §9.3. However, for the purposes of this section 9.2.1, we will simply assume they are the same, i.e.,

$$P(R) \simeq P_g(R), \tag{9.13}$$

so that substituting (9.12) into (9.10) gives

$$R\ddot{R} + \frac{3}{2}\dot{R}^2 = -\frac{1}{\rho_0}\left[P_0 - P_0 \left(\frac{R_0}{R} \right)^{3\kappa_p} \right] . \tag{9.14}$$

An equivalent of (9.14) was obtained by Rayleigh (1917).

There is now an ordinary differential equation, (9.14), in a single variable R that depends only on time. However, (9.14) is nonlinear, and moreover, recalling from §1.2.2 that κ_p is at least 1, and, for air, close to 1.4, the last term on the right-hand-side could become very large if $R(t)$ became small, suggesting rather wild behaviour. Indeed, when microbubbles are driven by ultrasound, (as will be described in §9.6.2), quite extreme behaviour has been observed in experiments - behaviour that is exploited in several industrial and medical technologies.

9.2.2 Natural frequencies of bubbles

Setting aside, for now, the possibilities of extremely nonlinear behaviour, it is instructive to linearise (9.14) now, since this will reveal the fundamental physics mentioned earlier that is at the core of bubble acoustics: the tendency for a bubble to oscillate like a mass on a spring, where the 'spring' is the compressibility of the gas in the bubble and the mass is that of the surrounding liquid. After considering this linear physics in this section 9.2.2, we will return to the nonlinear equation (9.14) in order to consider additional aspects of the physics neglected in Rayleigh's original analysis. For now, therefore, assume the bubble is oscillating volumetrically by only an infinitesimal fraction. Hence (9.14) can be linearised with

$$R(t) = R_0 + \delta(t), \tag{9.15}$$

where $\delta \ll R_0$. Thus, δ is a perturbation positive outwards from the bubble. The left-hand side of (9.14) reduces to $R_0\ddot{\delta}$. Meanwhile a Taylor-series expansion of the troublesome last term gives to first order

$$P_0(R_0/R)^{3\kappa_p} \simeq P_0\left(1 - (3\kappa_p/R_0)\delta\right); \tag{9.16}$$

substituting (9.16) into (9.14) gives

$$R_0\ddot{\delta} = -\frac{1}{\rho_0}P_0(3\kappa_p/R_0)\delta, \tag{9.17}$$

giving the equation for simple harmonic motion,

$$\ddot{\delta} = -{\omega_0}^2\delta, \tag{9.18}$$

in which the natural frequency of the bubble in radians per second, ω_0, is given by

$$\boxed{\omega_0 = \sqrt{\frac{3\kappa_p P_0}{\rho_0}\frac{1}{R_0}}}, \tag{9.19}$$

so that the natural frequency in Hertz is

$$\boxed{f_0 = \frac{1}{2\pi}\sqrt{\frac{3\kappa_p P_0}{\rho_0}\frac{1}{R_0}}}. \tag{9.20}$$

Although, as noted above, Rayleigh (1917) obtained the equivalent to (9.14). Rayleigh's interest was in the collapse of a 'cavity' (a bubble) with an initial pressure much less the equilibrium value P_0, and in the maximum pressure generated during this collapse; he did not directly linearise (9.14). Instead, (9.20) was independently derived by Marcel Minnaert in 1933; Minnaert was interested in the 'music' made by bubbles, and since music implies very low amplitudes, he began immediately with expressions for the kinetic energy of the liquid and the potential energy of the gas based on the sinusoidal oscillations expected from low-amplitude fluid waves; equating the two energies gives (9.20), which is now is called *Minnaert's equation.*

One might wonder about the validity of the assumption of spherical symmetry, when bubbles large enough to see are evidently not spherical in shape. The restoring force for the bubble-acoustic oscillation, the pressure inside the bubble, depends on the *volume* of the bubble, according to the Ideal Gas Law (9.11). Hence, variations in shape have little effect on f_0; it was shown by Strasberg (1953) that even quite large distortions shift the frequency by only a few percent.

For an air bubble ($\kappa_p = 1.4$) in still water at room temperature so the density is approximately 1000 kg m^{-3}, and not very deep, so that P_0 is approximately atmospheric pressure (the standard atmosphere has a pressure of 101 325 Pa), the natural frequency of the bubble in Hertz is approximately

$$\boxed{f_0 \simeq \frac{(3.3 \text{ m s}^{-1})}{R_0} \text{ Hz}}. \tag{9.21}$$

Thus, a bubble with a radius of 3.3 mm - the sort of bubble one might readily create, for example, on pouring water into a glass - has a natural frequency of about 3.3/0.0033 Hz, or 1 kHz. Meanwhile, a bubble with a radius of 1 mm, which might also be created in the glass at the same time, has a natural frequency of 3.3 kHz. These cases fall in the low-kilohertz frequency band where human hearing is at its best.

Consider a simple experiment, in which water is first poured gently into a glass such that the water surface is not broken and no bubbles are seen. This first test reveals that without bubble formation, water motion is an almost silent process. In the second test, the water is poured more vigorously, so that bubbles are formed; 'splashing' and 'gurgling' sounds are heard. The formation of the bubble, in particular, the pinch-off of the bubble from the surface, is a process that creates a volume of air subjected, at the point of pinch-off, to an extreme distortion from a spherical shape. As that shape relaxes to a sphere there is a small perturbation to the volume corresponding to a small perturbation in radius, δ, as assumed above. The consequence is that the bubble 'rings' like a bell.

In addition to the sounds created by pouring (the column of poured water is more formally called a *plunging jet*), individual drops of water, such as raindrops hitting a water surface, can entrain small bubbles, creating a 'plink' sound. When waves break in the ocean, a topic to be discussed in detail in chapter 10.2, immense numbers of bubbles are formed. These bubbles create the 'whitecap' appearance of breakers and ring with a multitude of superimposed frequencies called the 'roar' of the surf (to be discussed in §9.5.2 below); likewise with waterfalls. Thus, the sounds people instinctively associate with the motion of water are mostly due to the formation of bubbles millimetres in size.

A huge disparity in scales is apparent on comparing (9.21), $f_0 \simeq 3.3/R_0$, with the standard relation for all waves, $f = c/\lambda$. The speed of sound in water is about 1480 m s^{-1}, which is almost 450 times greater than the 3.3 m s^{-1} in (9.21). Therefore, when a bubble oscillates naturally on formation, generating a frequency f_0, the wavelength λ of the resulting sound waves will be hundreds of times larger than the bubble size; in the example given above, the 3.3 mm radius bubble would generate sound waves almost 1.5 m long. Thus, the effects of compressibility of the liquid are only significant over scales hundreds of times larger than the bubble, explaining why it was safe to neglect compressibility in deriving the basic relation (9.20). However, as will be seen in the following section, the radiation of sound waves away from the bubble not only alerts observers to bubble formation (or, more commonly, alerts them to the vigorous movement of water). The radiation of sound waves away from the bubble also contributes to the damping of the bubble, which causes the 'ringing' of this fluid-dynamical 'bell' to die out.

9.3 Rayleigh-Plesset equation

9.3.1 Surface tension, vapour and driving pressures

There were several simplifying assumptions in the derivation of the nonlinear equation (9.14), which affect the accuracy of Minneart's equation (9.20). In this section §9.3.1, improvements to the representation of pressure are included.

Firstly, the surface tension was neglected. The added pressure in the gas inside the bubble due to surface tension, p_σ, called *Laplace pressure*, can be derived from considerations of the force balance on a bubble and is given by (1.11), which for equal radii of curvature becomes

$$p_\sigma = \frac{2\sigma}{R}, \tag{9.22}$$

where σ for the surface tension between air and water at room temperature is about 0.072 kg s^{-2}, i.e. 0.072 N/m, (as detailed in §1.2.2.5, $\sigma = 0.07275 \pm 0.00036$ N m^{-1} at 20°C), but σ rises to about 0.076 kg s^{-2} at 0° C (Vargaftik et al., 1983). For example, the added pressure inside a 1 mm radius bubble is 144 kg m^{-1}s^{-2}, while for a 0.01 mm (i.e. 10 μm) radius bubble, $p_\sigma =$ 14 400 kg m^{-1}s^{-2} Thus, compared to the constant part of the total pressure, P_0, that is 10^5 kg m^{-1}s^{-2} even under a negligible depth of water, the added pressure due to surface tension is only significant for very small bubbles.

Secondly, vapour pressure, p_v, was neglected. Molecules of the liquid will have evaporated into the gas of the bubble, contributing to the total pressure inside the bubble, and the resulting partial pressure due to water vapour in the bubble will depend on the temperature and pressure. However, it is usually assumed that as the bubble expands and contracts, evaporation and condensation across the gas-liquid interface occurs instantly, so the vapour pressure stays constant. Therefore, the ideal gas law (9.11) only applies to the portion of the bubble contents that is gas and not vapour. The vapour pressure of water in the air is 2340 kg m^{-1}s^{-2}, which is only about 2% of the ambient pressure. (Even so, as will be explained in §9.6.2, vapour has an important effect on the life of the bubble.)

Including surface tension and vapour pressure results in a more accurate version of (9.12),

$$P_g(R) = \left(P_0 + \frac{2\sigma}{R} - p_v\right)\left(\frac{R_0}{R}\right)^{3\kappa_p}, \tag{9.23}$$

and since, at equilibrium (i.e. when $R(t) = R_0$), the pressure in the liquid just outside the bubble, $P(R)$, must equal the constant part of the total pressure, P_0, the liquid pressure should be given by

$$P(R) = \left(P_0 + \frac{2\sigma}{R} - p_v\right)\left(\frac{R_0}{R}\right)^{3\kappa_p} - \frac{2\sigma}{R} + p_v. \tag{9.24}$$

Thirdly, recall that in deriving (9.10), the pressure at an infinite distance from the bubble was set equal to the constant part of the total pressure, P_0. However, as noted earlier, the pressure at infinity could also include a time-varying pressure $p_\infty(t)$ due to fluid flows of any kind, such as water waves or sound waves from some other phenomenon. As noted in §9.2.2 above, the bubble is so much smaller than the wavelength of sound that an 'infinite' distance, in other words, a distance where fluid flows due to sound waves are dominant over fluid flows due to the bubble, need not be that far away. In many bubble-acoustic applications, this 'driving' pressure at infinity is the oscillating pressure due to applied ultrasound; in other bubble-acoustic applications, it is the pressure created as fluid flows through a contraction, causing a rapid pressure change. Including this possible $p_\infty(t)$, (9.14) now becomes

$$\begin{aligned} &R\ddot{R} + \frac{3}{2}\dot{R}^2 \\ &= -\frac{1}{\rho_0}\left[P_0 + p_\infty(t) - \left(P_0 + \frac{2\sigma}{R} - p_v\right)\left(\frac{R_0}{R}\right)^{3\kappa_p} + \frac{2\sigma}{R} - p_v\right]. \end{aligned} \tag{9.25}$$

9.3.2 Viscous dissipation

Including the viscous stress occurring in the liquid just outside the bubble as the bubble expands and contracts leads to an additional term inside the square brackets of (9.25), which now becomes

$$\boxed{\begin{aligned} &R\ddot{R} + \frac{3}{2}\dot{R}^2 \\ &= -\frac{1}{\rho_0}\left[\left(P_0 + \frac{2\sigma}{R} - p_v\right) - \left(P_0 + \frac{2\sigma}{R} - p_v\right)\left(\frac{R_0}{R}\right)^{3\kappa_p}\right. \\ &\qquad \left. + \frac{4\mu}{R}\dot{R} + p_\infty(t)\right] \end{aligned}} \tag{9.26}$$

This nonlinear equation for the bubble radius as a function of time, (9.26), is attributed to Rayleigh (for the original terms from Rayleigh (1917) noted in (9.14)) and to Plesset, Noltingk, Neppiras and Poritsky who were responsible for adding the other terms between 1949 and 1952. It is usually called the *Rayleigh-Plesset equation.*

Although (9.26) is useful for many problems in cavitation and ultrasound physics, it has been mostly superseded by more sophisticated equations that take into account liquid compressibility. Furthermore, the thermal damping of the bubble was not included in (9.26), and has a significant bearing on the time the 'ringing' of the freely-oscillating bubble takes to die out.

Unfortunately, the derivation of thermal damping, as well as compressibility effects, is already difficult when small-amplitude behaviour is assumed, and it would be challenging if no limitation on the amplitude were made. Thus, the usual approach in this book - beginning with the full, nonlinear equations that include all the physics and then simplifying them in a consistent manner to linear equations - will not be followed for bubble acoustics. Instead, we will jump to linearisation of the Rayleigh-Plesset equation (9.26), just as in §9.2.2 we linearised Rayleigh's original cavity-collapse equation (9.14) that neglected the physics of surface tension and vapour pressure, in order to immediately grasp the basic mass-on-a-spring physics of bubble acoustics. After linearisation, the additional terms representing thermal damping and compressibility will be added. As outlined by Leighton (1994), such work on damping was undertaken by Devin (1959) and by other researchers around that time. This process of stitching together the momentum equation, rather than cutting it from the one piece of cloth at the start, may not be mathematically satisfying, and raises the nagging doubt that some terms might have been misrepresented. However, it is practical and mirrors the history of development of bubble-acoustic theory, so it will be followed below in §9.4.2 and subsequent sections. Nearly two decades later, as detailed by Illesinghe (2007), a wholistic approach to the addition of viscous damping, thermal damping and compressibility was undertaken by Prosperetti (1977); this offered the advantage of mathematical consistency of the derivations, but in the end makes only a small difference to the resulting damping constants, with a maximum difference of about 7% between the theories for bubbles of about 10-micron radius (Illesinghe, 2007). Repeated experiments have shown that irrespective of the approach, the resulting damping constants lead to a very good model for many practical situations, particularly at low amplitudes.

9.4 Linear bubble acoustics

9.4.1 Linearised Rayleigh-Plesset equation

In order to apply the same linearisation $R(t) = R_0 + \delta(t)$ (9.15) to the Rayleigh-Plesset equation (9.26) as to the equation without surface tension, vapour pressure and viscosity that led to the equation of simple harmonic motion in §9.2.2, note the Taylor's series expansions to first order of $2\sigma/R(t)$ and $[4\mu/R(t)]\dot{R}(t)$ are

$$\frac{2\sigma}{R(t)} \simeq \frac{2\sigma}{R_0}\left(1 - \frac{1}{R_0}\delta(t)\right), \tag{9.27}$$

and

$$4\mu\frac{\dot{R}(t)}{R(t)} \simeq \frac{4\mu}{R_0}\dot{\delta}(t). \tag{9.28}$$

Thus, substituting (9.16), (9.27) and (9.28) into (9.26) and for consistency with linearisation on the basis of small δ, neglecting terms in δ^2, gives

$$R_0\ddot{\delta} = -\frac{1}{\rho_0}\left[\left(P_0 + \frac{2\sigma}{R_0} - p_v\right)\left(\frac{3\kappa_p}{R_0}\right)\delta + \frac{4\mu}{R_0}\dot{\delta}(t) + p_\infty(t)\right], \tag{9.29}$$

which can be re-arranged to give

$$\ddot{\delta} + 2\zeta_\mu\omega_0\dot{\delta} + \omega_0^2\delta = -\frac{p_\infty(t)}{\rho_0 R_0}, \tag{9.30}$$

and which for zero driving pressure $p_\infty(t)$ becomes the standard form of the damped harmonic oscillator equation (1.74), reproduced here for convenience,

$$\ddot{\delta} + 2\zeta_\mu\omega_0\dot{\delta} + \omega_0^2\delta = 0, \tag{9.31}$$

in which, for bubble acoustics,

$$\omega_0^2 = \left(P_0 + \frac{2\sigma}{R_0} - p_v\right)\left(\frac{3\kappa_p}{\rho_0 R_0^2}\right), \tag{9.32}$$

and

$$2\zeta_\mu\omega_0 = \frac{4\mu}{\rho_0 R_0^2}. \tag{9.33}$$

The damping ratio is written as ζ_μ with the subscript μ to emphasise that only the viscous damping has been included to far; in the next sections, the other forms of damping will be added, presuming, as noted in §9.3.2, that linear behaviour permits the various forms of damping to be super-imposed.

Clearly, if the bubble size is not very small, as already noted in §9.3.1, the term $2\sigma/R_0$ is negligible relative to P_0 and if the temperature is not very high, p_v is also very small compared to P_0, so (9.32) reduces to Minnaert's equation (9.19). If the bubble were forced by a driving pressure $p_\infty(t)$, (9.30), is applicable, and if $p_\infty(t)$ were sinusoidal with time and had a frequency ω, the bubble response could be calculated just as for any forced harmonic oscillator. (The response of a forced harmonic oscillator was derived in §1.5.5.)

9.4.2 Thermal damping

Following Devin (1959), the dimensionless damping ratio due to thermal damping, ζ_T, is given by

$$2\zeta_T = \frac{\left[\dfrac{\sinh(R_0/l_T) + \sin(R_0/l_T)}{\cosh(R_0/l_T) - \cos(R_0/l_T)} - \dfrac{1}{R_0/(2l_T)}\right]}{\left[\dfrac{R_0/l_T}{3(\gamma-1)} + \dfrac{\sinh(R_0/l_T) - \sin(R_0/l_T)}{\cosh(R_0/l_T) - \cos(R_0/l_T)}\right]}, \tag{9.34}$$

where γ is the value taken by κ_p when the gas is expanding and contracting adiabatically, and l_T is the thermal boundary-layer thickness, which, for a bubble oscillating at frequency ω, is given by $l_T = \sqrt{\alpha_T/(2\omega)}$, where α_T is the thermal diffusivity of the gas (for air, $\alpha_T = 2.239 \times 10^{-5}\mathrm{m^2s^{-1}}$ at 27°C and atmospheric pressure).

9.4.3 Radiation damping

An expression for the damping due to the radiation of sound waves from the bubble can be derived, as for the viscous and thermal damping expressions, by assuming linear behaviour. This leads (Leighton, 1994; Illesinghe, 2007) to the damping ratio due to radiation, ζ_r being given by

$$2\zeta_r = \frac{R_0 \omega_0}{c}, \tag{9.35}$$

where c is the speed of sound in the liquid. (The speed of sound was derived in §3.3.2.)

9.4.4 Linear damped bubble oscillator equation

Finally, (9.30), which included only viscous damping, can be now be improved to represent the missing physics, assuming, as did Devin (1959), that thermal and radiation damping can be superimposed, becoming

$$\boxed{\ddot{\delta} + 2\zeta\omega_0\dot{\delta} + \omega_0^2\delta = -\frac{p_\infty(t)}{\rho_0 R_0}}, \tag{9.36}$$

so that the standard-form oscillator equation without forcing is (1.74),

$$\ddot{\delta} + 2\zeta\omega_0\dot{\delta} + \omega_0^2\delta = 0, \tag{9.37}$$

and for bubble acoustics the overall damping ratio is given by

$$\boxed{\zeta = \zeta_\mu + \zeta_T + \zeta_r}, \tag{9.38}$$

with ζ_μ given by (9.33), ζ_T given by (9.34) and ζ_r given by (9.35). The damping ratio as a function of bubble radius for an air bubble in water is shown in figure 9.3. Note that figure 9.3 can be compared with figure 3.20 in the textbook by Leighton (1994) noting that Leighton's "dimensionless damping constant" is 2ζ. There is a slight difference with figure 3.20 in Leighton (1994) for the smallest bubbles, which appears to be because the added pressure due to surface tension is used in the expression for ω_0 in (9.32). In figure 9.4, ζ is shown for a limited range of bubble sizes of relevance to *passive bubble acoustics* (§9.5), in which bubbles are large enough to emit sounds in the audible-frequency range, while still small enough to remain as closed bubbles. If air bubbles in water in normal gravity are more than about a centimetre in size (i.e. about 5 mm in radius) they do not exist as 'closed' bubbles, instead forming *spherical-cap* bubbles. These dome-shaped bubbles rise trailing a 'skirt' of tiny bubbles, and look a bit like a jellyfish; alternatively, spherical-cap bubbles are readily broken up into smaller bubbles (Maxworthy et al., 1996).

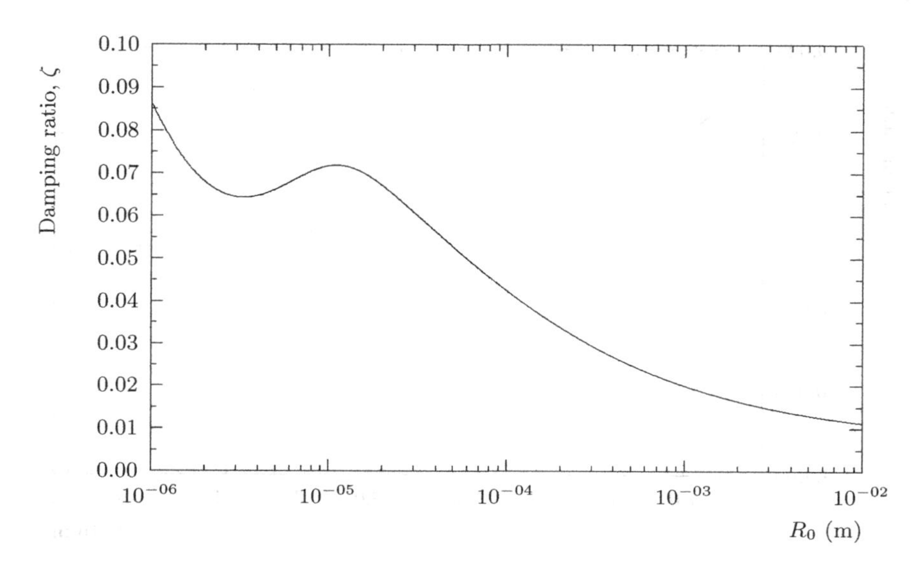

FIGURE 9.3
The overall damping ratio, ζ, given by (9.38), for air bubbles in water ranging from bubble a micron in radius to a bubble a centimetre in radius. Note that in reality an uncoated bubble a micron in radius would dissolve in a few seconds, while a bubble more than 5 mm in radius in normal gravity would either break up or rise as a 'spherical-cap bubble' shedding small bubbles as it rises.

Now, the standard solution to (9.37), originally given as (1.80) and repeated here for convenience,

$$\delta = A\mathrm{e}^{\mathrm{i}\left(\sqrt{1-\zeta^2}\right)\omega_0 t - \zeta\omega_0 t}, \tag{9.39}$$

shows that, like all damped oscillators, the bubble will oscillate with a frequency different from its undamped natural frequency ω_0, by a factor of $\sqrt{1-\zeta^2}$. The frequency is usually only slightly different to ω_0; as shown in figure 9.3, the values of ζ turn out to be much less than 0.1 for bubbles with radii larger than about 100 microns; and even for bubbles as small as a few microns, ζ is still less than 0.1, so the ζ^2 in (9.39) is very small. This alteration to the natural frequency due to the $\sqrt{1-\zeta^2}$ factor was derived in §1.5.5 as (1.79) and is called the *damped natural frequency*, and for clarity, it will be denoted $\omega_{0\zeta}$; from (9.39), it is given by

$$\omega_{0\zeta} = \left(\sqrt{1-\zeta^2}\right)\omega_0. \tag{9.40}$$

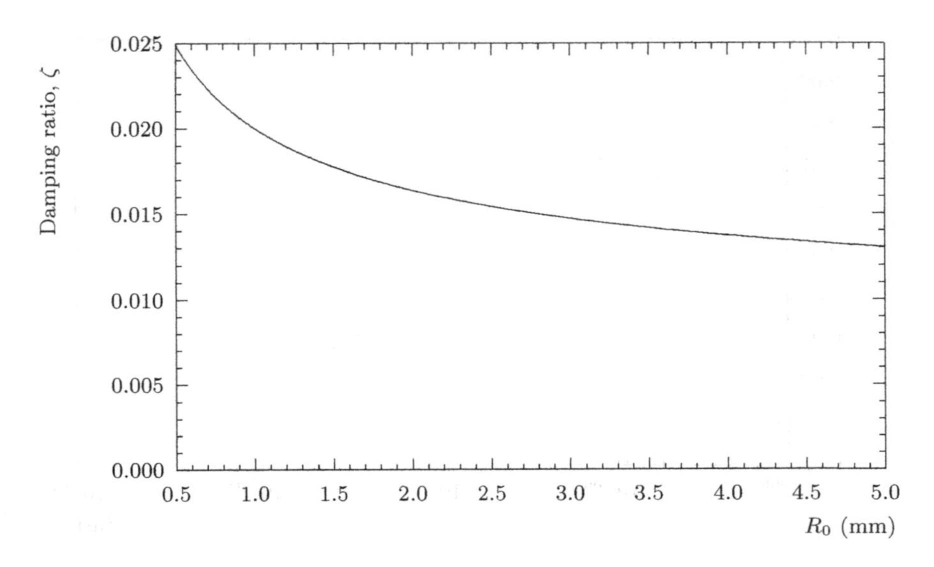

FIGURE 9.4
The overall damping ratio, ζ, given by (9.38), for air bubbles in water in the 'passive acoustic' range from 0.5 mm to 5.0 mm radius where bubbles are large enough to emit audible sounds, while small enough to exist as closed bubbles.

In practical cases where millimetre-sized bubbles are formed and the resulting oscillations agree very well with linear theory, δ, the perturbation to the radius, is extremely small. It is the resulting pressure perturbation in the liquid that propagates as audible sound. Thus it is useful to translate δ to the pressure perturbation that would be heard - and measured. Noting from (9.16) that when linearised, $(R_0/R)^{3\kappa_p} \simeq (1 - (3\kappa_p/R_0)\delta)$, means that (9.24) gives

$$P(t, R_0) = P_0 - \left(P_0 + \frac{2\sigma}{R_0} - p_v\right)\frac{3\kappa_p}{R_0}\,\delta(t). \tag{9.41}$$

Hence, defining (just as in §3.3.2) the pressure perturbation representing sound waves, p, with $p = P - P_0$, gives

$$p(t, R_0) = -\left(P_0 + \frac{2\sigma}{R_0} - p_v\right)\frac{3\kappa_p}{R_0}\,\delta(t). \tag{9.42}$$

The resulting pressure that propagates as sound is given at the bubble radius by substituting (9.39) into (9.42) and using (9.40) for convenience, yielding

$$p(t, R_0) = -A\left(P_0 + \frac{2\sigma}{R_0} - p_v\right)\frac{3\kappa_p}{R_0}\,\mathrm{e}^{\mathrm{i}\omega_{0\zeta}t - \zeta\omega_0 t}. \tag{9.43}$$

Finally, the sound waves will propagate outwards through the liquid from the bubble at the usual sound-wave speed in the liquid, c. The solution of the wave equation in the spherically-symmetric co-ordinate system used in this chapter 9.2 results in spherical Bessel functions of order zero in the radial direction r given by (3.35); these are simply sines and cosines in the radius but divided by r. Therefore, the pressure at any distance r in the liquid surrounding a gas bubble volumetrically oscillating with a small amplitude (so that linear theory is valid) is given by

$$\boxed{p(t,r) = -\frac{A}{r}\left(P_0 + \frac{2\sigma}{R_0} - p_v\right)\frac{3\kappa_p}{R_0}\,\mathrm{e}^{\mathrm{i}(\omega_{0\varsigma}t-\kappa r)-\zeta\omega_0 t}}, \tag{9.44}$$

where, as usual, the wavenumber is given by $\kappa = \omega_{0\varsigma}/c$.

9.5 Applications of linear bubble acoustics

9.5.1 Industrial measurements

The simple relation (9.20) suggests that a measurement of sound frequency will immediately yield a measurement of bubble size. To understand why the bubble size matters, first consider some industries in which bubbles are introduced. These include the chemical, pharmaceutical, minerals-processing, food and water industries (Pandit et al., 1992; Boyd and Varley, 2001; Manasseh and Ooi, 2009). Bubbles are typically pumped into liquids in many processes, usually to dissolve gas in the liquid. The most common situation is where air is pumped into liquid (a process called *sparging*), usually with the aim of dissolving oxygen in the liquid (*aeration*). Genetically-engineered cells in recombinant-DNA fermenters need oxygen, (incidentally, an example of this is given in §9.6.2), as do bacteria in wastewater treatment plants. Technologies under development to remove CO_2 from the atmosphere involve bubbling of CO_2 (e.g. Lobaccaro et al., 2016).

While chemical properties define the dissolution rate per unit area of the interface between gas and liquid, the total area clearly matters. In all these processes, for a given volume of gas injected, the size of the bubbles affects the ratio of surface area to volume, which is a powerful influence on gas-liquid *mass transfer* rate (Takemura and Yabe, 1998); mass-transfer concepts are briefly outlined in §10.3.2. The smaller the bubble, the better. Making small bubbles requires more energy and is problematic: fine holes may clog up or may not function as intended if there is a change in process conditions, resulting in much larger-than-expected bubbles.

Thus, the measurement of bubble size is an important aspect of process control, but industrial liquids may be acidic or caustic or very hot (including molten metals), which would damage delicate probes. The liquids could also be opaque, rendering cameras useless. The sound created by bubble formation, however, is immune to these issues. For millimetre-sized bubbles, the wavelengths of sound in most liquids are a metre or more, facilitating transmission of sound to hydrophones (detailed in §3.3.8.2) or even to accelerometers in the wall of the reaction vessel. These are passive-acoustic techniques, relying entirely on the sounds naturally emitted (figure 9.4).

However, measurements of sound spectra are difficult to translate unambiguously to bubble-size spectra. Many empirical and semi-empirical relations have been made from data measured in gas-liquid reaction vessels and similar complex industrial systems (Boyd and Varley, 2001; Manasseh et al., 2001), and complex signal-processing strategies have been applied (Al-Masry and Abdennour, 2006). There are two issues: bubbles interact acoustically, shifting the emitted frequencies (Manasseh and Ooi, 2009) so that (9.20) is no longer valid; and the amplitude emitted by different-sized bubbles may vary, biasing the spectrum. At present, following Pandit et al. (1992), it is assumed that all bubbles in a complex bubbly flow are perturbed by the same proportion, so that in (9.44), $|A|/R_0 = \text{const}$. On the frequency-shift issue, progress has been made in understanding the inter-bubble interactions (Leroy et al., 2005; Roshid and Manasseh, 2020), and on understanding some mechanisms for the sound-emission amplitude (Manasseh et al., 2008; Deane and Czerski, 2010), but there is still no comprehensive method of simply measuring sound and deducing the bubble sizes.

9.5.2 Sounds of ocean waves

The sounds made by waves breaking on a shore are familiar to coastal dwellers worldwide. There is a specific word for this sound in Greek, $\phi\lambda o\iota\sigma\beta o\varsigma$. However, as will be detailed in §10.3, it is over the 71% of the Earth's surface that is ocean where most wave-breaking occurs, and therefore, where most of the sounds due to wave breaking occur. This was of importance to submariners, who needed a detailed understanding of natural oceanic noise in order to conceal their own operations - such as the noise made by cavitating propellers. Studies since the mid-20th century showed that wave-breaking is a significant contributor to oceanic noise (Wenz, 1962). As waves break, bubbles are formed; details on these processes are in §10.3.4. Each bubble-formation event creates a brief pulse of sound, and the superimposition of a vast number of bubble-acoustic pulses forms the sound of each breaking wave. Meanwhile, bubbles bursting at the surface also create noise, but by a different mechanism: the bubble behaves as a *Helmholtz resonator* (Spiel, 1992), in which a small mass of air (in this case, air in the hole in the top of the 'dome' of the bubble) effectively 'bounces' on the 'spring' formed by the compressible volume of air in the bulk of the bubble.

Interest in the air-sea exchange problem (§10.3.2) prompted many studies in which the sounds of breaking waves were measured (Farmer and Vagle,

1988; Melville et al., 1988; Bass and Hay, 1997; Manasseh et al., 2006; Deane and Stokes, 2010), with an important aim, similar to that described above in §9.5.1, being to estimate the rate of gas-liquid mass transfer. Just as with the industrial measurements, these studies must be semi-empirical until a better understanding is achieved of the relation between the sounds of a complex bubbly flow and the distribution of bubble sizes within it.

9.5.3 Volcanic bubbles

Volcanoes can inject huge amounts of gas into the Earth's atmosphere. As detailed in §10.3.5, a key gas is CO_2, which can create particularly explosive eruptions. Furthermore, in §10.3.5, it is mentioned how the injection of CO_2 into the Earth's atmosphere by volcanoes is a part of the planet's long-term carbon cycle, in which carbon absorbed by oceanic life from the atmosphere (a process accelerated by breaking waves) and buried in marine sediments, is eventually returned to the atmosphere. Aside from their role in the Earth's atmospheric CO_2 levels and hence its climate, volcanoes have more immediate relevance in their threat to human life and property.

A particular form of volcanic eruption is the *Strombolic eruption*, named after Mount Stromboli. In such eruptions, a large gas bubble rises up the volcanic conduit through the molten magma, then bursts at the crater, ejecting magma over considerable distances. Acoustic sensors, together with seismic sensors placed on active volcanoes have been used to estimate the bubble size, and possibly also the magma viscosity.

In order to calculate a vibration frequency as the bubble rises up the conduit, a number of different phases of the eruption may be considered. When the bubble (which may be roughly cylindrical and tens of metres long and several metres across) is near the surface of the crater, its volume vibrates, transmitting sound into the atmosphere over the thin layer of magma overlying it. The analysis is similar that leading to the linearised bubble-acoustic model in §9.2.2, but considers the geometry of the thin film (Vergniolle and Brandeis, 1996). When the bubble bursts, it can be modelled as a Helmholtz resonator (Vergniolle and Caplan-Auerbach, 2004), rather like the small bubble bursting at the sea surface noted in §9.5.2 above. Matching the measured frequency and damping gives an estimate of the bubble size and the magma viscosity. The bubble size indicates the energy to be released during the eruption, and the magma viscosity also indicates the likely consequences.

9.6 Nonlinear bubble acoustics and applications

9.6.1 Sonochemistry

When ultrasound is applied to a liquid, any cavitation that may result is somewhat different to the growth and subsequent collapse of hydrodynamic cavitation bubbles described in §9.2.1. As with all cavitation events, it is initially presumed that there are some very small nuclei of gas, consisting of a small number of gas molecules that have formed a *nanobubble*, defined as a bubble smaller than a micron in size. Reasons for such nanobubbles to exist are unclear but may include cosmic rays (Brennen, 1995) (which are incidentally mentioned in §11.2.3). In hydrodynamic cavitation, because the pressure in the liquid is below the vapour pressure, vapour must evaporate into any nanobubble nuclei present, which then grow explosively until they become unstable and collapse.

When sound is applied, during the low-pressure part of the cycle, the nanobubble is expanded, and in the high-pressure part, it is contracted. One might think that the result would be neutral, with the bubble remaining the same equilibrium size. However, as explained in the derivation of (9.24), there is a vapour content of the gas in the bubble. When the bubble is expanded, more vapour will evaporate into the bubble, and when it is contracted, vapour will dissolve back into the liquid. These effects would cancel out if it were not for the fact that when the bubble is expanded, it has a larger surface area than when it is contracted; the oscillation is asymmetric with respect to the surface area. Since the rate of mass transfer depends on the surface area, greater evaporation of vapour into the bubble occurs than dissolution back into the liquid, as illustrated schematically in figure 9.5; the ratio of maximum to minimum radii shown is about $4^{1/3}$, so the surface area in figure 9.5(a) is about 4 times that in figure 9.5(b).

The consequence is that the bubble grows with each passing ultrasonic cycle (figure 9.5(c)), a process called *rectified diffusion*, which has been extensively studied (e.g. Leong et al., 2011). In the cartoons of figure 9.5(c), the bubble appears to be oscillating with the same proportional amplitude, which may be approximately true while the bubble radius remains much smaller than R_0. Recalling that with ultrasound there are millions of cycles per second, what was initially a nanobubble can rapidly grow towards the radius, R_0, that resonates at the applied frequency. As its radius gets closer to R_0, its amplitude of oscillation becomes larger and larger, until it becomes unstable. If it is near a surface, it may collapse violently, rather like the bubble on the ship's propeller described in §9.2.1, or otherwise fragment into many smaller bubbles (Brennen, 2002). These small 'daughter' bubbles will again be subject to rectified diffusion and will grow and collapse anew.

The foregoing explanation is oversimplified. Bubble nuclei are present in most liquids, and certainly in water. The simple explanation predicts that the

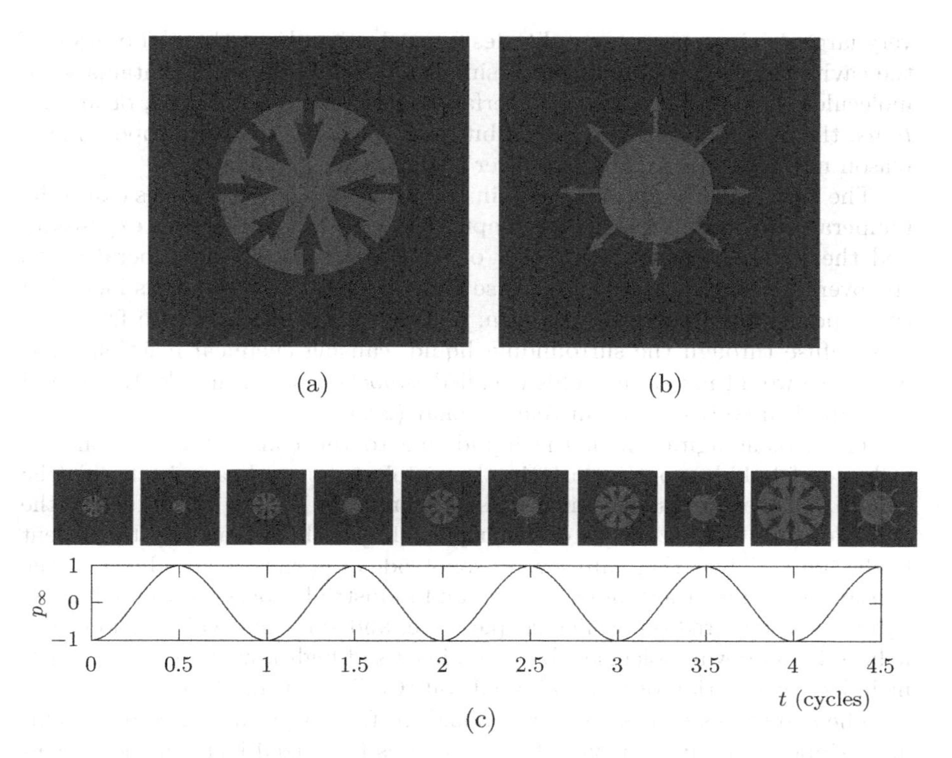

FIGURE 9.5
The principle of rectified diffusion; arrow thickness indicates amount of mass transferred over bubble surface. (a) Bubble at cycle-maximum radius (at $t = 0$, say), hence minimum pressure, causing evaporation of surrounding liquid into bubble. (b) Bubble at cycle-minimum radius ($t = \pi/\omega$), hence maximum pressure, causing dissolution of vapour into liquid. (c) Timeline of driving sound pressure together with cartoons of corresponding stages of bubble growth at each half-cycle. For example, pressure $p_\infty = 1$ maybe 1 bar (100 kPa) and one cycle may be 1/30 000 s.

gentlest of music, if repeated indefinitely, would eventually cause cavitation in a glass of water! That does not occur because the gases in the bubble, without any driving sound waves, will eventually dissolve in the liquid; the Laplace pressure due to surface tension (§9.3.1) imposes a higher pressure inside the bubble than outside, forcing gases to dissolve in the liquid. The pumping of vapour into the bubble by rectified diffusion has to overcome this bias towards dissolution. Thus, there are *cavitation inception* thresholds, detailed in Leighton (1994); the higher the frequency, the higher the pressure amplitude of sound waves needed to cause cavitation. The cavitation threshold also depends on the size of the initial nuclei: very small and very large nuclei require very large driving-pressure amplitudes for cavitation. Even this elaboration of the cavitation mechanism is an oversimplification: if the liquid contains some molecules that tend to coat the interfaces between gas and liquid, or *surfactants*, the mass transfer will be inhibited. Surfactants are also proposed as a reason nuclei persist in liquids, rather than rapidly dissolving.

The very high amplitudes found in ultrasonically-driven bubbles cause the temperature to rise during the compression and fall during the expansion, and the just-mentioned asymmetry of the cycle causes the temperature to rise overall. The temperature may rise high enough to cause the gas including the vapour molecules to break down, creating *free radicals.* These free radicals diffuse through the surrounding liquid, causing chemical reactions that otherwise would not occur. This is called *sonochemistry* (Suslick, 1990), and is comprehensively covered in Ashokkumar (2016).

The violent agitation of the liquid due to the constant formation and collapse of bubbles, particularly in the vicinity of the ultrasonic 'horn' (the *sonotrode*) that transmits the ultrasound into the liquid, is useful for the breakdown of biological cells or other particles or fibres that may be present in the liquid. Thus, there are several sonotrode or 'sonicator' products on the market used in a variety of laboratory and industrial processes, where intense agitation is required on a microscopic scale, and sonochemical reactions also help to break down molecules. Large-scale uses of high-power ultrasound have included remediation of contaminated soil (Collings et al., 2010).

The conditions can be extreme enough to form a plasma, in which atoms are stripped of their electrons. If the bubble is held fixed inside a special vessel, which is usually achieved by the use of an ultrasonic standing-wave field, the plasma can be observed to emit light. This phenomenon is called *sonoluminescence*, in which temperatures are thought to be at least 5 000 K and may be well over 10 000 K (Hilgenfeldt et al., 1999), hotter than the core of the Earth or the surface of the Sun. Sonoluminescence is not restricted to laboratories but can occur in any sufficiently-intense cavitation event; astonishingly, sonoluminescence is created by cavitation in the intense jet of water emanating from the claw of the snapping shrimp (Lohse et al., 2001), which uses the shock of the jet to stun its prey.

9.6.2 Medical ultrasound diagnostics

A *contrast agent* is any substance that is artificially introduced into the human body to make particular tissues of interest stand out on a scan. Contrast agents for X-ray imaging (iodine or barium) and for magnetic resonance imaging (gadolinium) are chemical elements that absorb or react strongly to the type of imaging energy, or, for positron emission tomography, emit radiation, clearly delineating that part of the body. However, ultrasound does not involve electromagnetic or ionizing radiation, rather, it is a fluid wave which is a mechanical oscillation. Therefore, a mechanical oscillator is needed in order to achieve a strong reaction to ultrasound.

Microbubbles, which are bubbles micrometres in size, have been introduced to the body by intravenous injections since the late 1960s to perform as *Ultrasound Contrast Agents* (UCAs) (Gramiak and Shah, 1968). They were initially made simply by agitating saline water, which will cause some gas to come out of solution, forming microbubbles. Since the 1980s, UCAs have been made by cavitating the water with sonotrodes (Keller et al. (1986); see §9.6.1) prior to injection. They are now made in more sophisticated ways, as will be detailed shortly.

Most medical ultrasound is at 1-10 MHz. Use the rule-of-thumb simplification to Minnaert's equation (9.21), bubbles that would resonate at $f_0 = 1$ MHz would have a diameter $2R_0$ of roughly 6-7 μm, while bubbles that would resonate at 10 MHz would be a tenth the size. A convenient coincidence of physics and biology is now evident. The narrowest blood vessels in the human body are *capillaries*, which connect the smallest arteries to the smallest veins, and capillaries are roughly 10 μm in diameter. Thus, bubbles that would resonate in the ultrasound scan will also pass through the entire circulatory system without causing a blockage.

When the microbubble resonates, it strongly absorbs energy from the ultrasonic wave, which on the scale of the microbubble is a plane wave. The microbubble then re-radiates sound in the form of spherical waves (i.e. it *scatters* the waves); sound can not coherently return to the scanner as would a simple reflection from a tissue boundary, for example. The consequence is that reflections from the tissues where the microbubbles are present are much weaker, just as if the tissue were very dense and had strongly scattered the ultrasound. Since some of the energy in the bubble vibration is lost to heat, owing to the damping described in §9.4.1, this is effectively a form of *inelastic scattering*. The scanner algorithms thus show a region with microbubbles as bright relative to a dark background, just as they would show up bone.

Use of Minnaert's equation, or the more sophisticated but still linear solutions to the Rayleigh-Plesset equation, implies that the behaviour of ultrasound contrast agents may be understood from linear behaviour. However, in practice, the behaviour of UCAs requires an understanding of nonlinear bubble behaviour. That is because the behaviour of microbubbles in the ultrasound-scanner beam is invariably nonlinear. In fact, the forced oscillation of the microbubble typically features a nonlinear harmonic, and many modern clinical scanners have a feature that filters out the base frequency of the scan, retaining only the expected harmonic frequency. Since only the bubbles should emit the nonlinear harmonic frequency, only the tissues in which microbubbles have collected show up once the filter is applied.

The Laplace pressure (§9.3.1) due to surface tension in a microbubble is large; as noted in §9.6.1, this causes an elevated pressure inside. A micron-sized bubble would dissolve in the surrounding liquid in a few seconds (Brennen, 1995). For that reason, gases with limited solubility are used, but the main innovation that makes UCAs practical is a shell, or coating, which may be a protein (Grinstaff and Suslick, 1991), polymer, or lipid

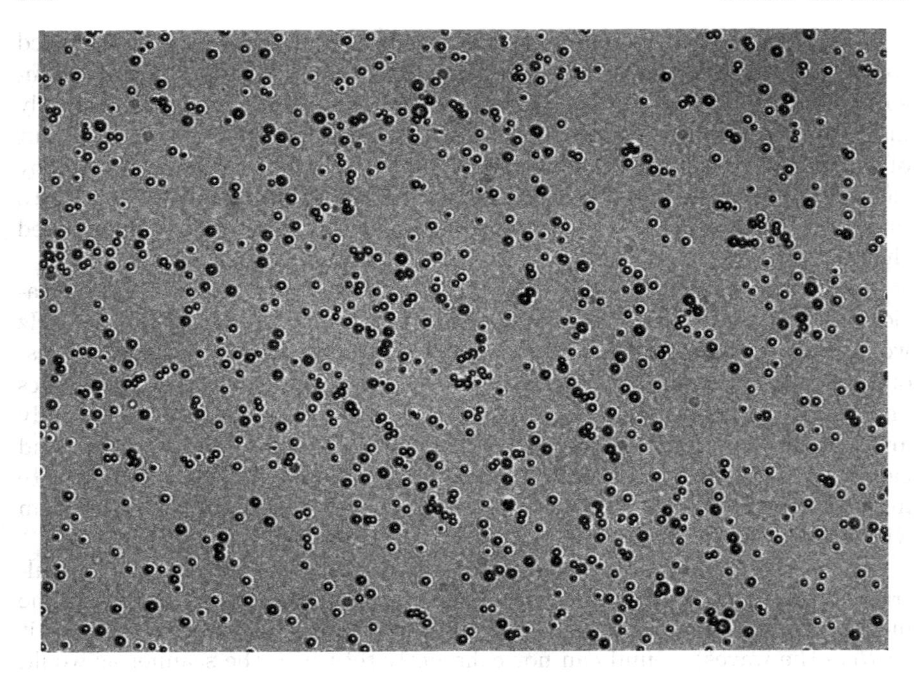

FIGURE 9.6
Microbubbles averaging about 4 μm in diameter, with shells made from a protein (bovine serum albumin; details in Threlfall et al. (2013)). Image reprinted under a CSIRO Licence to Publish.

(Sirsi and Borden, 2009). Figure 9.6 shows an example of protein-shelled bubbles. The shell greatly reduces the dissolution of gas into the liquid. The most common shell materials are phospholipids, which are the same class of molecules that form the membranes of biological cells. Nevertheless, the coating eventually ruptures, particularly for bubbles that have been subjected to the ultrasound beam, and then the microbubbles dissolve.

With or without microbubbles in the patient's system, the application of ultrasound to the human body raises the risk of cavitation. The *Mechanical Index* (MI), an empirical measure used by medical clinicians that is thought to represent the risk of cavitation damage, is given by

$$\mathrm{MI} = \frac{\hat{p}_-}{\sqrt{f}}, \tag{9.45}$$

where $\hat{p}_-$ is the peak negative pressure measured in MPa and f is the frequency in MHz. It is generally recommended the MI should be less than 0.3 for ultrasound diagnostic procedures; details are in ter Haar (2012).

Extensive efforts, academic, medical and commercial, have been devoted to the development of *targeted ultrasound contrast agents* (Lindner, 2004;

Klibanov, 2009). Here, the microbubble shell or coating has a targeting *ligand* attached to its outside. In this context, a ligand is a general term for a molecule that forms a 'bridge' between the bubble shell and some 'target' molecules. Most commonly, the target is a protein that identifies pathological tissue; the proteins may be markers of inflammation (Lau et al., 2020), or of cancer (Browning and Stride, 2018). If the target is a protein, the ligand is typically an *antibody* or antibody fragment. Antibodies are also proteins and have several types; the most commonly found in the body, are 'Y'-shaped molecules called immunoglobulin-G or IgG. The two tips of the 'Y' are specific to the protein they are meant to target, which in nature is a protein identifying a bacterium or virus that has invaded the bloodstream. The specificity has been described with a 'lock-and-key' analogy; if the target protein is a lock, the antibody tip is the key the uniquely fits it. Antibodies may be engineered: designed with computer software so they are specific to the desired target, then manufactured in industrial fermenters by genetically-modified bacteria or other cells. The base of the 'Y' may be attached to the bubble shell, with the two tips pointing outwards. When the tips bind to the target, the microbubble will be attached to it. A group of cells expressing the target protein will eventually collect several microbubbles on their surface. This means that the targeted tissue will show up brightly on the scan.

9.6.3 Medical ultrasound therapeutics

Since ultrasound contrast-agent microbubbles were first introduced, they have been used not just to passively delineate blood vessels and, via targeting, to identify pathological tissue, but to perform therapeutic roles as well. These have included targeting of pathological tissue, desirable damage to pathological tissue, the release of therapeutic molecules, the localised mixing of molecules, and the stimulation of subtle behaviours by cells, such as the uptake of drugs and genes (§13.3.2). It is clear that different aspects of physics, chemistry and biology may all operate together, to some extent making the strictly physics-based organisation of the chapters and sections of this book artificial from the perspective of medical therapeutic applications. Thus, reading of §13.3.2 together with the present section is recommended.

In addition to targeting the microbubbles to pathologies, the microbubble shells may be loaded with drugs. Increasing the ultrasound power, once it is clear that the microbubbles are attached to, or at least near to the target tissue, causes the microbubbles to rupture. This releases the drug in a high concentration locally, achieving a targeted treatment of, for example, cancer (Rapoport et al., 2009; Boissenot et al., 2016).

Cavitation transfers energy from the ultrasonic device into the fluid in which the cavitation bubbles form, heating the fluid locally. This heating, together with the intense localised stresses created by collapsing cavitation bubbles, can be used to destroy undesirable tissue. Importantly, sound waves, like all waves, can be focused. An ultrasound device can be applied to the outside of the human body (an *extracorporeal* treatment), and provided the focus point is known precisely, sound waves would pass through the body without causing damage until the focus region is reached, effectively achieving a form of surgery without ever cutting into the body. This technique is called *High Intensity Focused Ultrasound* and has been used to treat prostate cancers (Poissonnier et al., 2007) and several other conditions (Duc and Keserci, 2019).

An extracorporeal transmitter can also be used to create a focused, high-intensity pulse of sound aimed at the bladder. As it reaches the focus point, it creates a pressure lower than the vapour pressure on the surface of a kidney stone. The resulting cavitation bubbles, together with the mechanical shock of the focused wave, break kidney stones into fragments small enough to pass naturally, removing the need for surgery. This procedure is called *shock-wave lithotripsy* and has been an established clinical technique since the 1980s (Sackmann et al., 1988).

10

Surface-wave breaking in weather and climate

10.1 Summary of key points

- The **Stokes limiting steepness** of a wave in deep water is given by (10.6) on page 232,

$$\boxed{k\hat{\eta} = 0.444},$$

where k is the wavenumber ($k = 2\pi/\lambda$ where λ is the wavelength), $\hat{\eta}$ is the amplitude of the waves and $k\hat{\eta}$ is the wave steepness. (In oceanography, the symbol a is generally used instead of $\hat{\eta}$ for amplitude, so it is common to see ka used for steepness.)

- **Practical wave-breaking limits** from laboratory experiments and ocean data on deep water suggest breaking occurs within the ranges of steepnesses given by (10.7) on 232.

$$\boxed{0.20 < k\hat{\eta} < 0.55}.$$

- The nonlinear **wave height** is $\hat{H}$, so that it is better to replace the $\hat{\eta}$ from linear theory with $\hat{H}/2$; the wave height is equal to $2\hat{\eta}$ when the waves are linear, but as the waves become nonlinear, the trough depth and crest height may no longer be equal to $\hat{\eta}$.

- The three types of **breaking on beaches** can be estimated to occur, as noted on page 233, according to

Spilling breakers	$\mathrm{Ir} < 0.5$
Plunging breakers	$0.5 < \mathrm{Ir} < 3.3$
Surging or collapsing breakers	$\mathrm{Ir} > 3.3$

where Ir is the Iribarren number given by (10.8) on page 233,

$$\boxed{\mathrm{Ir} = \frac{\tan\alpha}{\sqrt{\hat{H}/\lambda}},}$$

DOI: 10.1201/9780429295263-10

FIGURE 10.1
A plunging breaker viewed parallel to the wave crest encloses air within its 'barrel' or 'tube' that can be fitted by an ellipse with a universal ratio of major to minor axes of $\sqrt{3}$:1. Photograph at Palm Beach, Gold Coast, Australia, copyright by and courtesy of Rio Harvey.

where α is the slope of the seabed.

- A useful **book** on wave-breaking is Babanin (2011).

10.2 Wave breaking and air-sea exchange

10.2.1 Criteria for wave breaking

Our experience of the sea almost invariably includes the experience of *wave breaking*. Even if wave breaking is not observed out to sea, all but ripples inevitably break on a beach (e.g. figure 10.1). Despite the ubiquity of wave-breaking phenomena, they have been remarkably difficult to model, since they are nonlinear processes. The breaking of waves out to sea is an active area of research (as reviewed by Babanin, 2011), and, as will be mentioned later, wave breaking over the Earth's oceans is of great importance to weather and climate. Meanwhile, owing to its importance in coastal engineering and the imperative to protect lives, coastal environments and infrastructure, the breaking of waves on beaches and structures has been modelled by many empirical formulae (as reviewed by Robertson et al., 2013). Increasing expenditures on coastal-protection infrastructure are forecast under climate-change scenarios (Hinkel et al., 2014).

In §2.3.12, breaking was described simplistically as occurring when the speed u with which the water is set into motion at the surface exceeds the phase speed c. In reality, more detailed criteria predict breaking, but it is instructive to begin with this simplest of criteria. It is clear that waves must have broken when the horizontal component of wave orbital velocity in the direction of wave propagation exceeds the phase speed. Under this condition, the water is moving so fast relative to the wave propagation speed that it overshoots the wave, destroying the sinusoidal pattern of the linear solution. Consider the linear solution for the horizontal x-direction component of water velocity, which from (2.63) is

$$u = -\mathrm{i}kA\frac{\mathrm{g}}{\omega}\frac{\cosh(kh+kz)}{\cosh(kh)}\mathrm{e}^{\mathrm{i}(-kx+\omega t)}. \tag{10.1}$$

At the surface ($z = 0$) and noting from §2.3.7 that the magnitude of A is simply $\hat{\eta}$, the maximum value of u is given by

$$\hat{u}|_{z=0} = k\hat{\eta}\frac{\mathrm{g}}{\omega}. \tag{10.2}$$

If we now insist that $\hat{u}$ at $z = 0$ equals the phase speed c, recalling that $c = \omega/k$ gives for this simplest and most extreme of breaking criteria, called the *kinematic limit* (Babanin, 2009),

$$\frac{\omega}{k} = k\hat{\eta}\frac{\mathrm{g}}{\omega}, \tag{10.3}$$

and using the dispersion relation (2.35), $\omega^2 = \mathrm{g}k\tanh kh$, (10.3) reduces to

$$\tanh kh = k\hat{\eta}. \tag{10.4}$$

In deep water, as noted in §2.3.9, $\tanh kh \simeq 1$ so (10.4) becomes simply

$$k\hat{\eta} = 1, \tag{10.5}$$

or $\hat{\eta}/\lambda = 1/(2\pi)$. The equivalent for the shallow-water limit, $\hat{\eta}/h = 1$, was derived in §2.3.12. Quite consistently with the destruction of the sinusoidal wave pattern, this extreme criterion (the kinematic limit) also means that the assumption of linearity on which the sinusoidal solution is based, that the wave amplitude $\hat{\eta}$ is very small compared with the wavelength λ, has been violated: it is no longer true that $\hat{\eta}/\lambda \ll 1$.

The parameter $k\hat{\eta}$ is usually called the *wave steepness*. In oceanography, the symbol a is normally used for the wave-elevation amplitude $\hat{\eta}$, so that it is very common to find results quoted as a function of ka.

The approach above of using the linear solution to predict its own demise is not particularly accurate. Observations of the sea surface show that as wave amplitudes increase, the steepening is represented by a gradual loss of the sinusoidal shape, well before breaking. Waves tend to become 'peaky' (as in figure 8.1), so that the sinusoidal shape tends to a triangular shape. This process was examined by George Gabriel Stokes, who re-developed Airy's solution for small but nonzero values of the steepness in 1847. The steepness was used as a parameter in a perturbation expansion, a similar approach to that undertaken in §6.5.1 for shallow-water waves that led to solitary-wave solutions. In §6.5.1, the small parameter was kh, but in Stokes's wave-steepening theory, as well as in the full solution to the soliton (or 'KdV') equation, (6.41), the small parameter was $k\hat{\eta}$. More details are in Whitham (1999). This allowed the derivation of the 'peaky' wave profile which does a remarkably good job of predicting the properties of surface gravity waves that are somewhat but not strongly nonlinear.

In 1880, Stokes found that the included angle under the wave crest reaches a minimum of 120° when the wave is at its steepest. The wave height $\hat{H}$ is no longer exactly twice the wave amplitude $\hat{\eta}$, because the crests of the waves become sharper than the troughs. The value of $\hat{H}/\lambda$ for waves in the deep-water approximation, calculated by increasingly sophisticated methods over the 20th century, reaches a maximum of about 0.1412, or approximately 1/7 (Schwartz and Fenton, 1982), so that breaking is theoretically predicted to occur approximately at $k\hat{\eta} = (2\pi/\lambda)\hat{H}/2 = \pi \times 0.1412$, i.e., at

$$\boxed{k\hat{\eta} = 0.444}. \tag{10.6}$$

This *Stokes limiting steepness* (10.6) predicts that breaking will occur at a much lower steepness than the kinematic limit (10.5), $k\hat{\eta} = 1$. However, laboratory experiments suggest far steeper waves than $k\hat{\eta} = 0.444$ can arise, and when ocean observations and laboratory experiments are combined, a limit of $k\hat{\eta} = 0.55$ is evident (Toffoli et al., 2010), while some laboratory waves can become almost as steep as the kinematic limit (Babanin, 2009). However, wave breaking far from coasts is not usually a total and catastrophic elimination of the wave, as is a wave breaking on a beach; laboratory observations suggest that some breaking occurs for $k\hat{\eta} > 0.20$ (Toffoli et al., 2010), representing the beginning of a 'brake' on the further growth of waves. Thus, empirical results suggest that waves will be breaking for

$$\boxed{0.20 < k\hat{\eta} < 0.55}. \tag{10.7}$$

The steepness at which waves break is very important since it limits the maximum amplitude to which waves can grow and thus the energy that waves can transport. If there were no breaking, the wind could continue to pump energy directly into waves (with a maximum wavelength λ such that the wave speed c equals the wind speed U), and the wave height would grow indefinitely. This knowledge of the limit on wave height due to breaking is needed for our prediction of the impacts of waves on coastal environments and infrastructure, as well as for our attempts to extract renewable energy from ocean waves (as discussed in chapter 8). Knowledge of the wind speed, which is relatively easy to obtain from meteorological data and models, enables an estimation of the maximum wavelength (and hence the minimum k) of waves (as outlined in §8.2.1). If we had a better estimate than (10.7), we would have a better estimate of wave energy, enabling us to address these practical needs.

As noted earlier, the breaking of waves on shores and on engineering structures are of importance owing to the potential damage that can be caused. Wave breaking on beaches can be classified using a nondimensional ratio of the breaker height to its wavelength,

$$\boxed{\mathrm{Ir} = \frac{\tan\alpha}{\sqrt{\hat{H}/\lambda}},} \tag{10.8}$$

where α is the angle of the seabed slope where breaking occurs; this ratio is called the *Iribarren number* (Ir) after Ramón Iribarren Cavanillas who proposed it in 1949 together with Nogales (Robertson et al., 2013). It is sometimes called the *surf similarity parameter* and examples of values taken by Ir for various types of breaking are given in §10.2.2 below. Empirical formulae for wave breaking on beaches are often given as a function of a *breaker index*, γ_b, given by

$$\gamma_b = \frac{\hat{H}_b}{h_b}, \tag{10.9}$$

where $\hat{H}_b$ is the height of waves at their point of breaking, and h_b is the depth of the water where this occurs (Robertson et al., 2013).

10.2.2 Types of wave breaking

A casual observation of breaking waves at sea will allow their division into two broad categories: *plunging breakers* and *spilling breakers* (Banner and Peregrine, 1993). Waves breaking on beaches can exhibit these two forms of self-destruction, plus a third form is found on very steep beaches, *surging breakers*, sometimes called *collapsing breakers*, where the wave abruptly disintegrates into turbulent foam. On beaches, spilling breakers are estimated to occur for $\mathrm{Ir} < 0.5$, whereas surging or 'collapsing' breakers are estimated to occur for $\mathrm{Ir} > 3.3$ (Robertson et al., 2013).

Plunging breakers (such as those shown in figure 2.15 on page 84 and figure 10.1) could be thought of as the classical breaking wave. The crest of the wave overshoots the body of the wave, forming a short-lived overhang. On beaches, they are estimated to occur for $0.5\ \mathrm{Ir} < 3.3$ (Robertson et al., 2013).

Even rather small waves can form plunging breakers. If the wave is large enough, the plunging breaker forms the 'barrel' or 'tube' capable of enclosing a person and much sought-after by surfers. Indeed, the development of surfing as a widespread recreation and sport in the second half of the 20th century led to a large number of high-quality photographs being taken of plunging breakers - in particular, photographs of the *profile* of the breaker, looking along the 'tube' (e.g. figure 3 in Peregrine (1983), and figure 10.1). Images of ocean breakers and laboratory experiments on the profile of a plunging breaker revealed a remarkable similarity that was explained by the early 1980s. Irrespective of the size of the plunging breaker, the cross-sectional profile of the 'tube' enclosed by the plunging breaker can be fitted by an ellipse with a ratio of major to minor axes of $\sqrt{3}$:1 (New, 1981). Despite the plunging breaker representing a severe departure from Airy's original potential-flow solution (§2.3.5), it is impressive that a complex-potential description of the free surface, in which gravity is neglected so that the water is effectively in free fall, is able to generate the $\sqrt{3}$:1 ellipse that represents the plunging-breaker profile very well (New, 1981). A cubic solution permits the free surface to intersect itself and also closely fits the $\sqrt{3}$:1 ellipse on the front face of the wave (Longuet-Higgins, 1982). The major axis of the ellipse, which is also the line of symmetry of the cubic solution, rotates with time as the plunging breaker develops, reaching a maximum angle to the horizontal of about 47° (Longuet-Higgins, 1982).

It seems likely that a plunging breaker will form if the phase speed of the wave is reduced rapidly with time, which would be expected as waves reach a steep beach or reef. However, the more common circumstance is that as the wave steepens the front face of the wave becomes unstable and water cascades down the front face, forming a spilling breaker. Over the global oceans, spilling breakers are far more common (Banner and Peregrine, 1993).

10.3 Global climate consequences of ocean-wave breaking

10.3.1 Energy transfer from whitecapping to microscale processes

While our most common experience of wave breaking is coastal surf, since 71% of the Earth's surface is ocean, waves breaking on the narrow strip of ocean bordering the land represent a tiny fraction of the waves breaking on the windswept reaches of the global oceans, where it is called *whitecapping*. Satellite data suggest that roughly 1-4% of the entire global ocean surface is covered by whitecaps at any given time, but whitecapping is highly variable geographically, with over 6% in some regions and close to zero in others (see Anguelova and Webster, 2006, figure 5a).

It is important to recall that breaking processes represent a loss of energy from waves. As mentioned in §10.2.1 above, knowledge of the energy *remaining* in the waves when they reach a coast, after they have lost energy from breaking on their journey to the coast, is important for both engineering and coastal-environment applications. However, some components of the energy *lost* from waves - and lost over the global ocean surface - have even greater significance.

The greater the height of a wave, the greater the energy lost as it breaks. The energy lost is eventually transformed into heat. However, the energy-transformation path from the energy of waves given by (8.2) to heat in the water and atmosphere is complex, and in outlining it below, aspects of ocean-wave breaking of critical relevance to climatic and biological sciences will be revealed.

Above the sea surface, some of the energy of breaking is transformed by the fragmentation of the wave crest. Characteristic 'claw-like' fragments of the crest can be seen in photographs taken with a sufficiently fast shutter speed (such as figure 2.15). These fragments break up further, transforming their kinetic and potential energy into kinetic and gravitational-potential energy or airborne filaments and water masses and drops (generically called *spume*), and into the surface-tension energy of these drops. Small droplets produced directly from the crest fragmentation may be carried away by the wind, and their fate will be mentioned shortly. Larger drops will hit the water surface; the impact of a liquid drop on a liquid surface can produce a crater, the rim of which in turn breaks up into a 'crown' of tiny droplets. This exquisite process was first captured in a high-speed photograph by Worthington (1883) and continues to be investigated well over a century later (e.g. Liow, 2001; Thoroddsen et al., 2008). The tiny droplets from the impact 'crown' may also be carried away by the wind.

Below the sea surface, some of the energy of breaking is transformed temporarily into the gravitational-potential energy of large air cavities and bubbles dragged underwater, and into the surface-tension energy of bubbles, before the bubbles eventually rise to the surface. Furthermore, drops of spume falling back to the sea can entrain bubbles from the collapse of the crater noted above (Morton et al., 2000). Laboratory experiments suggest that for spilling breakers, large (millimetre-sized) bubbles are mostly contained within the wave crest and hence are above the mean sea level, but for plunging breakers, large bubbles may be pushed below the mean sea level (Loewen et al., 2010). Once formed, however, bubbles may be carried much deeper by other processes, as will be outlined in §10.3.4. As they rise, bubbles transfer energy to the water via the drag force the bubbles exert on the water as well as the lateral motions of rising bubbles (Clift et al., 1978). The significance to the climate of the bubbles formed by breaking waves will be discussed in §10.3.4.

When bubbles burst at the surface the thin 'film' that is the rupturing bubble wall produces the smallest of all ocean droplets, which are typically less than 1 μm in size (Grythe et al., 2014). Thus, some surface-tension energy is temporarily transformed into the gravitational-potential energy of airborne tiny droplets. These tiny droplets, like the smallest droplets created by fragmentation of the crest above the surface and those ejected from the 'crown' due to drop impact, are carried away by the wind. The products of these droplets (and, in some cases, the droplets themselves) are called *aerosols* because they may stay aloft for a very long time.

As water evaporates, the seawater droplets transform into tiny crystals of salt. The oceans are estimated to inject an astonishing $9.0 \pm 0.6 \times 10^{12}$ kg per annum of aerosols less than 10 μm in diameter into the atmosphere (Grythe et al., 2014), or about 9 billion tonnes per annum. It should be noted that sea salt is not the only component of particulate aerosols. Bursting bubbles and evaporating droplets also release into the atmosphere a gas, dimethyl sulphide, that is produced in the seawater by planktonic life in the oceans, a topic to be mentioned in more detail in §10.3.4. Once in the atmosphere, dimethyl sulphide gas participates in a chain of chemical reactions leading to solid sulphate particles (Menzo et al., 2018). These particles may play a similar role to the salt-crystal aerosols produced from evaporating water droplets that will be discussed in §10.3.3 below.

The importance of oceanic life extends beyond well the addition of simple molecules like dimethyl sulphide to sea-spray aerosols. Most importantly, lifeforms extract carbon from the seawater, which is an essential part of the global carbon cycle to be outlined in §10.3.5. Even the physics of droplet and bubble formation is affected by the presence of oceanic life. As will be discussed in §10.3.4, complex organic molecules are present in seawater. Sea-spray droplets can also entrap live biological cells (Blanchard and Syzdek, 1970; Sultana et al., 2017), hence transporting them in the wind to the land, or to colonise distant seas.

10.3.2 Outline of air-sea exchange

The term *mass transfer* includes four key processes. In evaporating, liquids change phase into their gaseous (or *vapour*) form, requiring the absorption of heat; and the reverse process, *condensation*, is when vapour becomes liquid, giving up heat. Gases dissolved in the liquid may come out of solution, and pass into the atmosphere, a process called *evolution* or sometimes 'outgassing'; and the reverse, when gas is dissolved in liquid, is called *dissolution* or sometimes 'invasion'. The processes by which mass transfer between the ocean and atmosphere occurs are collectively called *air-sea exchange.*

Since the transfer of mass occurs across interfaces, the greater the surface area, the faster the exchange of mass between the air and sea. Therefore, the surface area created by wave breaking is very important, but the rate of mass transfer not the only factor. The time drops stay aloft affects the amount of mass transferred, and the time bubbles stay underwater affects the amount of mass transferred. Ocean experiments on air-sea exchange have found that mass transfer rates in seas marked by significant wave breaking were 50–200 times higher than those predicted without breaking (Wallace and Wirick, 1992); other studies have similarly suggested that air-sea exchange is two orders of magnitude higher during those times when the wind exceeds the 10 m s^{-1}, which is associated with breaking seas (Vagle et al., 2010).

It was noted in §10.3.1 that the formation of sea-spray droplets facilitates the transfer of a gas of biological origin, dimethyl sulphide, from the water to the air, which will be discussed in §10.3.3. The Earth's atmosphere is 78% nitrogen and 21% oxygen with the remaining one percent composed of a number of other *trace gases.* Oxygen is more soluble in water than nitrogen, and is moreover oxygen is produced by photosynthesising organisms and consumed by aerobic bacteria and by animal life, so exchanges of oxygen between the atmosphere and ocean, facilitated by breaking waves, are an essential part of life on Earth.

However, the one percent of trace gases in the atmosphere includes one particularly critical gas: carbon dioxide (CO_2). Prior to the Industrial Revolution of the mid-eighteenth century, CO_2 made up 0.028% of the Earth's atmosphere. Presently, CO_2 makes up about 0.042% of the Earth's atmosphere. As with other trace gases, the concentration is more commonly expressed in parts per million (ppm) so that the value at present is about 419 ppm. The role of CO_2 in the climate of planets was elucidated by Svante Arrhenius in 1896. The Earth (and similar planets) absorbs solar energy over the full spectrum of electromagnetic wavelengths emitted by the Sun: wavelengths that include the infrared wavelengths that is heat, but also every other wavelength, including, of course, visible light. Some of the incident solar energy that was not at infrared wavelengths, and hence was not incoming heat, is transformed into heat by the Earth and re-radiated back into space as heat. Carbon dioxide is one of the gases that absorb heat, and thus it absorbs some of the heat re-radiated by the Earth, preventing its loss to space and hence warming the Earth's atmosphere.

It is thought that roughly 25-30% of all human CO_2 emissions are absorbed by the ocean (DeVries, 2014; Le Quéré et al., 2018), and the influence of wave-breaking on this process will be discussed in §10.3.4 below. Since actual measurements of dissolution rates can only be locally determined from dedicated ocean experiments, the true global rate is unknown (Yang et al., 2011) and estimates vary by 20% (Khatiwala et al., 2013; Le Quéré et al., 2018). Since, as noted above, breaking waves increase the rate of gas dissolution by two orders of magnitude, understanding the percentage of the ocean over which breaking is occurring (e.g. Anguelova and Webster, 2006) and the *microscale processes* in breaking waves (outlined in §10.3.1) is very important to improving predictions of the oceanic sink for carbon.

Water vapour can make up a substantial amount of the one percent of trace gases, and also absorbs re-radiated heat. Since wave-breaking, as noted in §10.3.3, increases evaporation, waves are also significant in affecting the atmosphere's water-vapour concentration.

10.3.3 Influence of sea-spray aerosols on climate

Salty spray can be blown inland from an ocean that regularly features breaking waves, a phenomenon only too familiar to coastal residents whose metal window fittings and cars suffer from the resulting corrosion. Fine sea-spray aerosols may be transported hundreds of kilometres inland by the wind (May et al., 2018).

However, sea-spray has a much greater significance than corrosion. Fine sea-spray aerosols may stay in the atmosphere indefinitely, where they reflect sunlight back into space, cooling the Earth (Grythe et al., 2014). Considering *only* this reflective property, aerosols act as a negative feedback on climate change, reducing global warming.

Aerosols also act as cloud condensation nuclei, 'seeding' the formation of clouds (Grythe et al., 2014). Cloud formation is rapid, reacting to changes over timescales of days to weeks (Andrews et al., 2012). The formation of clouds is thought to be one of the least-understood feedback mechanisms affecting climate change (Stephens, 2005) and differences in the way clouds and their effects are represented in climate models is presently a major cause of differing predictions of the extent of climate change (Sherwood et al., 2020).

Before outlining the possible effects of clouds on the climate, it is important to note that aerosol seeding from breaking ocean waves is only one contributor of the genesis of clouds. The amount of moisture in the atmosphere, and the *relative humidity,* which indicates how close the atmosphere is to the point of condensation, strongly affect cloud formation. Higher ocean temperatures increase evaporation of water from the sea, providing more moisture for cloud formation, but higher atmospheric temperatures also permit the air to hold more moisture before condensation occurs, decreasing cloud formation. Thus, climate-change-induced changes in air and sea temperatures, in addition to any changes in wave-breaking derived aerosols, may alter the fraction of the Earth's surface covered in clouds, and, of course, the distribution of clouds in different geographical regions as well as in different heights in the atmosphere.

The effects of clouds on the climate are manifold and complex. Clouds, like the aerosols that seeded them, also reflect solar energy back into space, cooling the Earth. However, the water in the clouds also absorbs heat radiated from the Earth's surface, trapping it in the atmosphere analogously to the effect of CO_2 (Stephens, 2005). These opposing effects are generally familiar to residents of the Earth's temperate regions; we notice that clouds provide some shade from the sun during a hot day, but also act as a 'blanket', prevent nights from getting very cold. Clouds formed at low altitudes may exert different influences on the climate to clouds formed at higher altitudes (Stephens, 2005). Furthermore, whether clouds are composed of ice or of water droplets also shifts their influence (Tan et al., 2016).

Thus, clouds have the potential to provide both a negative and a positive feedback mechanism, either reducing or enhancing global warming. Importantly, these feedbacks occur in two broad ways: in the number and distributions of clouds; and in the nature of the effect exerted by clouds on the climate. Analyses over the last decade appear to be converging to the opinion that clouds provide a net positive feedback, increasing global warming (Zelinka et al., 2013; Tan et al., 2016; Sherwood et al., 2020).

Considering the long chain of processes and events outlined in §8.2.1 and earlier in this chapter 10 that lead from winds to waves and to the formation of sea-spray aerosols, many uncertainties may be identified. If climate change causes wind speeds over the ocean to increase, we might expect higher waves and thus more energetic wave breaking and more sea-spray aerosols, possibly slowing the rate of warming of the climate. Meanwhile, more sea-spray aerosols may seed increased cloudiness, which might perturb the climate either towards higher or towards lower temperatures, depending on other factors briefly outlined above. However, if climate change causes wind speeds to decrease, the negative feedback due to sea-spray would be weakened, possibly increasing the rate of warming. Analyses of satellite data indicate a slight overall increase of wave heights occurred from 1985 to 2018, with the largest increase in the Southern Ocean and greater increases in extremes (Ribal and Young, 2019).

10.3.4 Influence of bubbles from breaking waves on climate

As noted in §10.3.1, breaking waves inevitably create bubbles. Like the spume and aerosols created above the surface, bubbles greatly increase the surface area between the sea and the air.

Bubbles larger than about a millimetre do not rise in a straight line, but zig-zag or spiral, prolonging their time underwater (Maxworthy et al., 1996). Although, as noted in §10.3.1, bubbles are formed by breaking waves at a depth that is mostly within the wave's amplitude, other processes may carry bubbles much deeper, to a depth of 10 m or greater (Vagle et al., 2010). The behaviour of bubbles and the gases they contain at such depth in the sea is rather different to the behaviour of droplets in the atmosphere. In particular, since water has approximately 800 times the density of air, there is a large increase in hydrostatic pressure with depth. A further aspect of physical chemistry now comes into play: the solubility of gases in liquids is strongly dependent on pressure, and the higher the pressure (and the lower the temperature), the greater the mass of gas that may be absorbed by seawater. This applies to any component of the atmospheric air that is trapped in the bubble. Carbon dioxide is already very soluble in water, and, as depth increases, solubility increases significantly. If the solubility is greater, the rate at which the gas dissolves is greater, and the amount of gas dissolved during the bubble's journey from the surface to the depths and back to the surface is greater. Thus, the dissolution of gases from bubbles into seawater, particularly bubbles injected deep into the sea by breaking waves, is an asymmetric process compared to the evolution of gases from droplets in the atmosphere.

Large breaking waves can thus be thought of as mechanical pumps, on balance pressurising and dissolving atmospheric gases and particularly CO_2

into the ocean faster via injected bubbles than those gases are released from spume at the surface. Large breaking waves are more likely where winds are strong and steady, and this is the case in the mid-latitudes, as explained in §8.2.1, and these mid-latitude waters are also cold, further promoting CO_2 absorption. Nevertheless, if strong steady winds less common, and this may occur in the tropics, breakers generate only near-surface bubbles and droplets; and if the ocean temperature is high enough, CO_2 may be returned to the atmosphere.

Understanding of these processes requires measurement, and one technique for measuring bubble formation is the acoustic emissions of breaking waves which were outlined in §9.5.2.

10.3.5 Outline of the oceanic part of the carbon cycle

A brief review of the oceanic part of the Earth's carbon cycle is now necessary. The aim is to put the physics of waves and wave breaking that is our primary focus into a context where calculations on waves and wave breaking could lead to future calculations on carbon dioxide and climate.

In summary, over hundreds of millions of years, carbon dioxide is removed from the atmosphere as rocks are weathered by rain, depositing calcium carbonate at the seabed. Photosynthetic ocean organisms also remove carbon dioxide that has dissolved in seawater, and, on dying, they also deposit carbon at the seabed. Eventually, tectonic processes drag these ocean-floor carbon deposits deeper into the Earth, where heat causes carbon dioxide gas to form again; the CO_2 is returned to the atmosphere in volcanic eruptions. This process has resulted in long cycles of extreme climate change. However, over the last 250 million years or so, ocean organisms evolved that result in the storage of dissolved carbonates in deep cold seawater, from which CO_2 can be returned to the atmosphere in the tropics. This enables a 'rapid' adjustment of the Earth's CO_2 levels, so that climate changes may be modified over approximately 10 000 year timescales. Thus, oceanic air-sea exchange, mediated by breaking waves and intimately tied to oceanic chemistry and biology, has controlled the Earth's climate by controlling the atmospheric CO_2 level. More details are in Rohling (2017) and included references.

Finally, the microbes in the ocean produce organic surfactants (briefly introduced in §9.6.1 in the context of ultrasonic cavitation), which promote the formation and longevity of bubbles, and thence the 'foaminess' of seawater, increasing the propensity of bubbles to both trap atmospheric gases and also to release droplets to the atmosphere from bursting bubbles. Meanwhile, organic surfactants may also act as a barrier, somewhat reducing the rate of mass transfer from the bubble into the water.

11

Rotating-fluid waves in space and planetary systems

11.1 Summary of key points

- Rotation is inherently present in astrophysical and planetary systems (§11.2.1).
- Rotation-induced inertial waves occur in **stars** (§11.2.2), inside **planets** that have fluid interiors such as the liquid outer core of the Earth, and possibly in **moons** with liquid water oceans (§11.2.3).
- Magnetic fields protect life from dangerous stellar radiation and thus enhance the habitability of planets (§11.2.3).
- Rotation and inertial waves may be relevant to the **generation of magnetic fields** (§11.2.4).
- Inertial waves occur in **rotating spacecraft carrying fluids**, affecting their stability (§11.3.1) and possibly habitability (§11.3.2).
- A useful **book** on the fundamental theory is Greenspan (1968).

11.2 Rotating-fluid waves in stellar and planetary physics

11.2.1 The origin of rotation

The Coriolis force introduced in §5 may be a fictitious force that represents the effects of rotation in the rotating frame of reference, but what it represents - the tendency of a particle to continue in a straight line - exerts a powerful influence on many systems. Rotating systems have been utilised since the beginning of spaceflight, and continue to be proposed (e.g. figure 5.1, and §11.3 below). The largest and most significant fluid systems are astrophysical,

DOI: 10.1201/9780429295263-11

FIGURE 11.1
An aurora over Mawson Station, Antarctica. Auroras form owing to the Earth's magnetic field, in which waves due to rotation are speculated to play a role. Further details in §13.4. Image copyright by and courtesy of Lydia Jean Dobromilsky.

in which rotation is almost inevitable, and may affect many phenomena (e.g. figure 11.1).

Space is an inhospitable vacuum, so even the individual molecules of gas, particles of dust or larger bodies that are occasionally encountered are spaced very far apart. Thus, it may be surprising that systems of sparsely-separated molecules and particles of dust might be considered a 'fluid'. However, in some cases, the volumes over which motion is considered are so large that there are a sufficiently large number of entities (gas molecules, particles, etc) within an 'infinitesimal' element of volume. Therefore, we might consider continuum properties such as density, pressure and fluid velocity to exist. Thus, for some astrophysical systems, the continuum hypothesis introduced in §1.2.1 is still applicable. Therefore these systems could be considered fluids, and fluid waves could occur, with restoring forces being compressibility, gravity, and the Coriolis force due to rotation. The continuum assumption has long been applied to the formation of stars and planets to be discussed shortly. The continuum assumption has even been applied to the vastly greater scale of entire galaxies, where rotation and waves are also thought significant (e.g. Lin and Shu, 1966; Baba et al., 2016). Nevertheless, at the galactic scale, continuum models may be limited to the fraction of the galaxy that is gas, excluding the intense concentrations of mass such as stars (which, as just noted, may be modelled as continua on a much smaller scale), and excluding aspects of physics that are still unclear, such as the dynamics of dark matter (Vogelsberger et al., 2020).

Individual stars and planets rotate because the parent body of gas and dust from which they formed (the *accretion disc*) was rotating, as outlined below. Black holes, which were originally stars, also rotate, although for black holes, the laws of mechanics established by Newton that are the basis of this book are no longer valid, requiring both relativistic and quantum-mechanical modifications.

It is supposed that a shock of material expanding from a distant exploding star (a *supernova*) compresses interstellar gas, dust or the material from an earlier supernova. Once such a concentration of mass is achieved, gravitational attraction causes the mass to fall inwards towards the centre of gravity of the mass concentration. Inevitably, the distribution of mass around the centre of gravity and their initial velocities are not precisely spherically symmetric, so that falling particles will be pulled slightly off the direct path to the centre of gravity by the gravitational attraction of other nearby masses. The consequence is that the cloud of material must rotate as it falls inwards upon itself. Once rotation has commenced, angular momentum is conserved, so that rotation is largest towards the centre. Eventually, a quasi-equilibrium is reached in which molecules and particles of collective mass M have concentrated at the centre of gravity. If this rotating mass eventually reaches sufficient mass and density to form a star, it will be a rotating star. The Sun's rotation period is slightly more than 25 Earth days, although the rotation rate varies with latitude over the Sun's surface as well as with radius, as will be discussed in §11.2.2 below.

Meanwhile, some particles attain a sufficient angular velocity that their centripetal acceleration is exactly that due to gravity, i.e., $\Omega^2 r = GM/r^2$ where Ω is the particle's rotation rate around the centre of gravity in radians per second, G is the universal gravitational constant of value 6.6743×10^{-11} m^3 kg^{-1}s^{-2} and M is the mass of the centre of gravity. These particles are said to be in *orbit* around the centre of gravity; the relation just-quoted was formulated by Johannes Kepler in 1619 and is known as *Kepler's Third Law*.

Despite the great distances between particles and even between individual molecules in the accretion disc, the scales over which the disc's overall behaviour are of interest are so large that it is one of the systems for which the continuum hypothesis still holds. Therefore, the accretion disc itself is typically modelled as a fluid (e.g. Bondi and Hoyle, 1944; Alexander, 2008). It is a compressible fluid with gravity providing an attractive force *within* the fluid, rather than an external body force applied to the whole fluid, as in the rest of this book. Magnetic forces may also play a role in accretion disc dynamics (Papaloizou and Lin, 1995). Since the disc fluid is compressible, it has a speed of sound, just like any fluid considered in chapter 3, and early theories of particle concentration to form larger particles supposed that this *accretion* occurred when the orbital velocity equalled the speed of sound (Bondi and Hoyle, 1944).

When a star has formed, the accretion disc can be termed a *protoplanetary disc.* Inevitable local inhomogeneities in the distribution of particles of the accretion disc lead to further concentrations at various radii from the star at the disc's centre. Some of the local inhomogeneities will become large enough to become planets, which are defined to be spherical bodies that are large enough to have swept their orbits clear of other objects. Again owing to conservation of angular momentum, material slightly farther out than the planet's orbit must orbit the star slightly faster than the planet as it moves inwards to the planet, whilst the material slightly closer to the star must orbit slower as it moves outwards to the planet. This differential rotation imparts a spin to the planet as well, and it is, therefore, a rotation in the same sense as the rotation of the protoplanetary disc and the star. On a smaller scale, moons, which are bodies that orbit planets, also rotate.

Thus, rotation and hence the Coriolis force is an inherent feature of the physics of stars, planets and moons. Provided these celestial bodies contain fluids, which is certainly the case with stars, waves made possible by the Coriolis force may influence the behaviour of these fluids. It is worth noting that the Coriolis force is important in accretion discs and protoplanetary discs as well. However the rotation rate of the discs, Ω, is not a constant, as in chapter 5; instead, it varies continuously with radius.

11.2.2 Inertia waves in stars

Fluid flows in a star feature a complex interaction of forces due to thermal convection and magnetic fields as well as the Coriolis force. Oscillation modes in the Sun were theoretically predicted in the early 20th century, taking into account both compressibility and gravity as restoring forces permitting oscillations. Rotation was known to be significant in modifying those modes (Cowling and Newing, 1949). In addition to theory, oscillation modes can be observed on the surface of the Sun (Leighton et al., 1962; Gough et al., 1992).

Stars like our Sun also known have a rotation rate that varies with radius (Duvall et al., 1984), with a large increase in the rotation rate over its value at the surface occurring at the innermost 10-20% of the Sun's radius (Duvall et al., 1984). Nevertheless, the presumption that the star rotates as a rigid fluid-filled body was behind Poincaré's original derivation of the governing equation for the Coriolis-dominated oscillations known as inertia waves, (5.11). Stars are not perfect spheres; they are spheroidal since rotation causes a 'bulge' at the equator, and solutions to Poincaré's equation (5.11) for such geometries have been found (Greenspan, 1968). (Any rotating spherical celestial body, including the Earth, also has an equatorial bulge.)

Inertial oscillations in stars like our Sun may be significant to the explanation of the *solar dynamo* (Ulrich, 2001), the process by which the movement of the electrically conducting fluid in the Sun generates a self-sustaining magnetic field. Magnetism in the Sun is very important; it affects many phenomena such as *solar flares* and *coronal mass ejections.* These solar events can disrupt communications on Earth and endanger space travellers, as briefly discussed in §11.2.3. Over the long term, solar variations may affect the Earth's climate.

11.2.3 Magnetism and life on planets and moons

Rocky planets like the Earth were initially molten on formation, and some like the Earth still contain molten interiors where the Coriolis force can generate waves (to be discussed below). Other planets are composed of gas, which being a fluid may also admit waves due to the Coriolis force. Some moons, such as Jupiter's moon Europa, are thought to have liquid-water oceans beneath an ice crust (Kivelson et al., 2000), again potentially admitting waves due to rotation (Rovira-Navarro et al., 2019).

The Earth remains the planet of most importance to us, and one of the most vital but oft-forgotten features of our planet is its magnetic field. We will shortly see that some researchers speculate that inertial waves may play a role in the magnetic fields of the Earth and similar planets. It is not surprising that stars, which are full of electrically charged, rapidly flowing gases, possess powerful magnetic fields, even if full details of how the solar dynamo operates remain to be elucidated. The gas-giant planets Jupiter and Saturn are known to have a magnetic field, with Jupiter's field being very powerful (Hide, 1980). For rocky planets like Earth, the mechanism by which they generate a magnetic field is also not perfectly understood. Moreover, for some planets, it is not clear why they have a magnetic field at all.

The Earth's magnetic field is dominated by a dipolar component, apparently like a giant bar magnet, with its south pole located in the Arctic and its north pole in the Antarctic. Thus the magnetic poles are close to the Earth's rotation axis but do not coincide with it. The magnetic south pole is actually nearest to the Earth's North Pole. Furthermore, the magnetic poles have been measured 'wandering' over the Earth's surface over the last century, and geological data show that the polarity of the field has actually reversed several times over the Earth's history (Roberts and Glatzmaier, 2000). These observations are clues that the Earth's magnetism is not fixed, like that of a bar magnet. Moreover, Joseph Larmor pointed out in 1919 that the Earth's interior is too hot for permanent magnetism to exist, so some dynamic process must generate the magnetic field (Hollerbach, 1995).

Our planet's magnetic field is important to life and therefore we might speculate that magnetic fields of other planets may be equally important. The magnetic field safely deflects the *solar wind*, a stream of electrically charged high-energy particles from the Sun that creates the auroras at high latitudes (figure 11.1), as well as *cosmic rays*, similar particles of galactic origin. These represent ionizing radiation well known to damage biological molecules such as DNA (e.g. Okayasu, 2012). The region over which these dangerous particles is deflected is called the Earth's *magnetosphere.* Without the magnetosphere, significant radiation damage may occur to many organisms, and atmospheric chemistry alterations may occur. Magnetic field reversals can occur quite rapidly on a geological timescale and during the time the reversal is taking place, the Earth's surface may lose protection from radiation. The last reversal, about 42 000 years ago, is thought to be associated with widespread environmental and human impacts (Cooper et al., 2021).

During a solar event, space travellers outside the Earth's magnetosphere may be subjected to a radiation dose sufficient to cause acute radiation sickness (Kim et al., 2009) or death (Townsend et al., 1989). To date, the only humans who ventured well outside the magnetosphere were those travelling to the Moon on the Apollo missions from 1969 to 1972. On the 4th of August, 1972, an enormous solar event occurred (Vette, 1973) capable of delivering a life-threatening radiation dose (Townsend et al., 1989); fortunately this occurred between the Apollo 16 and Apollo 17 missions. It has been estimated that the risk of cancer death to long-term space travellers or residents of the Moon or Mars maybe two to four orders of magnitude higher than that from the terrestrial benchmark, a chest X-ray (Durante and Cucinotta, 2008).

The solar wind will gradually strip away the atmosphere of a planet without a magnetosphere, a process that has well advanced in the case of Mars, which has no global magnetic field but may once have had one (Jones, 2011). A magnetosphere would make Earth-like life more tenable by reducing dangerous radiation doses. An analysis of 10 recently discovered *exoplanets* (planets in other solar systems) suggests that if an Earth-like magnetic field were present, radiation doses under an Earth-like atmosphere would be an order of magnitude less (Atri, 2019).

Thus, the determination of the origin and dynamics of the magnetic fields of potentially habitable planets and moons is of great interest. In order to appreciate theories on the rotational origin of the magnetic field, and ultimately to assess theories on the role of inertial waves in the magnetic field, a brief survey of the Earth's internal composition is necessary.

The Earth's surface is the *crust*, (which is composed of the continental crust on which we live and the oceanic crust). Beneath the crust, which is roughly 10 km thick, is the *mantle*, composed of rocks heated sufficiently that they can flow extremely slowly, taking many millions of years for a significant movement to occur. The crust is divided into *tectonic plates*, each of which might comprise both continental and oceanic zones. Slow movements of the tectonic plates, driven by flows of the mantle, build mountains and recycle oceanic rocks (including the carbonate minerals that are part of the carbon cycle detailed in §10.3.5). The mantle's immense viscosity is not that of a Newtonian fluid, as defined in chapter 1, since the relation between shear stress and the mantle's rate of strain is nonlinear. Thus, from the perspective of fluid waves and the generation of the Earth's magnetic field, the mantle might be assumed to be a solid, and also an electrically-insulating solid (Roberts and Glatzmaier, 2000).

The interior of the Earth is hot mostly owing to heat leftover from the violent impacts that formed the Earth, and in part because of the radioactive decay of heavy elements. The temperature, which tends to melt materials, increases towards the centre of the Earth, but pressure also increases, and pressure tends to maintain materials in a solid state. At the Earth's centre, pressure wins. The *inner core* of the Earth is a solid ball believed to be composed of iron and nickel, but the *outer core*, extending from a radius of about 1220 km to about 3480 km, is a fluid, composed of liquid iron and nickel. The core is so hot that this molten metal has a low viscosity and can flow relatively rapidly. Incidentally, the properties of fluid waves permitted this diagnosis of the Earth's internal structure. Since a fluid deforms indefinitely in response to shear stress (§1.2.1), in the absence of an interface (such as the interfaces between the sea and atmosphere considered in chapter 2 or the interfaces between different fluids considered in chapter 4), shear waves (or *s-waves*), where the restoring force is shear elasticity, can pass through a solid, but not through a fluid. Meanwhile longitudinal waves (or *p-waves*) - effectively, sound waves - can pass through a fluid as well as a solid. Seismic waves of both s- and p-type, generated by earthquakes and tracked across the globe, revealed the Earth's fluid outer core.

11.2.4 Geodynamo mechanisms

As noted earlier, the Earth's core is far too hot for permanent magnetism, and the movement of the magnetic poles is further evidence for a dynamo. In 1919, Larmor not only pointed out that permanent magnetism was not possible, as noted earlier; he recognised that flow of an electrically conducting fluid is required to generate a magnetic field, a process is called the *geodynamo* (Hollerbach, 1995). Estimates of the energy lost through electrical resistance and viscosity show that any geodynamo capable of generating the observed magnetic field requires a sustained energy source; if there were no energy source, fluid flows, electric currents and the resulting magnetic field would have disappeared long ago in Earth's history (Roberts and Glatzmaier, 2000).

Most opinions are that the energy source sustaining the geodynamo is the heat in the core, driving convection currents (Jones, 2011), although, as will be detailed shortly, there are dissenting opinions (Tilgner, 2007; Andrault et al., 2016). Certainly, heat from the Earth's interior is slowly escaping to the surface and thence to space, providing a thermal gradient that should drive convection currents in the fluid outer core. However, the geometry of the flows created by thermal convection *alone* does not seem consistent with the simplest models one might imagine for a geodynamo (Hollerbach, 1995). That is because convective flows must be predominantly radial in direction, in order to carry heat to the surface. A magnetic field with the observed dipolar structure would not be generated by radial convective flows alone. To create the required electric currents and dipolar magnetic field, a fluid flow is required that is always at right angles to the axis of rotation, or *geostrophic*

(Hollerbach, 1995). However, convection alone could not create this geostrophic flow. Instead, rotation, coupled with convection, is invoked; the Coriolis force, as noted in §5.3.2, deflects radial motions. This is thought to result in convection currents spiralling along cylindrical surfaces with axes parallel to the Earth's rotation axis (Jones, 2011); and these spiralling flows, in turn, could create the required electric currents and the observed dipolar field. Columnar flows parallel to the rotation axis are Taylor columns mentioned in §5.3.2 that are a direct consequence of the Taylor-Proudman theorem arising from (5.4).

Indeed, numerical simulations in which *all* the physics is carefully included, comprising viscous fluid flow in a rotating reference frame, and electric and magnetic fields, does simulate 'Taylor-column'-like flows that carry heat to the surface along spiral paths. The results predict a predominately dipolar magnetic field, and even its periodic reversals (Roberts and Glatzmaier, 2000). However, limitations to computational power mean that key parameters in the simulations are much larger than in any planetary system (Jones, 2011), so it is not certain if the flows observed in simulations exist. A further concern with the convection-dominated geodynamo is that the heat flux estimated over the Earth's history may be insufficient to explain the geodynamo (Andrault et al., 2016).

Moreover, the question remains of what mechanisms are both necessary and sufficient. The answer to this question may enable predictions of the next, potentially hazardous reversal of the Earth's magnetic field. Possibly of even greater significance is the understanding of what features of exoplanets might be supportive of life on these worlds.

If convection were not the primary driver of the magnetic field, what would be? It has long been recognised, based on order-of-magnitude estimates, that *precession* of the planet's core relative to its mantle, in which the rotation axis itself oscillates over time, has the potential to drive a dynamo (Bullard, 1949). Malkus (1968) suggested that precession of the Earth creates a fluid flow in the outer core that drives the geodynamo, and significant research continues to explore this theory (e.g. Vanyo, 1991; Giesecke et al., 2019). The precession by itself would not create the required geostrophic flow, so the idea was that precession-driven flows would somehow break down to create a turbulent geostrophic flow. And inertial waves are a precession-driven flow that is known to break down (Manasseh, 1992).

Inertial waves in the Earth's outer core were identified in seismic data by Aldridge and Lumb (1987). Tidal flows in the outer core, driven by the Moon, could also contribute (Andrault et al., 2016). These flows in the fluid outer core, be they tidal or inertial waves, can be predicted using linear theory (as in §5.3). Critically, however, being wavelike, tidal oscillations and inertial waves also cannot directly generate the required geostrophic flow, because, like all wavelike flows, they are constantly reversing. To generate a steady flow from waves, nonlinear phenomena (such as those outlined in §6.4) are required, and nonlinear phenomena were originally noted by Malkus (1968) to generate turbulence in systems dominated by rotation-driven waves. Once there is turbulence, it is supposed there could be a rapid transfer of energy to the geostrophic flow required by the geodynamo, specifically, a circulation in the azimuthal direction. This is an energy-based argument: turbulence causes a rapid dissipation of energy. Since that energy is being removed from the rotational kinetic energy of the fluid, the energy loss must be represented by a steady flow in the opposite direction to the planet's rotation.

At this point, we will leave the discussion of the geodynamo, and by extension, the question of the origin of the magnetic fields of other planets and celestial bodies. That will require a deeper examination of the processes by which a steady current could be generated by inertial waves, which will be detailed in §13.4.2.

11.3 Engineering of rotating spacecraft

11.3.1 Rotation for attitude control

Rotating spacecraft containing large mass-fractions of liquid fuels are subject to stability issues owing to inertial waves (Gerrits and Veldman, 2003), a problem originally noticed for military projectiles such as artillery shells containing liquids (Rogers et al., 2013). The problem became acute in the 1980s as larger communications satellites were built that were *spin-stabilised* (Hill, 1985; Bao and Pascal, 1997).

Spin-stabilisation utilises the *gyroscopic* effect as part of the *attitude control* strategy that ensures the spacecraft points in the correct direction. Early communications satellites, such as Telstar 1, launched in 1962 and still in orbit today, were spin-stabilised throughout their intended mission. Interplanetary probes such as the Pioneer 10 and Pioneer 11 probes launched in 1972 and 1973 respectively, and now on interstellar journeys, are also spin-stabilised. The complexities of orienting an antenna while keeping the body of the satellite spinning, and the availability of greater computer power to manage control systems, caused a gradual shift to three-axis stabilisation, which requires significant amounts of liquid fuel to be held on board the spacecraft throughout its operational life. Nevertheless, spin stabilisation was required during the launch process. Once the satellite has been released into low Earth orbit at an altitude of several hundred kilometres, which is the limit of travel of most rockets and reusable launch vehicles, a transfer orbit is required to deliver it to *geostationary orbit* at an altitude of about 35 800 km. Prior to the firing of the rocket injecting the spacecraft into the transfer orbit, the spacecraft was spun to achieve gyroscopic stability. Experiments showed that spacecraft with full fuel tanks became unstable; the spin axis began to *precess*, destroying the benefit of spin-stabilisation and risking the loss of the spacecraft (Pocha,

1987). The empirically found solution was to fit baffle plates in the tanks, eventually shown to be plates that disrupt the lowest-order mode of inertial waves (Manasseh, 1993). Thus, the resonance of inertial waves was speculated to be responsible for destabilising the entire spacecraft.

The *sloshing* of liquids in spacecraft continues to cause stability issues (Veldman et al., 2007; Hahn et al., 2018), which can become very complex when the fuel tanks are partially drained, as eventually, they must be, and when the absence of significant gravity means that other forces can dominate. Furthermore, even without spin stabilisation, spacecraft rotation is required during orbital manoeuvres (Xu et al., 2019).

11.3.2 Rotation for artificial gravity

If humans are to undertake lengthy space voyages, 'Artificial Gravity' (AG) of some sort is required to prevent unacceptable health problems arising from a very prolonged or permanent experience of gravity much less than that of Earth (Hall, 1999; Chen et al., 2020). The most commonly proposed form of AG is that due to rotation, in which part of the spaceship or all of it is subject to rotation. The concept of artificial gravity via rotation has been attributed to Konstantin Tsiolkovsky in 1920 (Braddock, 2017) and has been subjected to very many studies and plans over the past century. Rotational AG systems have been tested on mice in space with promising results (e.g. Shiba et al., 2017). To design an AG system, the simple equating of centripetal acceleration and the desired acceleration due to gravity, which here we will call g_a, gives,

$$\Omega^2 r = g_a, \tag{11.1}$$

where Ω is the rotation rate as before and where r is the radius. For example, to match the acceleration due to gravity on Mars, which is about 38% of that on Earth so that $g_a = 3.71$ m s^{-2}, given a value of Ω corresponding to 2 rotations per minute, i.e. $\Omega = 2\pi \times 2/60$ Hz, the habitable parts of the spaceship must be located at a radius of 85 m from the rotation axis, demanding a very large structure. In some designs (e.g. Rousek, 2010, figure 5.1) a considerable radius is achieved not by having an impractically large spaceship but by having the habitation module and a module of equal mass located at either end of a long cable or truss, with propulsion modules on the axis of rotation. Clearly, increasing Ω would permit a more compact vehicle; Taraba et al. (2006) propose a Mars-mission habitation radius of 30 m for a rotation rate of 3.8 revolutions per minute. However, at higher values of Ω, the Coriolis force will begin to cause complex issues that are still poorly understood. For example, Gupta et al. (2004) found that thermal convection under AG was fundamentally different from those under terrestrial gravity. These differences in convection due to rotation are a microcosm of the physics discussed in §11.2.3 in the context of the flows occurring inside planets. At higher rotation rates, human-health issues emerge, such as nausea due to the Coriolis force (Sandler, 1995). Since all of the fluid systems subjected to AG have boundaries completely enclosing the fluid, inertia-wave modes are possible according to the theory derived in §5.3.4. The relatively low rotation rates found compatible with human health means that the criterion for inertial waves to exist, (5.12), reproduced here as

$$|\omega| < 2\Omega, \tag{11.2}$$

is most likely to be satisfied by mechanical oscillations ω generated at the rotation rate, i.e. $\omega = \Omega$. For example, a central, non-rotating hub in a rotating air-filled habitat has the potential to generate inertial waves.

12

Nonlinear environmental waves

12.1 Summary of key points

- **Rogue ocean waves** may occur owing to a nonlinear mechanism, modulational instability (§12.2).
- **Tsunamis** set the entire ocean depth into motion and can be modelled as solitary waves (solitons), in which nonlinearity and dispersion balance (§12.3).
- **Atmospheric solitons** occur over large scales (§12.4.1).
- **Gravity currents** (§12.4.2) created by thunderstorms may impact a stratified layer, creating solitons that may be damaging (§12.4.3). The speed of the nose of a gravity current is given by (12.7) on page 260,

  $$\boxed{u_C = C\sqrt{g'_c h}},$$

  where C is an empirical constant between 1 and $\sqrt{2}$, h is the depth of the current and

  $$g'_c = \frac{\rho_C - \rho_0}{\rho_C}\mathrm{g},$$

  with ρ_C being the density of the fluid in the current, ρ_0 being the ambient-fluid density and g the acceleration due to gravity.
- **Oceanic internal solitons** may be important to oceanic mixing (§12.4.4).
- Useful **books** include Whitham (1999) on the fundamental theory, while Simpson (1997) covers gravity currents.

12.2 Rogue waves

As long as mariners have put to sea, there have been anecdotes of individual 'freak' or 'rogue' waves reaching enormous heights that would not be possible if

DOI: 10.1201/9780429295263-12

FIGURE 12.1
The beach at Mahabalipuram, Tamil Nadu, India. On 26 December 2004, a magnitude 9 earthquake in Sumatra, over 1,800 km away, created a tsunami that caused an estimated 5 m-high run-up on this coast (run-up is defined in §7.3); over 8000 people were killed in the state of Tamil Nadu alone (Sheth et al., 2006). Photograph by Richard Manasseh.

their heights were limited by the Stokes limiting steepness, (10.6), $k\hat{\eta} = 0.444$, where, as usual, k is the wavenumber, i.e. $k = 2\pi/\lambda$ where λ is the wavelength, and $\hat{\eta}$ is the surface elevation amplitude due to the waves. A brief search for 'rogue wave' will uncover dramatic images and video recordings, mostly, unfortunately, without reliable estimates of wavelength or height. Benjamin and Feir (1967) showed that a nonlinear mechanism, *modulational instability*, can result in waves transferring energy to 'side-bands' differing slightly in frequency from the primary wave frequency. Modulational instability has been implicated in the development of laboratory extreme waves (Onorato et al., 2009; Toffoli et al., 2017).

12.3 The tsunami: an ocean-surface soliton

Cultures on earthquake-prone coasts have experienced large, rare waves that are completely unlike any waves due to ocean swell or even the most severe of storms. The oldest continuous record-keeping of such events is in Japan. Fishermen working far out to sea might have felt only an unusual and gradual rise and fall in sea level over a minute or so, but on returning home to port, they discovered devastation that survivors reported was due to an enormous wave or set of waves. Since the wave had not been recognised out to sea, it may have seemed as if the wave had arisen in the harbour only. This is one speculation on the origin of the term created by combining the Japanese words for 'harbour' (*tsu*) and 'wave' (*nami*); the term has been in use since 1611, although records of earthquake-generated waves date to 684 in Japan and 47 BCE in China (Cartwright and Nakamura, 2008). Records from the 17th century indicate that inhabitants of the Japanese archipelago already understood the association between earthquakes and tsunamis, since they ran for higher ground on feeling an earthquake (Cartwright and Nakamura, 2008).

Earthquakes result from the movement of the tectonic plates of the Earth's crust (the internal structure of the Earth was described in §11.2.3), or from volcanic activity, but it is tectonic-plate earthquakes that cause the most frequent and devastating tsunamis. A sudden vertical displacement of a large region of the seabed may occur over a region many kilometres or tens of kilometres in length. The earthquake thus disturbs the bottom of the sea: the *opposite* boundary to winds blowing at the surface that create conventional water-surface waves. Thus, if it is to create a surface wave at all, the earthquake-generated disturbance must be large enough to influence the entire depth, from the bottom to the surface. Therefore, the shallow-water approximation discussed in §2.3.11 must be more appropriate than even the full solution, since the full solution still has the maximum displacement at the surface, not the bottom. Even if the seabed is the deep ocean floor, four kilometres down, if the disturbance is to affect the surface at all, the resulting wave must, by definition, be a shallow-water wave. As far as the tsunami is concerned, even the deep ocean is shallow.

We now know that tsunamis can be described as solitary waves, where nonlinearity and wave dispersion balance in shallow-water waves, allowing the wave to travel over long distances unchanged in shape. Tsunamis have been modelled by soliton solutions to the weakly-nonlinear, weakly-dispersive equations for shallow-water surface waves, such as the Korteweg-de Vries equation, (6.41) derived in §6.5.1 and repeated here as

$$\frac{\partial \mathsf{u}}{\partial \mathsf{t}} - 6\mathsf{u}\frac{\partial \mathsf{u}}{\partial \mathsf{x}} + \frac{\partial^3 \mathsf{u}}{\partial \mathsf{x}^3} = 0, \tag{12.1}$$

where $\mathsf{u} = u/c_0 - 1$, u being the horizontal component of water velocity, c_0 is the linear shallow-water speed given by $\sqrt{gh}$ where h is the water depth, and the independent variables are related to the time t and horizontal distance x by $t = (6h/c_0)\mathsf{t}$ and $x = h\mathsf{x}$. It is remarkable that a simple analytic solution to (12.1) exists, (6.42), reproduced here as

$$\mathsf{u} = a_{\mathsf{u}} \operatorname{sech}^2\left(\frac{\mathsf{x}}{\mathcal{L}}\right), \tag{12.2}$$

where the wavelength $\mathcal{L}$, unlike its linear equivalent, depends on the wave amplitude a_{u}, according to (6.43), repeated here as

$$\mathcal{L}^2 = \frac{B}{a_{\mathsf{u}}}, \tag{12.3}$$

where B is a parameter that depends on the initial conditions. A graph of this solution was shown in figure 6.3. Depending on the nature of the initial seabed disturbance, the leading part of the tsunami may be a trough or a crest, and soliton solutions can have several crests and troughs travelling together, with the first crest not necessarily the largest. Similarly to linear shallow-water waves, recalling (2.84) in §2.3.12, as the tsunami propagates into shallower water such as continental-shelf waters - and, eventually, into a harbour or onto a beach - conservation of mass requires that the wave becomes higher and higher as the water becomes shallower and shallower. While that occurs for wind-driven waves too, with the tsunami, the entire ocean depth, rather than just the surface waters, has been set into motion. Thus the mass of water set into motion is immense, resulting in waves that may reach disastrous heights as the tsunami reaches shore. Should a trough precede the crest, the sea may appear to recede for a great distance, unusually revealing the seabed.

Meanwhile, out at sea, the tsunami's height may be comparatively modest. The very large wavelength also makes the steepness gentle out at sea, so that the passage of a tsunami out at sea may not even be noticed by the crew of a vessel. However, the very large wavelength may also create a flood penetrating many kilometres inland. This may bear a superficial resemblance to the incoming tide in some parts of the world where tides travel as a 'bore' with a sharp front, leading to the misleading term 'tidal wave' formerly used in the English-speaking world. Astronomical tides, however, are routine and almost perfectly predictable, whereas the earthquakes that generate tsunamis remain unpredictable.

One of the most severe tsunami situations is thought to be created by the Cascadia fault off the Pacific North-West Coast of North America. The last magnitude-nine Cascadia earthquake occurred on 26 January 1700 and the resulting tsunami travelled all the way across the Pacific to Japan (Satake, 2003). On 26 December 2004, a magnitude-nine earthquake off Sumatra made a tsunami that travelled across the Indian Ocean from Sumatra to India and Sri Lanka (figure 12.1), killing an estimated 230 000 people (Suppasri et al., 2012).

12.4 Internal solitons

12.4.1 Mesoscale atmospheric solitons

Under some circumstances, the part of the atmosphere overlying the ground could be modelled by the shallow-water approximation of §2.3.11, which, in deference to meteorologists and atmospheric physicists, is best called the *long-wave approximation* rather than referring to the watery origin of the surface-wave theory. The shallow-water approximation is really an assumption that the wavelength is long relative to the depth of the fluid layer set into motion.

It is quite common to observe clouds that are aligned in periodically-spaced patterns. In general, clouds form when moist air is lifted to a sufficient altitude that the moisture condenses to form ice crystals or water droplets.

(a) (b)

FIGURE 12.2
Periodic arrangements of clouds (a) seen from above (tropical North Pacific) and (b) seen from below (Port Phillip Bay, Australia). Periodic arrangements of clouds may be caused by convection cells, alone, or by shear-layer instabilities alone, rather than internal gravity waves; or they may be caused by these phenomena in combination with internal gravity waves. Photographs by Richard Manasseh.

It is worth noting in passing that a precondition for cloud formation is that there should be some nuclei 'seeding' the condensation, which might come from atmospheric dust or aerosols from ocean wave breaking (as discussed in §10.3.3).

Thus, periodically-spaced rows of clouds can represent phenomena where a lower layer of air is subject to some wavelike disturbance, and the amplitude of the waves and the altitude of the layer is such that the crests of the disturbance are above the altitude where clouds form. It is possible that waves propagating horizontally on the interface between two layers in the atmosphere (as derived in §4.3.2) are responsible for these observations, but in a continuously stratified atmosphere, waves could be propagating with a vertical component as well as horizontally (as derived in §4.3.3) and what is observed could be the region where the waves happen to pass through the altitude of cloud formation.

It is also possible that the periodic disturbances causing clouds are not due to stratification, but are partially or wholly caused by *shear-layer instabilities* due to differences in wind speed across a narrow layer. A common version of a shear-layer instability is the *Kelvin-Helmholtz* instability, which creates a series of periodically-spaced billows (Drazin and Reid, 1981; Sutherland, 2010). It is also possible that atmospheric convection cells could form periodic patterns that could also cause these phenomena. Some examples are shown in figure 12.2.

While clouds in rows are often observed, somewhat more rare is the observation of a large, spectacular *roll cloud*, and sometimes a set of parallel

FIGURE 12.3
'Morning Glory' internal solitary wave train near Burketown, northern Australia. Note the 'skirt' or 'apron' of the cloud appears to delineate a $\mathrm{sech}^2 x$ profile as predicted by solutions to the KdV equation, (12.2) and shown in figure 6.3, although in practice other equations also admitting soliton solutions are more appropriate (Christie, 1989). Photograph copyright by and courtesy of Michael Zupanc, mike@zupy.net.

roll clouds. They appear to propagate across the sky almost as if one were sitting on the seabed, looking up and observing a large shallow-water wave on the sea surface, with its crest delineated by a whitecap. Unlike any common row clouds that might be caused by waves, these clouds are due to nonlinear waves. An approach can be taken similar to that in §6.5, in which nonlinear wave-steepening balances the slight dispersion that occurs when waves are in reality not infinitely long compared to the layer depth. This leads to a prediction of solitary waves in the atmosphere, or indeed any stratified fluid with a large extent above the stratified zone, like the atmosphere, or a large extent below the stratified zone, like the deep ocean (Benjamin, 1967). These can be thought of as a sort of atmospheric analogue of the tsunami. Although they need not be as damaging in general as tsunamis, in §12.4.3 below the potential danger atmospheric solitary waves may pose is discussed.

The best known of the roll-cloud phenomena is the spectacular 'Morning Glory' of northern Australia (figure 12.3), which typically travels at 10 m s^{-1} (Clarke et al., 1981). The 'Morning Glory' can be modelled using the long-wave theory that leads to solitary-wave solutions (Christie, 1989; Rottman and Einaudi, 1993; Porter and Smyth, 2002), as detailed in §6.5.

Similar spectacular roll clouds have been analysed in North America (Coleman et al., 2009) and cloud lines elsewhere have also been attributed to solitary waves (e.g. Birch et al., 2014). These are *mesoscale* atmospheric phenomena, i.e. flows on scales from several kilometres up to several hundred kilometres and thus larger than the flows due to individual storms, but smaller than the *synoptic-scale* high and low-pressure systems that control the broad features of the weather (the synoptic-scale was introduced in §5.3.3). The 'Morning Glory' is thought to be due to a sea breeze generated on the previous evening that travels from the Coral Sea over Cape York towards the west (Christie, 1989); a sea breeze travelling from the east may collide with this westwards breeze, creating a large disturbance in a stratified layer over the Gulf of Carpentaria (Goler and Reeder, 2004).

While rare, these roll clouds offer an opportunity to understand atmospheric solitary-wave phenomena that are otherwise invisible. The sea breezes just mentioned in the context of the 'Morning Glory' are a typical initiator of large atmospheric gravity waves; more generally, sea breezes are *gravity currents* (Simpson, 1994). In order to understand nonlinear waves created by gravity currents, a brief explanation of gravity currents is required.

12.4.2 Gravity currents

In this sub-section §12.4.2, we will digress from the theme of fluid waves to introduce gravity currents; although these are definitely first-order currents and not waves, they can have a close connection with internal waves and internal solitary waves in particular, as we will see in §12.4.3 below. A thorough exposition of gravity currents is given by Simpson (1997).

The genesis of a gravity current can be understood by a 'dam-break' thought experiment. A reservoir of dense fluid of density ρ_C is very large in the horizontal (x) direction. It is of depth h and is held behind a wall. Imagine that elsewhere is light fluid (such as atmospheric air) of density ρ_0. The wall disappears, and we assume that the bottom of the dense fluid at the base of the just-vanished wall floods outwards at a constant speed u_C. Equating the initial potential energy to the kinetic energy of the current gives

$$u_C = \sqrt{2g'_c h}, \tag{12.4}$$

where

$$g'_c = \frac{\rho_C - \rho_0}{\rho_C}\mathrm{g} \tag{12.5}$$

is a reduced gravity similar to the g' defined in (4.7). It might be noticed that u_C given by (12.4) has a similar form to the linear shallow-water wave speed, c_0, given by (2.81), reproduced here as

$$c_0 = \sqrt{g'h}, \tag{12.6}$$

for waves on the interface between the denser and lighter fluid. If $\rho_C \gg \rho_0$, as in the case of water and air, $g'_c \simeq g' \simeq \mathrm{g}$ and (12.6) becomes the same as (2.81).

Since $u_C > c_0$ irrespective of the relative values of the fluid densities, the dense fluid always rushes horizontally faster than any waves can travel on its surface. The dam-break results in a supercritical flow: that is a gravity current. It is definitely a current and not a wave, since it transports mass. In practice, the gravity current suffers considerable turbulent dissipation within it, as well as friction from the ground over which it flows. A more realistic form of (12.4) that recognises the highly dissipative nature of a gravity current is

$$\boxed{u_C = C\sqrt{g_c' h}}, \tag{12.7}$$

where C is some empirical constant that depends on many factors such as the roughness of the surface over which the current flows. It is clear that C lies between 1 and $\sqrt{2}$; if C were less than 1, shallow-water waves emanating from the reservoir would outpace the current's 'nose', break at the nose and by projecting fluid beyond the nose, effectively, increase its speed. Meanwhile, as mentioned above, $C = \sqrt{2}$ represents the unrealistic case where the current is propagating without any turbulent or frictional losses. Another version of (12.7) can be derived in which the dense fluid spreads out radially, rather than in a two-dimensional Cartesian system as above; for a radial gravity current, conservation of mass demands that h must be a function of radius r.

While we digressed into gravity currents in order to seek an explanation for what might drive nonlinear gravity waves, it is worth noting that gravity currents are of great importance in geophysical and environmental fluid dynamics, and in many aspects of engineering such as fire and industrial safety (more details are in Simpson, 1997). Many gases used in industry are denser than air, and their accidental release from storage tanks would create invisible and possibly hazardous gravity currents. The 'dense' fluid need not be more dense than its surroundings and thus travelling on top of the ground; it could be less dense and travelling under a ceiling or roof; indeed this may happen when hot air from a fire reaches a ceiling, providing a dangerous new path for fire propagation inside a building.

A particular form of gravity current occurs when the dense fluid obtains its added density from suspended particles, or where suspended particles add to the density of an already denser fluid. These could be dust in the case of dust storms, smoke from forest fires, snow in the case of avalanches, or sediment in the case of undersea landslides. Such particle-driven currents are more properly called *turbidity currents*, which exhibit complex and interesting phenomena. For example, a turbidity current propagating over a bed of similar particles loses turbulent energy to that surface, just as all gravity currents lose energy to friction with the surface over which they travel. However, the turbulence can re-suspend the particles in the bed, locally increasing the density of the current and endowing it with greater longevity; effectively, some of the kinetic energy transferred to turbulence is transferred back into potential energy rather than being lost immediately.

12.4.3 Thunderstorm solitons and aviation

As mentioned above, sea breezes are gravity currents that could initiate large nonlinear waves in the atmosphere such as the 'Morning Glory' roll-cloud phenomena. While these are large phenomena occurring over the mesoscale, they are not particularly intense or hazardous. Before outlining the actual mechanism of wave generation, another gravity-current phenomenon should be mentioned. These are gravity currents created by thunderstorms, and they could be very damaging. In order to understand how thunderstorms could create gravity currents, and thence waves, a brief explanation of thunderstorms is necessary.

A thunderstorm is an extraordinarily complex fluid-dynamical and thermodynamical system, involving three-dimensional, turbulent flows of air, moisture and heat, as well as potentially all of the phase transitions between water vapour, water and ice, all occurring within relatively small scale of a few tens of cubic kilometres. Enormous though such a scale is compared to a person, and aircraft or even a city, thunderstorms are nevertheless classified as *microscale weather* systems, since they are small compared to the mesoscale and very small compared to the synoptic scale of the atmosphere's circulation. While the detailed behaviour of individual thunderstorms is chaotic and unpredictable, the basic principles of thunderstorm evolution are well known and detailed in a number of textbooks (e.g. Kessler, 1982; Cotton and Anthes, 1992).

Intense vorticity between the thunderstorm updraft and downdraft, and turbulent entrainment, can suddenly draw in a large mass of mid-altitude air external to the storm. This air has lower humidity than the air inside the storm and evaporates the precipitation inside. Evaporation cools the air suddenly, making it dense. The result may be an enormous mass of air, a kilometre or so across, that 'falls out' of the thunderstorm. This is sometimes called a *downburst*, or, consistently with the labelling of a thunderstorm as a microscale process (and somewhat belying its enormity relative to human scale), it can be called a *microburst* (Fujita, 1990). Since it is generated by the evaporation of precipitation, the microburst air itself may be invisible, and hence is called a dry microburst. Occasionally, it drags some precipitation with it, becoming visible, when it is called a wet microburst.

The flow structure of a microburst is that of any *thermal*, which is a finite volume of fluid of different density that is released suddenly. A thermal forms a *ring vortex* which if made visible by smoke in air or dye in water forms a 'mushroom cloud' (Linden and Simpson, 1985). Although mushroom clouds are commonly associated with large explosions in the atmosphere, all thermals are characterised by the ring-vortex and can appear on any scale, for example, if a dense ink drop is released in water. When the thermal fluid is denser than its surrounds, as in the case of the microburst, the ring vortex hits the ground and expands radially outwards. Occasionally, wet microbursts are captured on video revealing the ring-vortex structure, but the dry microburst remains invisible. The dense fluid travels outwards as a gravity current.

The ring vortex maybe hundreds of metres or a kilometre or more high, and an aircraft landing or taking off across a microburst may be in danger (Fujita, 1986). The aircraft may be accelerated in its direction of travel initially, causing the pilot to throttle back to maintain the correct ground speed. When the aircraft enters the centre of the ring vortex, the horizontal component of the wind suddenly drops to zero. Simultaneously, the wind becomes vertically downwards, which is a very dangerous condition.

Thunderstorm-generated disturbances, sometimes called 'gust fronts', have been recorded far from the storm itself (Doviak and Ge, 1984), and stable stratification of the atmosphere has been implicated in this phenomenon (Doviak et al., 1991). If the atmosphere is stably stratified, it may be stratified into an approximately two-layer system, as analysed in §4.3.2, in which cool air lies above the ground with warmer air above an 'interface' that may be a kilometre or two above the ground. Circumstances like this are usually temperature inversions, (mentioned in §3.3.4 as causing sound wave trapping), but they can be a waveguide trapping internal gravity waves too. Furthermore, a stable stratification, which might occur if a cloudless, cool night follows a warm day, may assist the initiation of thunderstorms by 'trapping' and thus concentrating warm humid air close to the ground.

When the gravity current from the microburst (or a similar downburst) travels along the ground, it pushes into the stable layer, creating a large disturbance in the layer. Since the layer can support wave propagation, a variety of possible outcomes could follow. For the microburst air, g'_c is likely to be different to g', but owing to the extreme nature of the thunderstorm, it is likely that, at least initially, $g'_c > g'$, and since, as noted above, $u_C > c_0$ even if $g'_c = g'$, the microburst fluid will initially propagate supercritically into the stable layer. However, at some point, dissipation and radial spreading will cause u_C to drop, such that $u_C \simeq c_0(1 + \mathsf{u})$ where u is given by the solution to a nonlinear solitary-wave equation, such as the solution (6.42).

The 'head' of the gravity current, which may still be a rotating vortex from the microburst ring-vortex, then detaches and travels as a trapped mass of air contained within the crest of a solitary wave (Manasseh et al., 1998; Lamb, 2002). At this point, the original concept of a wave has broken down. Waves may not transport mass, but a sufficiently nonlinear solitary wave generated in the manner outlined above may indeed transport mass. It is not truly a wave, but because the 'head' has outrun the body of the microburst fluid, it is not a current either.

12.4.4 Oceanic internal solitons

Internal solitons transporting mass, sometimes described as waves with 'trapped cores', are found in the ocean as well as the atmosphere (Lamb, 2002; King et al., 2011), and are thought to be important for oceanic mixing, particularly since they can transport mass. The transport of mass within the ocean is important for planktonic ocean organisms such as those forming calcium-carbonate cell-walls or shells, which, as detailed in §10.3.5, are intimately connected to the capture of carbon dioxide from the Earth's atmosphere and hence regulation of the Earth's climate. The mass-transporting ability of nonlinear internal solitary waves is thought to be significant in transporting the larvae of plankton towards the shore where they can settle and reproduce (Scotti and Pineda, 2007). Just as gravity currents propagating over a bed of particles may resuspend these particles, internal solitary waves may also resuspend particles in the ocean (Boegman and Stastna, 2019).

Internal waves, whether they are solitary waves or not, can break. Just as there are empirical criteria for surface-wave breaking, there are empirical criteria for internal wave breaking, and for internal-wave breaking on a sloping seabed, the criteria are based on a version of the Iribarren number introduced in §10.2.1, Ir, given by (10.8), here reproduced as

$$\mathrm{Ir} = \frac{\tan\alpha}{\sqrt{\hat{H}/\lambda}}, \tag{12.8}$$

where as before α is the slope of the seabed, $\hat{H}$ is the wave height and λ is the wavelength.

13

Streaming in medicine, industry and geophysics

13.1 Summary of key points

- The **acoustic streaming speed** is proportional to the square of the sound-wave particle-velocity amplitude (and therefore proportional to the power of the applied sound), and is proportional to the gradient in the amplitude with distance; it has medical-ultrasound uses (§13.2).
- **Acoustic (or cavitation) microstreaming** occurs where there are large sound-field gradients over small distances (§13.3.1); like acoustic streaming, its speed is proportional to the square of the local particle-velocity amplitude.
- Owing to bubble resonant amplification, the local particle-velocity amplitude in acoustic microstreaming may be much higher than the sound-wave amplitude.
- **Inner vortices in acoustic microstreaming** have a thickness δ_ω that scales with the Stokes-layer thickness, given by (13.1) on page 268,

$$\delta_\omega = \sqrt{2\frac{\nu}{\omega}},$$

  where $\nu = \mu/\rho_0$, where μ is the dynamic viscosity and ρ_0 the density, and ω is the radian frequency.
- Microstreaming has been implicated in **medical therapies** using ultrasound (§13.3.2).
- Streaming in rotating flows should be proportional to the square of the inertial wave amplitude. However, in a precessing rotating fluid, it is possible for the streaming to be directly proportional to the wave amplitude (§13.4.2).
- A useful **reference** for streaming flows is the review by Riley (2001).

DOI: 10.1201/9780429295263-13

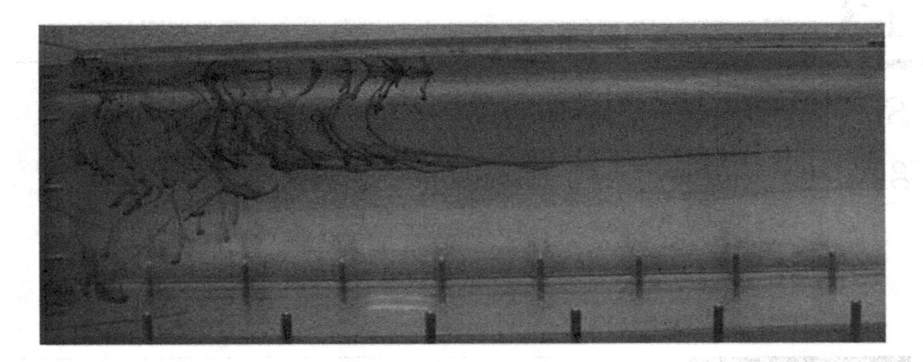

FIGURE 13.1
Acoustic streaming in water. A 2.25 MHz ultrasound transducer at the left-hand edge of the image, driven by a continuous-wave voltage of 20 V peak-to-peak, creates a 'jet' owing to acoustic streaming, visualised by drops of dye falling through the jet. The speed of the jet is about 0.02 m s^{-1}. Image courtesy of Pauline Lai; further details of this experiment are in Lai (2008).

13.2 Acoustic streaming in medicine

Acoustic streaming was discovered in the 1920s when ultrasonic transducers were developed from quartz crystals, giving sufficient power to cause what was termed the 'quartz wind'. Acoustic streaming is a phenomenon in which sound waves drive a mean flow, owing to the quadratic nonlinearity in the momentum equation, as outlined in §7.3. Basic principles of acoustic streaming are outlined in Manasseh (2016) and theoretical derivations are reviewed in Wu (2018). Acoustic streaming, like any nonlinear streaming, is significant whenever the nonlinearity in the momentum equation is large enough when averaged in time and space. Given that this term is $\boldsymbol{u} \cdot \boldsymbol{\nabla} \boldsymbol{u}$, there are two ways in which this term could become large. The first way is that $|\boldsymbol{u}|$ could be large, which means the sound power is high. The second way is that $|\boldsymbol{\nabla} \boldsymbol{u}|$ is large, which will be discussed in §13.3 below. Even if $|\boldsymbol{\nabla} \boldsymbol{u}|$ is not particularly large, it must still be nonzero, otherwise the entire nonlinear term would be zero. That is why, as explained in §7.3, something must cause a gradient in the sound power with distance, and most often that is viscosity. The streaming speed is proportional to the square of the wave amplitude, and proportional to the gradient in the wave amplitude with distance.

The situation in an effectively-unbounded fluid was analysed by Eckart (1948), who noticed that viscous stress was necessary to balance the radiation stress; unbounded acoustic streaming is now called *Eckart streaming.* An example is shown in figure 13.1; Eckart streaming takes the form of a jet transporting fluid in the direction the sound is propagating. However, acoustic streaming had been analysed much earlier, by Rayleigh (1884) in the context of sound waves between parallel plates; in this case, the streaming pattern, rather than a jet, forms cells between the plates in order to satisfy the boundary conditions of the confined geometry. This is called *Rayleigh streaming*. A further type of streaming arises in the fluid-dynamical boundary layer, causing a vortex in the boundary layer. This was analysed by Schlichting in 1932 (Boluriaan and Morris, 2003) and is called *Schlichting streaming*.

Ultrasound scanners have the ability to generate acoustic Eckart streaming, since all the required criteria of a gradient in the sound-wave amplitude and viscous dissipation exist in the human body. However, this is not usually noticed. Firstly, solid tissue, not unbounded fluid, makes up the bulk of the body. Secondly, if fluids are scanned, the fluids are usually blood that is flowing anyway, and the scanner is often set up in Doppler mode (the Doppler shift was derived in §3.3.10) where measurements are made of the speed of blood flow. Inevitably, the direction of sound propagation and thus the direction in which streaming would be generated is not the same as the direction of blood flow, and moreover, the speed of blood flow is usually much faster than the streaming speed.

Nevertheless, there are ultrasound scans of fluid-filled cavities in the human body in which there is no background flow. These are typically cysts, which are 'lesions' or abnormalities in the body. They may be asymptomatic, or benign yet painful, and in some cases may be malignant. While the 'gold standard' is a biopsy, in which tissue is removed for histological examination, clinicians use ultrasound to gain confidence on a diagnosis. An ideal development would render the invasive process of a biopsy unnecessary.

Betheras (1990) suggested that the observation of acoustic streaming inside a cyst could be a useful diagnostic criterion, since fluid-filled cysts may be less likely to be malignant (Edwards et al., 2003). More details are in Manasseh (2016). The acoustic-streaming criterion was applied to endometriomas (Clarke et al., 2004), as well as to lesions in breast tissue (Nightingale et al., 1999). However, larger clinical studies found that acoustic streaming could not discriminate between types of fluid-filled cysts (Van Holsbeke et al., 2010). More research is needed before ultrasound, in general, can replace surgical investigations for female reproductive-system lesions (Nisenblat et al., 2016).

The term $\boldsymbol{u} \cdot \boldsymbol{\nabla} \boldsymbol{u}$ exists in the momentum equation of solid mechanics as well as that of fluid mechanics, and thus the acoustic radiation force can cause even solid tissue to move, as long as it is elastic, which all parts of the human body are to some extent. Since the early 2000s, this phenomenon has led to the development of *elastography*, in which properties of human tissue are measured using ultrasound, facilitating the identification of abnormalities (Sigrist et al., 2017).

13.3 Acoustic microstreaming

13.3.1 Microstreaming principles

As mentioned above, the second way in which the nonlinearity in the momentum equation is large enough to cause streaming is when the gradient of the sound field, $|\nabla \boldsymbol{u}|$, is large. A bubble that is small compared to the wavelength of sound will satisfy this criterion, particularly if it is resonating, as detailed in §9.4. Basic principles of acoustic microstreaming and the history of research in this area were outlined in Manasseh (2016); a thorough review of the theoretical derivations is given by Wu (2018).

The streaming-flow pattern that develops depends on whether the bubble is oscillating volumetrically, or oscillating laterally (i.e. oscillating in translation). While the physics of bubble volumetric oscillation derived in §9.2 is purely spherically-symmetric, variations in the acoustic field driving the bubble, or the presence of nearby surfaces, or of other bubbles nearby, could cause lateral oscillations as well, and lead to a further form of steady radiation force called the secondary Bjerknes force (e.g. Mettin et al., 1997; Jiao et al., 2015). Bubbles oscillating in various combinations of lateral and volumetric motions create a great variety of streaming patterns, first observed by Elder (1959) and catalogued in detail by Tho et al. (2007). Theoretical derivations by Lane (1955) predicted what is now a commonly-observed flow structure for a spherical object in purely-lateral oscillations: there is a set of inner vortices in the *Stokes boundary layer* next to the object. The Stokes boundary layer is a layer of thickness δ_ω, given by (1.30), here reproduced as

$$\delta_\omega = \sqrt{2\frac{\nu}{\omega}}, \tag{13.1}$$

where ν is the kinematic viscosity of the liquid around the bubble ($\nu = \mu/\rho_0$ where μ is the dynamic viscosity and ρ_0 the density) and ω is the radian frequency; this boundary layer is found from an exact solution to the incompressible Navier-Stokes equation when it is linearised and there is a zero pressure gradient. These inner vortices in turn drive a set of secondary vortices. The secondary vortices are often what is observed since they cover a larger area. In fact, this basic structure of inner and outer vortices is found whenever any object with circular geometry is oscillated in a fluid (or has a fluid oscillated relative to it). For example, one of the vertical piles or columns supporting a pier or offshore structure will have such a streaming pattern around it as waves go past.

Meanwhile, a theory by Longuet-Higgins (1998) predicted that if the bubble is oscillating volumetrically only, and with no other bubbles or boundaries nearby, it would produce a 'dipolar' microstreaming pattern around it. This is indeed what is observed experimentally (figure 13.2), even if a wall is nearby,

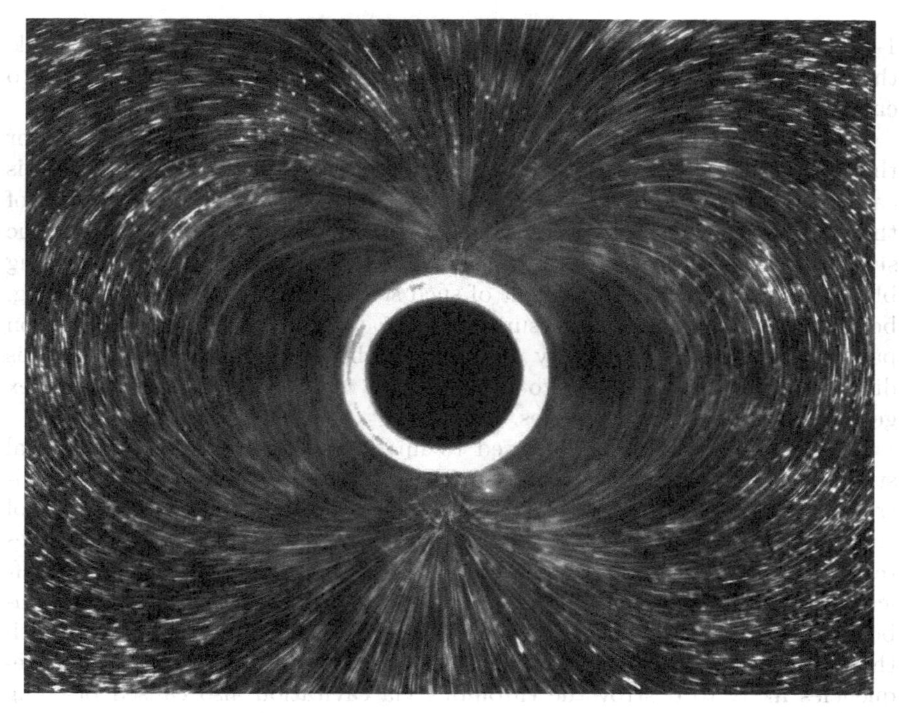

FIGURE 13.2
Experimental streak image around a bubble of radius 271 ± 4 μm attached to a micro-chamber wall and driven at 8.658 kHz such that the bubble oscillated volumetrically only. From Tho et al. (2007). The resonant frequency of a bubble this size on a rigid boundary is just under 10 kHz (Manasseh and Ooi, 2009). Image reprinted under a CSIRO Licence to Publish.

provided the frequency is close to the resonant frequency for a bubble on a wall (given in Manasseh and Ooi, 2009). The volumetric resonance is significant. Owing to the resonant amplification factor (1.83), the local velocity amplitude in acoustic microstreaming may be much higher than the sound-wave particle-velocity amplitude.

Some roles of microstreaming in microfluidic and engineering applications were reviewed by Jalal and Leong (2018); in the following, we concentrate on medical applications.

13.3.2 Microbubble microstreaming in medicine

The medical applications of acoustic microstreaming arise from the use of ultrasound contrast agents, which, as described in §9.6.2, are microbubbles

1-10 μm in diameter administered via a simple intravenous injection. While their use was initially as a diagnostic agent, interesting indications began to emerge that they might have a therapeutic ability as well.

Ultrasound had been used since the 1970s to break down blood clots, or thrombi (Trübestein et al., 1976), and since the break down of blood clots is called *thrombolysis* the use of ultrasound was called *sonothrombolysis*. One of the most interesting findings emerged from the clinical treatment of ischemic stroke. An ischemic stroke occurs when a blood clot blocks an artery supplying blood to the brain. The treatment of such strokes is particularly challenging, because the equivalent of the surgical interventions routinely performed on patients with blocked coronary arteries (which can lead to heart attacks) is difficult in the brain, owing to the presence of the skull and the complex geometry of the cerebral arteries.

Clinical ultrasound can be used to image the brain, and many clinical studies on stroke patients, backed by laboratory experiments, found that ultrasound applied with normal clinical scanners accelerates the dissolution of thrombi, particularly in the presence of intravenously-injected microbubbles (e.g. Molina et al., 2006). A meta-analysis of 105 studies found that significant improvement in clearing of the artery occurs with the use sonothrombolysis (Chen et al., 2019). However, this process only seemed effective with the megahertz frequencies of clinical ultrasound; attempts to use the low frequencies likely to destroy the thrombus via cavitation also caused bleeding (Daffertshofer et al., 2005), which was highly undesirable.

Streaming flows, in general, were proposed as a mechanism capable of gently breaking down the thrombus and mixing the thrombolytic drugs that can be administered after stroke into the affected region (Perren et al., 2008); however, there was no clear mechanism.

Further intriguing observations began to be made of the effect of ultrasound and microbubbles on biological cells. The cells were found to take in large molecules, such as useful drugs that cells normally would not absorb. Moreover, this phenomenon was not one in which the cells were crudely damaged, as one would expect from cavitation; rather, the 'opening' of the cell membrane was temporary (Wu et al., 2002). This is of particular value in the brain, where the *blood-brain barrier* prevents diffusive processes from passively transporting drugs circulating in the blood into the tissues behind the blood-vessel wall. Microbubble-mediated ultrasonic delivery of molecules into the brain was demonstrated in animals (Choi et al., 2007; Raymond et al., 2008) and ultrasonic opening of the blood-brain barrier has now been demonstrated in human patients (Lipsman et al., 2018). In all parts of the body, microbubbles may assist the uptake of drugs via microstreaming. The large molecules artificially introduced into cells by sonoporation have included DNA, enabling the process of *gene transfection* (Ferrara et al., 2007; Delalande et al., 2010; Ogawa et al., 2018), where genes with therapeutic properties can be incorporated into cells to remedy genetic disorders, or to promote the synthesis of therapeutic molecules by the body itself. All these observations of temporary and non-destructive uptake of molecules by cells were termed *sonoporation.*

The leading-order oscillations created by oscillating bubbles, as derived in §9.4, reverse every cycle, and although these would create stresses on nearby cell membranes (Doinikov and Bouakaz, 2010), at megahertz frequencies, the stresses reverse in a fraction of a millionth of a second, whereas biological processes such as protein expression take minutes. Collis et al. (2010) elaborated on the concept that microstreaming, being a steady flow, could cause both sonothrombolysis and sonoporation; they demonstrated that microstreaming on surfaces could generate a divergence in the flow over the surface (a cell wall, for example), that could gently stretch the cell wall. It is thought that this promotes a controlled and temporary uptake of the liquid surrounding the cell and any molecules in it (Qin et al., 2018), called *endocytosis.*

There are many other bioeffects of ultrasound on biological cells, reviewed by Rubin et al. (2018); not all effects are in the presence of microbubbles. Cells are well known to respond to mechanical stresses (Shah et al., 2014), but the detailed biological pathways for many effects remained unexplained.

13.4 Streaming in rotating fluids and planetary physics

13.4.1 Observations of streaming flow in rotating fluids

It has been recognised, as mentioned in §11.2.3, that precession of a rotating fluid has the potential to drive a geodynamo (Bullard, 1949; Malkus, 1968), creating the magnetic field that protects the Earth's biosphere from lethal solar events. This led to significant research to uncover a mechanism by which this might be possible. Interest in whether exoplanets are suitable for life also motivates interest on how widespread dynamo mechanisms might be. Recall from §11.2.3 that thermal convection by itself would not create the required 'geostrophic' fluid flow (circulation in the azimuthal direction) consistent with the dipolar magnetic field possessed by the Earth. Thermal convection, together with rotation, is at present the explanation preferred by the majority of researchers for the generation of the dipolar magnetic field. However, it was noted in §11.2.3 that convection may not provide sufficient energy to create the observed magnetic field (Andrault et al., 2016).

Tidal and inertial oscillations in the Earth's fluid outer core could not, according to linear theory, create the azimuthal circulation consistent with the dipolar magnetic field. This is simply because, as with all waves, and as noted in §11.2.3, the flow reverses every cycle. These oscillations would have to somehow generate a mean-streaming flow: a flow that is steady, or at least a flow with a steady component, such as some of the 'mean flows' described in chapter 7. Moreover, the flow would have to have a steady component roughly at right angles to the Earth's rotation axis, since the rotation axis is close to the magnetic axis.

The Earth's rotation axis is displaced from the normal to the plane in which it orbits the Sun by an angle of about 23.4°. Owing to the combined gravitational pull of the Moon and the Sun on the Earth's equatorial bulge, the axis itself rotates, i.e. precesses. (In addition, the plane of the Earth's orbit, established by Sun's original protoplanetary disc, mentioned in §11.2.1 and called the *plane of the ecliptic*, also precesses.) The result is a precession of the Earth's axis with a period of about 26 000 years, usually called *luni-solar precession*. Recalling the details of the Earth's structure outlined in §11.2.3, the precession of the mantle and crust differs from that of the solid inner core, providing a driving of inertial waves in the fluid outer core.

Laboratory experiments on rotating fluids as well as numerical simulations have provided many observations of mean-streaming flows, which are predominantly at right angles to the rotation axis as required. These studies, being in laboratories on Earth, cannot synthesise a gravitational acceleration towards the centre of the rotating fluid, as is the case with a planet or other celestial body; thus, thermal convection inside such a three-dimensional body cannot be included. There has been progress with experiments on rotating magnetic fluids (Stefani et al., 2019), but there are formidable experimental difficulties with appropriate magnetic fluids. Thus, magnetism is also not included in most studies, and the following studies are purely hydrodynamic.

A mean-streaming flow was reported in experiments in which inertial waves were generated (Fultz, 1959) as well as by Malkus (1968) who, as just-mentioned, had proposed the relevance of the streaming flow to the geodynamo. McEwan (1970), Kobine (1996), Meunier et al. (2008) and Horimoto et al. (2018) all reported a “zonal flow”, azimuthal circulation or “global circulation”. Some of these laboratory rotating-fluid systems used precessing equipment, which is much harder to set up than a simple rotating fluid. Others did not apply precession, confounding interpretations; Fultz (1959) used a disc oscillating along the axis to excite axisymmetric inertial modes, and McEwan (1970) used a differentially-rotating lid that provided a viscous torque also able to drive a mean flow. Furthermore, the geometry used by Fultz (1959), McEwan (1970) and Kobine (1996) and Meunier et al. (2008) was cylindrical. Malkus (1968) and Horimoto et al. (2018) used the spheroidal geometry more appropriate to a celestial body; even so, the solid inner core was not included.

Numerical simulations, which have fewer constraints than experiments, but struggle to reach realistic Reynolds numbers, have found mean-streaming flows in contained precessing fluids; Kong et al. (2015) showed, consistently with the experiments of Kobine (1996) and Horimoto et al. (2018), that an axisymmetric streaming flow occurs when a precessing cylinder is tuned to resonances of inertial modes, the formulae for which are given in §5.3.4.

13.4.2 Possible mechanisms for streaming flows in rotating fluids

Here, we consider some wave-interaction mechanisms that could drive an azimuthal streaming flow and hence, potentially, drive the geodynamo. There are three general classes of mechanisms that have been proposed: viscous boundary-layer mechanisms, inviscid interactions of inertia waves, and turbulent energy-loss mechanisms. While these three are nonlinear mechanisms, a fourth, linear mechanism is briefly outlined at the end of this section.

Recall that the wave equation is a hyperbolic equation, and Poincaré's equation for rotating fluids, (5.11), is also a hyperbolic equation, provided the oscillation frequency is less than twice the rotation rate. The solution thus exhibits characteristic surfaces, as mentioned in §1.5.3. Wood (1966) analysed these characteristics and found that a small amount of viscosity organised them into internal shear layers, sometimes called 'wave beams'. Where the characteristics intersect with the boundary of the rotating fluid in a precessing spherical shell (modelling the Earth's fluid outer core bounded by the Earth's mantle), Busse (1968) showed that the boundary layer would 'erupt' into the interior along with the characteristics; this would give a mean streaming flow. Tilgner (2007) showed that the nonlinear self-interaction of internal shear layers can also drive streaming flows. Most importantly, this mechanism requires viscosity.

The inviscid interactions of waves, which were explained in general in §7.4, form a second possible mechanism, and one that does not require viscosity in principle. Recall from §7.3 that the quadratic nonlinearity in the momentum equation permits time-averaged solutions at second order, driven by waves at leading order. The time-averaged or 'radiation-stress' mean flow outlined in §7.3 requires a gradient in the amplitude of waves. The gradient could be provided, for example, by friction with the seabed or wave breaking in the case of water waves, or energy loss to heat in the case of sound waves. The gradient could also be due to simple geometric spreading (§3.3.9) for any waves, which does not require viscosity. However, the inertia-wave modes formed in the fluid interior of a celestial body, being modes, fit entirely within the volume the body and do not suffer, at least not at leading order, the gradient in amplitude that would generate a time-averaged mean flow. Thus, the interactions of multiple waves of different frequencies, a weakly-nonlinear interaction described in §7.4, needs to be considered. An analysis of the inertia-wave modes by Greenspan (1969) concluded that the inviscid inertia-wave modes - solutions such as those given by (5.32) - could not interact using the weakly-nonlinear mechanism outlined in §7.4 to produce a 'geostrophic' mean flow. However, this did not rule out azimuthal circulations that varied in the axial direction, which would have the required ability to generate a dipolar magnetic field. The triadic interactions of waves described in §7.4 could combine two modes with a third that may be consistent with an azimuthal streaming flow (Meunier et al., 2008). The two modes need to be distinct, i.e. they could not be a mode interacting with itself, since that can be shown to be a zero-strength triad (Kerswell, 1999). It is also possible that the azimuthal mean flow could have elliptical rather than circular streamlines, making a whole class of triad-like instabilities called *elliptical-flow instabilities* possible (Waleffe, 1990; Kerswell, 1993). This possibility is interesting since tidal forcing generates oscillations with an azimuthal wavenumber $m = 2$, which may couple efficiently with an elliptical flow.

The third mechanism relies on a global energy-balance argument. Experiments since the 1960s had shown catastrophic transitions to turbulence occurring in contained fluid systems (Malkus, 1968). The systems were forced such that the inertia-wave modes with the lowest wavenumbers would resonate; these were called 'resonant collapses' (McEwan, 1970; Manasseh, 1992). More subtle collapses, which appeared to show two or three modes interacting (Manasseh, 1994), were shown to be caused by resonant triads (Albrecht et al., 2015). The most violent collapses were eventually shown to also be caused by resonant triads (Albrecht et al., 2018). Once some energy has been lost to turbulence, it must have been extracted from the background rotation rate, and this is represented by an azimuthal mean streaming. This mechanism is not deterministic, like the previous two mechanisms.

A fourth mechanism was proposed by Albrecht et al. (2021). The Coriolis term, which from (5.2) is $2\boldsymbol{\Omega} \times \boldsymbol{u}$, offers the possibility of driving a time-mean and azimuthally-mean flow via a linear mechanism. This works only in the special case of precession, in which the Coriolis term varies with time and in azimuth. Crucially, the variation in the Coriolis term due to precession will have the same temporal frequency and the same azimuthal wavenumber ($m =$ 1) as the linear response to precession, $\boldsymbol{u}$, since precession causes the same $m = 1$ variation in the Coriolis term. Hence, when the averaging operations detailed in §7.3 are applied, the averaged Coriolis term survives. One might imagine that this could be balanced by an averaged pressure-gradient term or a viscous term, but a pressure gradient cannot exist in the azimuthal direction after azimuthal averaging. Thus, the time- and azimuthally-averaged Coriolis term can only be balanced by a viscous mean flow. This mechanism is perfectly general, and linear. It implies that an azimuthal mean-streaming flow must exist in any precessing fluid.

In summary, waves excited the fluid interior or celestial bodies by precession or tides, or possibly some combination, could create an azimuthal mean streaming that has the appropriate structure to generate a dipolar magnetic field.

Bibliography

Al-Masry, W. and Abdennour, A. (2006). Gas hold-up estimation in bubble columns using passive acoustic waveforms with neural networks. *J. Chem. Tech. Biotech.*, 81(6):951–957.

Albrecht, T., Blackburn, H. M., Lopez, J. M., Manasseh, R., and Meunier, P. (2015). Triadic resonances in precessing rapidly rotating cylinder flows. *J. Fluid Mech.*, 778:R1.

Albrecht, T., Blackburn, H. M., Lopez, J. M., Manasseh, R., and Meunier, P. (2018). On triadic resonance as an instability mechanism in precessing cylinder flow. *J. Fluid Mech.*, 841:R3.

Albrecht, T., Blackburn, H. M., Lopez, J. M., Manasseh, R., and Meunier, P. (2021). On the origins of steady streaming in precessing fluids. *J. Fluid Mech.*, 910:A51–1–27.

Aldridge, K. D. and Lumb, L. I. (1987). Inertial waves identified in the earth's fluid outer core. *Nature*, 325:421–423.

Alexander, R. (2008). From discs to planetesimals: Evolution of gas and dust discs. *New Astron. Rev.*, 52(2-5):60–77.

Andrault, D., Monteux, J., Le Bars, M., and Samuel, H. (2016). The deep earth may not be cooling down. *Earth Planet. Sci. Lett.*, 443:195–203.

Andrews, D. G. and McIntyre, M. (1978). An exact theory of nonlinear waves on a Lagrangian-mean flow. *J. Fluid Mech.*, 89(4):609–646.

Andrews, T., Gregory, J. M., Forster, P. M., and Webb, M. J. (2012). Cloud adjustment and its role in CO2 radiative forcing and climate sensitivity: A review. *Surveys in Geophysics*, 33(3-4):619–635.

Anguelova, M. D. and Webster, F. (2006). Whitecap coverage from satellite measurements: A first step toward modeling the variability of oceanic whitecaps. *J. Geophys. Res.*, 111(C03017):1–23.

Ashokkumar, M. (2016). *Handbook of Ultrasonics and Sonochemistry.* Springer, Singapore.

Atri, D. (2019). Stellar proton event-induced surface radiation dose as a constraint on the habitability of terrestrial exoplanets. *Mon. Not. Roy. Astron. Soc.*, 492(1):1745–3925.

Baba, J., Morokuma-Matsui, K., Miyamoto, Y., Egusa, F., and Kuno, N. (2016). Gas velocity patterns in simulated galaxies: observational diagnostics of spiral structure theories. *Mon. Not. Roy. Astron. Soc.*, 460(3):2472–2481.

Babanin, A. (2011). *Breaking and Dissipation of Ocean Surface Waves.* Cambridge University Press.

Babanin, A. V. (2009). Breaking of ocean surface waves. *Acta Physica Slovaca*, 59(4):305–535.

Banner, M. L. and Peregrine, D. H. (1993). Wave breaking in deep water. *Annu. Rev. Fluid Mech.*, 25:373–397.

Bao, G. W. and Pascal, M. (1997). Stability of a spinning liquid-filled spacecraft. *Archive Applied Mech. (Ingenieur-Archiv.)*, 67(6):407–421.

Bass, S. J. and Hay, A. E. (1997). Ambient noise in the natural surf zone: wave breaking frequencies. *IEEE J. Ocean. Eng.*, 22:411–424.

Batchelor, G. K. (1973). *An introduction to fluid dynamics.* Cambridge University Press.

Beeby, S. P., Tudor, M. J., and White, N. M. (2006). Energy harvesting vibration sources for microsystems applications. *Meas. Sci. Technol.*, 17:R175–R195.

Benjamin, T. and Feir, J. (1967). The disintegration of wave trains on deep water. part 1. theory. *J. Fluid Mech.*, 27(3):417–430.

Benjamin, T. B. (1967). Internal waves of permanent form in fluids of great depth. *J. Fluid Mech.*, 29:559–592.

Berry, J. D., Neeson, M. J., Dagastine, R. R., Chan, D. Y., and Tabor, R. F. (2015). Measurement of surface and interfacial tension using pendant drop tensiometry. *J. Colloid Interface Sci.*, 454:226–237.

Betheras, F. R. (1990). Acoustic radiation force as a diagnostic modality. In *Proc. 20th Annu. Meeting ASUM*, page 69. Australian Society for Ultrasound in Medicine.

Birch, C. E., Reeder, M. J., and Berry, G. J. (2014). Wave-cloud lines over the Arabian Sea. *J. Geophys. Res.-Atmospheres*, 119(8):4447–4457.

Blanchard, D. C. and Syzdek, L. (1970). Mechanism for the water-to-air transfer and concentration of bacteria. *Science*, 170(3958):626–628.

Boegman, L. and Stastna, M. (2019). Sediment resuspension and transport by internal solitary waves. *Annu. Rev. Fluid Mech.*, 51:129–154.

Boissenot, T., Bordat, A., Fattal, E., and Tsapis, N. (2016). Ultrasound-triggered drug delivery for cancer treatment using drug delivery systems: From theoretical considerations to practical applications. *J. Control. Release*, 241:144–163.

Boluriaan, S. and Morris, P. J. (2003). Acoustic streaming: From Rayleigh to today. *Int. J. Aeroacoustics.*, 2(3):255–292.

Bondi, H. and Hoyle, F. (1944). On the mechanism of accretion by stars. *Mon. Not. Roy. Astron. Soc.*, 104:273.

Boon, W., Petkovic-Duran, K., White, K., Tucker, E., Albiston, A., Manasseh, R., Horne, M., and Aumann, T. (2011). Optical detection and quantification of micromixing of single-cell quantities of RNA. *BioTechniques*, 50(2):116–119.

Boyd, J. W. R. and Varley, J. (2001). The uses of passive measurement of acoustic emissions from chemical engineering processes. *Chem. Eng. Sci.*, 56:1749–1767.

Boyd, R., Ruming, K., Goodwin, I., Sandstrom, M., and Schröder-Adams, C. (2008). Highstand transport of coastal sand to the deep ocean: A case study from fraser island, southeast australia. *Geology*, 36(1):15–18.

Braddock, M. (2017). Artificial gravity: small steps on the journey to the giant leap. *J Space Explor*, 6(3):137.

Brennen, C. E. (1995). *Cavitation and Bubble Dynamics.* Oxford University Press.

Brennen, C. E. (2002). Fission of collapsing cavitation bubbles. *J. Fluid Mech.*, 472:153–166.

Browning, R. J. and Stride, E. (2018). Microbubble-mediated delivery for cancer therapy. *Fluids*, 3(4):74.

Bullard, E. C. (1949). The magnetic field within the Earth. *Proc. Royal Soc. A*, 197:433–453.

Bureau international des poids et measures, Paris (2006). *The International System of Units (SI).* Organisation Intergouvernementale de la Convention du Mètre, 8th edition.

Busse, F. H. (1968). Steady fluid flow in a precessing spherical shell. *J. Fluid Mech.*, 33:739–751.

Cartwright, J. H. E. and Nakamura, H. (2008). Tsunami: A history of the term and of scientific understanding of the phenomenon in Japanese and Western culture. *Not. Rec. Roy. Soc. Lond.*, 62(2):151–166.

Chen, M., Goyal, R., Majji, M., and Skelton, R. E. (2020). Design and analysis of a growable artificial gravity space habitat. *Aerospace Sci. Tech.*, 106:106147.

Chen, Z., Xue, T., Huang, H., Xu, J., Shankar, S., Yu, H., Wang, Z., and Toscano, M. (2019). Efficacy and safety of sonothombolysis versus non-sonothombolysis in patients with acute ischemic stroke: A meta-analysis of randomized controlled trials. *PloS one.*, 14(1):e0210516.

Choi, J. J., Pernot, M., Brown, T. R., Small, S. A., and Konofagou, E. E. (2007). Spatiotemporal analysis of molecular delivery through the blood-brain barrier using focused ultrasound. *Phys. Med. Biol.*, 52(18):5509–5530.

Christie, D. R. (1989). Long nonlinear waves in the lower atmosphere. *J. Atmos. Sci.*, 46(11):1462–1491.

Clarke, L., Edwards, A., and Graham, E. (2004). Acoustic streaming: an *in vitro* study. *Ultrasound Med. Biol.*, 30(4):559–562.

Clarke, R. H., Smith, R. K., and Reid, R. K. (1981). The Morning Glory of the Gulf of Carpentaria: An atmospheric undular bore. *Mon. Weather Rev.*, 109:1726–1750.

Clift, R., Grace, J. R., and Weber, M. E. (1978). *Bubbles, drops and particles*. Academic Press, London.

Coleman, T. A., Knupp, K. R., and Herzmann, D. (2009). The spectacular undular bore in Iowa on 2 October 2007. *Mon. Weather Rev.*, 137(1):495–503.

Collings, A., Gwan, P., and Sosa Pintos, A. (2010). Large scale environmental applications of high power ultrasound. *Ultrason. Sonochem.*, 17(6):1049–1053.

Collis, J., Manasseh, R., Liovic, P., Tho, P., Ooi, A., Petkovic-Duran, K., and Zhu, Y. (2010). Cavitation microstreaming and stress fields created by microbubbles. *Ultrasonics*, 50(2):273–279.

Cooper, A., Turney, C. S., Palmer, J., Hogg, A., McGlone, M., Wilmshurst, J., Lorrey, A. M., J, H. T., Russell, J. M., McCracken, K., et al. (2021). A global environmental crisis 42,000 years ago. *Science*, 371(6531):811–818.

Cotton, W. R. and Anthes, R. A. (1992). *Storm and cloud dynamics*. Academic Press.

Cowling, T. G. and Newing, R. A. (1949). The oscillations of a rotating star. *Astrophys. J.*, 109:149–158.

Craik, A. D. (1988). *Wave interactions and fluid flows*. Cambridge University Press.

Daffertshofer, M., Gass, A., Ringleb, P., Sitzer, M., Sliwka, U., Els, T., Sedlaczek, O., Koroshetz, W., and Hennerici, M. (2005). Transcranial low-frequency ultrasound-mediated thrombolysis in brain ischemia - Increased risk of hemorrhage with combined ultrasound and tissue plasminogen activator - Results of a phase II clinical trial. *Stroke*, 36(7):1441–1446.

Deane, G. B. and Czerski, H. (2010). Contributions to the acoustic excitation of bubbles released from a nozzle. *J. Acous. Soc. Am.*, 128:2625–2634.

Deane, G. B. and Stokes, M. D. (2010). Model calculations of the underwater noise of breaking waves and comparison with experiment. *J. Acous. Soc. Am.*, 127(6):3394–3410.

Delalande, A., Bureau, M.-F., Midoux, P., Bouakaz, A., and Pichon, C. (2010). Ultrasound-assisted microbubbles gene transfer in tendons for gene therapy. *Ultrasonics*, 50(2):269 – 272. Selected Papers from ICU 2009.

Devin, C. R. (1959). Survey of thermal, radiation and viscous damping of pulsating air bubbles in water. *J. Acous. Soc. Am.*, 31(12):1654–1667.

DeVries, T. (2014). The oceanic anthropogenic co2 sink: Storage, air-sea fluxes, and transports over the industrial era. *Global Biogeochem. Cycles*, 28(7):631–647.

Doinikov, A. and Bouakaz, A. (2010). Theoretical investigation of shear stress generated by a contrast microbubble on the cell membrane as a mechanism for sonoporation. *J. Acous. Soc. Am.*, 128(1):11–19.

Doviak, R. and Ge, R. (1984). An atmospheric solitary gust observed with a Doppler radar, a tall tower and a surface network. *J. Atmos. Sci.*, 41(17):2559–2573.

Doviak, R. J., Chen, S. S., and Christie, D. R. (1991). A thunderstorm-generated solitary wave observation compared with theory for nonlinear waves in a sheared atmosphere. *J. Atmos. Sci.*, 48(1):87–111.

Drazin, P. G. and Reid, W. H. (1981). *Hydrodynamic stability*. Cambridge University Press.

Duc, N. M. and Keserci, B. (2019). Emerging clinical applications of high-intensity focused ultrasound. *Diagn. Intervent. Radiol..*, 25(5):398–409.

Durante, M. and Cucinotta, F. A. (2008). Heavy ion carcinogenesis and human space exploration. *Nat. Rev.*, 8(6):465–472.

Duvall, T., Dziembowski, W., Goode, P., Gough, D. O., Harvey, J. W., and Leibacher, J. W. (1984). Internal rotation of the Sun. *Nature*, 310:22–25.

Eckart, C. (1948). Vortices and streams caused by sound waves. *Phys. Rev.*, 73(1):68–76.

Edwards, A., Clarke, L., Piessens, S., Graham, E., and Shekleton, P. (2003). Acoustic streaming: a new technique for assessing adnexal cysts. *Ultrasound Obstet. Gynecol.*, 22:74–78.

Elder, S. (1959). Cavitation microstreaming. *J. Acous. Soc. Am.*, 31:54–64.

Falcão, A. F. O. (2010). Wave energy utilization: A review of the technologies. *Renew. Sust. Energy Rev.*, 14:899–918.

Farmer, D. M. and Vagle, S. (1988). On the determination of breaking surface wave distribution. *J. Geophys. Res.*, C93:3591–3600.

Ferrara, K., Pollard, R., and Borden, M. (2007). Ultrasound microbubble contrast agents: Fundamentals and application to gene and drug delivery. *Annu. Rev. Biomed. Eng.*, 9:415–447.

Fine, R. A. and Millero, F. J. (1973). Compressibility of water as a function of temperature and pressure. *J. Chem. Phys.*, 59(10):5529–5536.

Fujita, T. T. (1986). DFW (Dallas-Ft. Worth) microburst on August 2, 1985. *University of Chicago*, page 63pp.

Fujita, T. T. (1990). Downbursts: meteorological features and wind field characteristics. *J. Wind Eng. Indust. Aerodyn.*, 36:75–86.

Fultz, D. (1959). A note on overstability and the elastoid-inertia oscillations of Kelvin, Solberg, and Bjerknes. *J. Meteorology*, 16:199–208.

Gerrits, J. and Veldman, A. (2003). Dynamics of liquid-filled spacecraft. *J. Eng. Math.*, 45(1):21–38.

Giesecke, A., Vogt, T., Gundrum, T., and Stefani, F. (2019). Kinematic dynamo action of a precession-driven flow based on the results of water experiments and hydrodynamic simulations. *Geophys. Astrophys. Fluid Dyn.*, 113(1-2):235–255.

Gill, A. E. (1982). *Atmosphere-Ocean Dynamics*. Academic Press, London.

Goler, R. A. and Reeder, M. J. (2004). The generation of the morning glory. *J. Atmos. Sci..*, 61(12):1360–1376.

Gough, D. O., Kosovichev, A. G., Toomre, J., Anderson, E., Antia, H. M., Basu, S., Chaboyer, B., Chitre, S. M., Christensen-Dalsgaard, J., Dziembowski, W. A., Eff-Darwich, A., Elliott, J. R., Giles, P. M., Goode, P. R., Guzik, J. A., Harvey, J. W., Hill, F., Leibacher, J. W., Monteiro, M. J. P. F. G., Richard, O., Sekii, T., Shibahashi, H., Takata, M., Thompson, M. J., Vauclair, S., and Vorontsov, S. V. (1992). The seismic structure of the Sun. *Science*, 272(5266):1296–1300.

Gramiak, R. and Shah, P. M. (1968). Echocardiography of the aortic root. *Invest. Radiol.*, 3(5):356–366.

Greenspan, H. P. (1968). *The theory of rotating fluids*. Cambridge University Press.

Greenspan, H. P. (1969). On the non-linear interaction of inertial modes. *J. Fluid Mech.*, 36:257–264.

Grinstaff, M. W. and Suslick, K. S. (1991). Air-Filled Proteinaceous Microbubbles - Synthesis Of An Echo-Contrast Agent. *Proc. Nat. Acad. Sci. USA*, 88(17):7708–7710.

Grythe, H., Strom, J., Krejci, R., Quinn, P., and Stohl, A. (2014). A review of sea-spray aerosol source functions using a large global set of sea salt aerosol concentration measurements. *Atm. Chem. Phys.*, 14(3):1277–1297.

Gupta, A., Baker, J., and Sharif, M. (2004). Numerical analysis of natural convection in an enclosure with rotationally produced artificial gravity. *Numer. Heat Transf. Part A,*, 46(2):131–145.

Hahn, M., Adami, S., and Förstner, R. (2018). Computational modeling of nonlinear propellant sloshing for spacecraft aocs applications. *CEAS Space J.*, 10:441–451.

Hall, T. W. (1999). Artificial gravity and the architecture of orbital habitats. *J. British Interplanetary Soc.*, 52(7/8):290–300.

Hide, R. (1980). Jupiter and saturn: Giant magnetic rotating fluid planets. *The Observatory*, 100:182.

Hilgenfeldt, S., Grossmann, S., and Lohse, D. (1999). A simple explanation of light emission in sonoluminescence. *Nature*, 398(6726):402–405.

Hill, D. E. (1985). *Dynamics and control of spin-stabilized spacecraft with sloshing fluid stores*. PhD thesis, Iowa State University.

Hinkel, J., Lincke, D., Vafeidis, A., Perrette, M., Nicholls, R., Tol, R., Marzeion, B., Fettweis, X., Ionescu, C., and Levermann, A. (2014). Coastal flood damage and adaptation costs under 21st century sea-level rise. *Proc. Nat. Acad. Sci. USA*, 111(9):3292–3297.

Hollerbach, R. (1995). On the theory of the geodynamo. *Phys. Earth Planet. Interiors*, 98:163–185.

Horimoto, Y., Simonet-Davin, G., Katayama, A., and Goto, S. (2018). Impact of a small ellipticity on the sustainability condition of developed turbulence in a precessing spheroid. *Phys. Rev. Fluids*, 3(4).

Illesinghe, S. (2007). Acoustical properties of multiple bubbles attached to a solid boundary. Master's thesis, Department of Mechanical Engineering, University of Melbourne.

Jalal, J. and Leong, T. S. (2018). Microstreaming and its role in applications: A mini-review. *Fluids*, 3(4):93.

Jiao, J., He, Y., Kentish, S. E., Ashokkumar, M., Manasseh, R., and Lee, J. (2015). Experimental and theoretical analysis of secondary bjerknes forces between two bubbles in a standing wave. *Ultrasonics*, 58:35–42.

Jones, C. A. (2011). Planetary magnetic fields and fluid dynamos. *Ann. Rev. Fluid Mech.*, 43(1):583–614.

Keller, M. W., Feinstein, S. B., Briller, R. A., and Powsner, S. M. (1986). Automated production and analysis of echo contrast agents. *J. Ultrasound Med.*, 5(9):493–498.

Kerswell, R. (1993). Elliptical instabilities of stratified, hydromagnetic waves. *Geophys. Astrophys. Fluid Dyn.*, 71(1-4):105–143.

Kerswell, R. R. (1999). Secondary instabilities in rapidly rotating fluids: inertial wave breakdown. *J. Fluid Mech.*, 382:283–306.

Kessler, E. (1982). *Thunderstorm morphology and dynamics*, volume 2. U. S. Department of Commerce, National Oceanic and Atmospheric Administration.

Khatiwala, S., Tanhua, T., Mikaloff Fletcher, S., Gerber, M., Doney, S. C., Graven, H. D., Gruber, N., McKinley, G. A., Murata, A., Ríos, A. F., and Sabine, C. L. (2013). Global ocean storage of anthropogenic carbon. *Biogeosciences*, 10(4):2169–2191.

Kim, M.-H. Y., Hayat, M. J., Feiveson, A. H., and Cucinotta, F. A. (2009). Prediction of frequency and exposure level of solar particle events. *Health Phys.*, 97(1):68–81.

King, S. E., Carr, M., and Dritschel, D. G. (2011). The steady-state form of large-amplitude internal solitary waves. *J. Fluid Mech.*, 666:477–505.

Kinsler, L. E. and Frey, A. R. (1962). *Fundamentals of Acoustics*. Wiley, New York.

Kivelson, M., Khurana, K., Russell, C., Volwerk, M., Walker, R., and Zimmer, C. (2000). Galileo magnetometer measurements: A stronger case for a subsurface ocean at Europa. *Science*, 289:1340–1343.

Klibanov, A. (2009). Preparation of targeted microbubbles: ultrasound contrast agents for molecular imaging. *Med. Biol. Eng. Comput.*, 47 (8):875–882.

Kobine, J. J. (1996). Azimuthal flow associated with inertial wave resonance in a precessing cylinder. *J. Fluid Mech.*, 319:387–406.

Kong, D., Cui, Z., Liao, X., and Zhang, K. (2015). On the transition from the laminar to disordered flow in a precessing spherical-like cylinder. *Geophys. Astrophys. Fluid Dyn.*, 109:62–83.

Lai, P. (2008). Acoustic streaming for medical diagnostics. Master's thesis, University of Melbourne, Department of Mechanical and Manufacturing Engineering.

Lamb, H. (1932). *Hydrodynamics*. Dover, 6th Edn.

Lamb, K. G. (2002). A numerical investigation of solitary internal waves with trapped cores formed via shoaling. *J. Fluid Mech.*, 451:109.

Lane, C. (1955). Acooustical streaming in the vicinity of a sphere. *J. Acous. Soc. Am.*, 27(6):1082–1086.

Lau, C., Hess, P., Shreves, T., and Lee, M.-S. (2020). Scoping review of targeted ultrasound contrast agents in the detection of myocardial ischemia. *J. Diagn. Med. Sonogr.*, 36(5):479–487.

Le Quéré, C., Andrew, R. M., Friedlingstein, P., Sitch, S., Hauck, J., Pongratz, J., Pickers, P. A., Korsbakken, J. I., Peters, G. P., Canadell, J. G., Arneth, A., Arora, V. K., Barbero, L., Bastos, A., Bopp, L., Chevallier, F., Chini, L. P., Ciais, P., Doney, S. C., and Gkritzalis, T. (2018). Global carbon budget 2018. *Earth Syst. Sci. Dat.*, 10(4):2141–2194.

Leighton, R. B., Noyes, R. W., and Simon, G. W. (1962). Velocity fields in the solar atmosphere. I. Preliminary report. *Astrophys. J.*, 135:474.

Leighton, T. G. (1994). *The Acoustic Bubble.* Academic Press, London.

Leong, T., Collis, J., Manasseh, R., Ooi, A., Novell, A., Bouakaz, A., Ashokkumar, M., and Kentish, S. (2011). The role of surfactant head group, chain length and cavitation microstreaming on the growth of bubbles by rectified diffusion. *J. Phys. Chem.*, 115(49):24310–24316.

Leroy, V., Devaud, M., Hocquet, T., and Bacri, J.-C. (2005). The bubble cloud as an n-degree of freedom harmonic oscillator. *Eur. Phys. J. E*, 17:189–198.

Li, X. and Götze, H.-J. (2001). Ellipsoid, geoid, gravity, geodesy, and geophysics. *Geophysics*, 66(6):1660–1668.

Lighthill, M. J. (1978). *Waves in Fluids.* Cambridge University Press.

Lin, C. and Shu, F. H. (1966). On the spiral structure of disk galaxies, ii. outline of a theory of density waves. *Proc. Nat. Acad. Sci. USA*, 55(2):229–234.

Linden, P. and Simpson, J. E. (1985). Microbursts: a hazard to aircraft. *Nature*, 317:601–347.

Lindner, J. R. (2004). Microbubbles in medical imaging: current applications and future directions. *Nat. Rev. Drug Discov.*, 3(6):527–532.

Liow, J. L. (2001). Splash formation by spherical drops. *J. Fluid Mech.*, 427:73–105.

Lipsman, N., Meng, Y., Bethune, A. J., Huang, Y., Lam, B., Masellis, M., Herrmann, N., Heyn, C., Aubert, I., Boutet, A., Smith, G. S., Hynynen, K., and Black, S. E. (2018). Blood-brain barrier opening in alzheimer's disease using mr-guided focused ultrasound. *Nat. Comm.*, 9(1):1.

Lobaccaro, P., Singh, M. R., Clark, E. L., Kwon, Y., Bell, A. T., and Ager, J. W. (2016). Effects of temperature and gas–liquid mass transfer on the operation of small electrochemical cells for the quantitative evaluation of CO2 reduction electrocatalysts. *Phys. Chem. Chem. Phys.*, 18(38):26777–26785.

Loewen, M. R., Rojas, G., and Loewen, M. R. (2010). Void fraction measurements beneath plunging and spilling breaking waves. *J. Geophys. Res.*, 115(C8).

Lohse, D., Schmitz, B., and Versluis, M. (2001). Snapping shrimp make flashing bubbles. *Nature*, 413(6855):477–478.

Longuet-Higgins, M. (1998). Viscous streaming from an oscillating spherical bubble. *Proc. Roy. Soc. Lond. A*, 454:725–742.

Longuet-Higgins, M. S. (1982). Parametric solutions for breaking waves. *J. Fluid Mech.*, 121:403–424.

Malkus, W. V. R. (1968). Precession of the Earth as the cause of geomegnetism. *Science*, 160:259–264.

Manasseh, R. (1992). Breakdown regimes of inertia waves in a precessing cylinder. *J. Fluid Mech.*, 243:261–296.

Manasseh, R. (1993). Visualization of the flows in precessing tanks with internal baffles. *Am. Inst. Aeronaut. Astron. J.*, 31(2):312–318.

Manasseh, R. (1994). Distortions of inertia waves in a rotating fluid cylinder forced near its fundamental mode resonance. *J. Fluid Mech.*, 265:345–370.

Manasseh, R. (1996). Nonlinear behaviour of contained inertia waves. *J. Fluid Mech.*, 315:151–173.

Manasseh, R. (2016). Acoustic bubbles, acoustic streaming, and cavitation microstreaming. In Yasui, K. and Ashokkumar, M., editors, *Handbook of Ultrasonics and Sonochemistry*, chapter 1, pages 1–37. Springer-Meteor.

Manasseh, R., Babanin, A., Forbes, C., Rickards, K., Bobevski, I., and Ooi, A. (2006). Passive acoustic determination of wave-breaking events and their severity across the spectrum. *J. Atmos. Ocean Tech.*, 23(4):599–618.

Manasseh, R., Ching, C. Y., and Fernando, J. S. (1998). The transition from density-driven to wave-dominated isolated flows. *J. Fluid Mech.*, 361:253–274.

Manasseh, R., LaFontaine, R. F., Davy, J., Shepherd, I. C., and Zhu, Y. (2001). Passive acoustic bubble sizing in sparged systems. *Exp. Fluids*, 30(6):672–682.

Manasseh, R., McInnes, K., and Hemer, M. (2017a). Pioneering developments of marine renewable energy in Australia. *Int. J. Ocean Clim. Syst.*, 8(1):50–67.

Manasseh, R. and Ooi, A. (2009). Frequencies of acoustically interacting bubbles. *Bub. Sci., Eng. Tech.*, 1(1-2):58–74 (Invited review).

Manasseh, R., Riboux, G., and Risso, F. (2008). Sound generation on bubble coalescence following detachment. *Int. J. Multiphase Flows*, 34:938–949.

Manasseh, R., Sannasiraj, S. A., McInnes, K. L., Sundar, V., and Jalihal, P. (2017b). Integration of marine renewable energy with the needs of coastal societies. *Int. J. Ocean Clim. Syst.*, 8(1):19–36.

Maxworthy, T., Gnann, C., Kürten, M., and Durst, F. (1996). Experiments on the rise of air bubbles in clean viscous liquids. *J. Fluid Mech.*, 321:421–441.

May, N. W., Gunsch, M. J., Olson, N. E., Bondy, A. L., Kirpes, R. M., Bertman, S. B., China, S., Laskin, A., Hopke, P. K., Ault, A. P., et al. (2018). Unexpected contributions of sea spray and lake spray aerosol to inland particulate matter. *Environ. Sci. Tech. Lett.*, 5(7):405–412.

McCormick, M. E. (2013). *Ocean wave energy conversion.* Courier Corporation.

McEwan, A. D. (1970). Inertial oscillations in a rotating fluid cylinder. *J. Fluid Mech.*, 40(3):603–640.

Melville, W. K., Loewen, M., Felizardo, F., Jessup, A., and Buckingham, M. (1988). Acoustic and microwave signatures of breaking waves. *Nature*, 336:54–56.

Menzo, Z., Elliott, S., Hartin, C., Hoffman, F., and Wang, S. (2018). Climate change impacts on natural sulfur production: Ocean acidification and community shifts. *Atmosphere*, 9(167):1–21.

Mettin, R., Akhatov, I., Parlitz, U., Ohl, C., and Lauterborn, W. (1997). Bjerknes forces between small cavitation bubbles in a strong acoustic field. *Phys. Rev. E*, 56:2924–2931.

Meunier, P., Eloy, C., Lagrange, R., and Nadal, F. (2008). A rotating fluid cylinder subject to weak precession. *J. Fluid Mech.*, 599:405–440.

Minnaert, M. (1933). On musical air bubbles and the sound of running water. *Phil. Mag.*, 16:235–248.

Molina, C., Ribo, M., Rubiera, M., Montaner, J., Santamarina, E., Delgado-Mederos, R., Arenillas, J., Huertas, R., Purroy, F., Delgado, P., et al. (2006). Microbubble administration accelerates clot lysis during continuous 2-mhz ultrasound monitoring in stroke patients treated with intravenous tissue plasminogen activator. *Stroke*, 37(2):425–429.

Moritz, H. (2000). Geodetic Reference System 1980. *J. Geodesy*, 74(1):128–133.

Morton, D., Rudman, M., and Liow, J.-L. (2000). An investigation of the flow regimes resulting from splashing drops. *Phys. Fluids*, 12(4):747–763.

Moulton, M., Dusek, G., Elgar, S., and Raubenheimer, B. (2017). Comparison of rip current hazard likelihood forecasts with observed rip current speeds. *Weather Forecast.*, 32(4):1659–1666.

New, A. L. (1981). A class of elliptical free-surface flows. *J. Fluid Mech.*, 130:219–239.

Nightingale, K. R., Kornguth, P. J., and Trahey, G. E. (1999). The use of acoustic streaming in breast lesion diagnosis: a clinical study. *Ultrasound Med. Biol.*, 25(1):75–87.

Nisenblat, V., Bossuyt, P., Farquhar, C., Johnson, N., and Hull, M. (2016). Imaging modalities for the non-invasive diagnosis of endometriosis. *Cochrane Db. Syst. Rev.*, (2).

Ogawa, K., Fuchigami, Y., Hagimori, M., Fumoto, S., Miura, Y., and Kawakami, S. (2018). Efficient gene transfection to the brain with ultrasound irradiation in mice using stabilized bubble lipopolyplexes prepared by the surface charge regulation method. *Int. J. Nanomed.*, 13:2309.

Okayasu, R. (2012). Repair of DNA damage induced by accelerated heavy ions - a mini review. *Int. J. Cancer*, 130(5):991–1000.

Onorato, M., Waseda, T., Toffoli, A., Cavaleri, L., Gramstad, O., Janssen, P. A. E. M., Kinoshita, T., Monbaliu, J., Mori, N., Osborne, A. R., Serio, M., Stansberg, C. T., Tamura, H., and Trulsen, K. (2009). Statistical properties of directional ocean waves: The role of the modulational instability in the formation of extreme events. *Phys. Rev. Lett.*, 102(11).

Pandit, A. B., J. Varley, J., Thorpe, R. B., and Davidson, J. F. (1992). Measurement of bubble size distribution: an acoustic technique. *Chem. Eng. Sci.*, 47(5):1079–1089.

Papaloizou, J. C. B. and Lin, D. N. C. (1995). Theory of accretion disks I: Angular momentum transport processes. *Ann. Rev. Astron. Astrophys.*, 33(1):505–540.

Pedlosky, J. (1987). *Geophysical fluid dynamics.* Springer, 2 edition.

Peregrine, D. H. (1983). Breaking waves on beaches. *Annu. Rev. Fluid Mech.*, 15:149–178.

Perren, F., Loulidi, J., Poglia, D., Landis, T., and Sztajzel, R. (2008). Microbubble potentiated transcranial duplex ultrasound enhances Iv thrombolysis in acute stroke. *J. Thromb. Thrombolys.*, 25(2):219–223.

Phillips, O. (1960). On the dynamics of unsteady gravity waves of finite amplitude part 1. the elementary interactions. *J. Fluid Mech.*, 9(2):193–217.

Pocha, J. J. (1987). An experimental investigation of spacecraft sloshing. *Space Comm. Broadcasting*, 4:323–332.

Poissonnier, L., Chapelon, J.-Y., Rouvire, O., Curiel, L., Bouvier, R., Martin, X., Dubernard, J. M., and Gelet, A. (2007). Control of prostate cancer by transrectal HIFU in 227 patients. *Eur. Urol.*, 51(2):381–387.

Porter, A. and Smyth, N. F. (2002). Modelling the Morning Glory of the Gulf of Carpentaria. *J. Fluid Mech.*, 454:1.

Prosperetti, A. (1977). Thermal effects and damping mechanism in the forced radial oscillations of gas bubbles in liquids. *J. Acous. Soc. Am.*, 61(1):17–27.

Qin, P., Han, T., Yu, A. C., and Xu, L. (2018). Mechanistic understanding the bioeffects of ultrasound-driven microbubbles to enhance macromolecule delivery. *J. Control. Release.*, 272:169–181.

Rakerd, B., Hunter, E. J., Berardi, M., and Bottalico, P. (2018). Assessing the acoustic characteristics of rooms: A tutorial with examples. *Perspectives of the ASHA special interest groups*, 3(19):8–24.

Rapoport, N. Y., Kennedy, A. M., Shea, J. E., Scaife, C. L., and Nam, K.-H. (2009). Controlled and targeted tumor chemotherapy by ultrasound-activated nanoemulsions/microbubbles. *J. Control. Release*, 138(3):268–276.

Rayleigh (1917). On the pressure developed in a liquid during the collapse of a spherical cavity. *Phil. Mag.*, 34:94–98.

Rayleigh, L. (1884). On the circulation of air observed in Kundt's tubes, and on some allied acoustical problems. *Phil. Trans. Roy. Soc. Lond.*, 175:1–21.

Raymond, S. B., Treat, L. H., Dewey, J. D., Mcdannold, N. J., Hynynen, K., and Bacskai, B. J. (2008). Ultrasound enhanced delivery of molecular imaging and therapeutic agents in alzheimer's disease mouse models. *Plos One*, 3:e2175.

Ribal, A. and Young, I. R. (2019). Multiplatform evaluation of global trends in wind speed and wave height. *Science*, 364(6440):548–552.

Riley, N. (2001). Steady streaming. *Annu. Rev. Fluid Mech.*, 33:43–65.

Roberts, P. H. and Glatzmaier, G. A. (2000). Geodynamo theory and simulations. *Rev. Mod. Phys.*, 72(4):1081–1123.

Robertson, B., Hall, K., Zytner, R., and Nistor, I. (2013). Breaking waves: Review of characteristic relationships. *Coastal Eng. J.*, 55(1):1–40.

Rogers, J., Costello, M., and Cooper, G. (2013). Design considerations for stability of liquid payload projectiles. *J. Spacecraft Rockets*, 50(1):169–178.

Rohling, E. J. (2017). *The oceans: A deep history.* Princeton University Press.

Roshid, M. M. and Manasseh, R. (2020). Extraction of bubble size and number data from an acoustically-excited bubble chain. *J. Acous. Soc. Am.*, 147(2):921–940.

Rottman, J. W. and Einaudi, F. (1993). Solitary waves in the atmosphere. *J. Atmos. Sci.*, 50(14):2116–2136.

Rousek, T. (2010). Artificial gravity systems concepts. Project report, International Space University, 50pp, August 2010, URL http://www.artificial-gravity.com/ISU-2010-Rousek.pdf, downloaded 26 September 2018.

Rovira-Navarro, M., Rieutord, M., Gerkema, T., and Maas, L. R. (2019). Do tidally-generated inertial waves heat the subsurface oceans of europa and enceladus? *Icarus*, 321:126–140.

Rubin, D. M., Anderton, N., Smalberger, C., Polliack, J., Nathan, M., and Postema, M. (2018). On the behaviour of living cells under the influence of ultrasound. *Fluids*, 3(4):82.

Sackmann, M., Delius, M., Sauerbruch, T., Holl, J., Weber, W., Ippisch, E., Hagelauer, U., Wess, O., Hepp, W., Brendel, W., and Paumgartner, G. (1988). Shock-wave lithotripsy of gallbladder stones. *New Engl. J. Med.*, 318(7):393–397.

Sandler, H. (1995). Artificial gravity. *Acta Astronaut.*, 35(4-5):363–372.

Satake, K. (2003). Fault slip and seismic moment of the 1700 cascadia earthquake inferred from japanese tsunami descriptions. *J. Geophys. Res.*, 108(B11).

Schwartz, L. W. and Fenton, J. D. (1982). Strongly nonlinear waves. *Annu. Rev. Fluid Mech.*, 14(1):39–60.

Scotti, A. and Pineda, J. (2007). Plankton accumulation and transport in propagating nonlinear internal fronts. *J. Mar. Res.*, 65(1):117–145.

Shah, N., Morsi, Y., and Manasseh, R. (2014). From mechanical stimulation to biological pathways in the regulation of stem cell fate. *Cell Biochem. Funct.*, 32(4):309–325.

Sherwood, S., Webb, M. J., Annan, J. D., Armour, K. C., Forster, P. M., Hargreaves, J. C., Hegerl, G., Klein, S. A., Marvel, K. D., Rorhling, E. J., Watanabe, M., Andrews, T., Braconnot, P., Bretherton, C. S., Foster, G. L., Hausfather, Z., Heydt, A. S. v. d., Knutti, R., Mauritsen, T., and Norris, J. R. (2020). An assessment of earth's climate sensitivity using multiple lines of evidence. *Rev. Geophys.* 58(4): e2019RG000678.

Sheth, A., Sanyal, S., Jaiswal, A., and Gandhi, P. (2006). Effects of the December 2004 Indian Ocean tsunami on the Indian mainland. *Earthq. Spectra*, 22(3_suppl):435–473.

Shiba, D., Mizuno, H., Yumoto, A., Shimomura, M., Kobayashi, H., Morita, H., Shimbo, M., Hamada, M., Kudo, T., Shinohara, M., Asahara, H., Shirakawa, M., and Takahashi, S. (2017). Development of new experimental platform 'mars' - multiple artificial-gravity research system - to elucidate the impacts of micro/partial gravity on mice. *Sci. Rep.*, 7(1):1.

Sigrist, R. M. S., Liau, J., Kaffas, A. E., Chammas, M. C., and Willmann, J. K. (2017). Ultrasound elastography: Review of techniques and clinical applications. *Theranostics.*, 7(5):1303–1329.

Simpson, J. E. (1994). *Sea breeze and local winds.* Cambridge University Press.

Simpson, J. E. (1997). *Gravity currents: In the environment and the laboratory.* Cambridge University Press, 2 edition.

Sirsi, S. and Borden, M. (2009). Microbubble compositions, properties and biomedical applications. *Bub. Sci. Eng. Tech.*, 1(1-2):3.

Spiel, D. E. (1992). Acoustical measurements of air bubbles bursting at a water surface: Bursting bubbles as Helmholtz resonators. *J. Geophys. Res.: Oceans*, 97(C7):11443–11452.

Stefani, F., Gailitis, A., Gerbeth, G., Giesecke, A., Gundrum, T., Rüdiger, G., Seilmayer, M., and Vogt, T. (2019). The DRESDYN project: liquid metal experiments on dynamo action and magnetorotational instability. *Geophys. Astrophys. Fluid Dyn.*, 113(1-2):51–70.

Stephens, G. L. (2005). Cloud feedbacks in the climate system: A critical review. *J. Climate*, 18:237–273.

Strasberg, M. (1953). The pulsation frequency of nonspherical gas bubbles in liquid. *J. Acous. Soc. Am.*, 25(3):536–537.

Streeter, V. L. and Wylie, E. B. (1979). *Fluid Mechanics.* McGraw-Hill International, 7th edition.

Sultana, C. M., Al-Mashat, H., and Prather, K. A. (2017). Expanding single particle mass spectrometer analyses for the identification of microbe signatures in sea spray aerosol. *Anal. Chem.*, 89(19):10162–10170.

Suppasri, A., Muhari, A., Ranasinghe, P., Mas, E., Shuto, N., Imamura, F., and Koshimura, S. (2012). Damage and reconstruction after the 2004 Indian Ocean tsunami and the 2011 Great East Japan tsunami. *J. Nat. Disast. Sci.*, 34(1):19–39.

Suslick, K. S. (1990). Sonochemistry. *Science*, 247(4949):1439–1445.

Sutherland, B. R. (2010). *Internal gravity waves.* Cambridge University Press.

Takemura, F. and Yabe, A. (1998). Gas dissolution process of spherical rising gas bubbles. *Chem. Eng. Sci*, 53 (15):2691–2699.

Tan, I., Storelvmo, T., and Zelinka, M. D. (2016). Observational constraints on mixed-phase clouds imply higher climate sensitivity. *Science*, 352(6282):224–227.

Taraba, M., Zwintz, K., Bombardelli, C., Lasue, J., Rogler, P., Ruelle, V., Schlutz, J., Schüssler, M., O'Sullivan, S., Sinzig, B., Treffer, M., Valavanoglou, A., Van Quickelberghe, M., Walpole, M., and Wessels, L. (2006). Project M^3 - a study for a manned Mars mission in 2031. *Acta Astronaut.*, 58(2):88–104.

ter Haar, G. (2012). Guidelines and recommendations for the safe use of diagnostic ultrasound: the user's responsibilities. In ter Haar, G., editor, *The Safe Use of Ultrasound in Medical Diagnosis*, page 166pp. 3 edition.

Tho, P., Manasseh, R., and Ooi, A. (2007). Cavitation microstreaming in single and multiple bubble systems. *J. Fluid Mech.*, 576:191–233.

Thoroddsen, S., Etoh, T., and Takehara, K. (2008). High-speed imaging of drops and bubbles. *Annu. Rev. Fluid Mech.*, 40:257–285.

Threlfall, G., Wu, H. J., Li, K., Aldham, B., Scoble, J., Šutalo, I. D., Raicevic, A., Pontes-Braz, L., Lee, B., Schneider-Kolsky, M., Ooi, A., and Manasseh, R. (2013). Quantitative guidelines for the prediction of ultrasound contrast agent destruction during injection. *Ultrasound Med. Biol.*, 39(10):1838–1847.

Tilgner, A. (2007). Kinematic dynamos with precession driven flow in a sphere. *Geophys. Astrophys. Fluid Dyn.*, 101(1):1–9.

Toffoli, A., Babanin, A., Onorato, M., and Waseda, T. (2010). Maximum steepness of oceanic waves: Field and laboratory experiments. *Geophys. Res. Lett.*, 37:L05603.

Toffoli, A., Proment, D., Salman, H., Monbaliu, J., Frascoli, F., Dafilis, M., Stramignoni, E., Forza, R., Manfrin, M., and Onorato, M. (2017). Wind generated rogue waves in an annular wave flume. *Phys. Rev. Lett.*, 118(14).

Townsend, L. W., Nealy, J. E., Wilson, J. W., and Atwell, W. (1989). Large solar flare radiation shielding requirements for manned interplanetary missions. *J. Spacecraft Rockets*, 26(2):126–128.

Trübestein, G., Engel, C., Etzel, F., Sobbe, A., Cremer, H., and Stumpff, U. (1976). Thrombolysis by ultrasound. *Clin. Sci. Mol. Med.*, 51(s3):697s–698s.

Turner, J. S. (1973). *Buoyancy effects in fluids*. Cambridge University Press.

Ulrich, R. K. (2001). Very long lived wave patterns detected in the solar surface velocity signal. *Astrophys. J.*, 560(1):466–475.

Vagle, S., McNeil, C., and Steiner, N. (2010). Upper ocean bubble measurements from the ne pacific and estimates of their role in air-sea gas transfer of the weakly soluble gases nitrogen and oxygen. *J. Geophys. Res.-Oceans*, 115.

Van Holsbeke, C., Zhang, J., Van Belle, V., Paladini, D., Guerriero, S., Czekierdowski, A., Muggah, H., Ombelet, W., Jurkovic, D., Testa, A. C., Valentin, L., Van Huffel, S., Bourne, T., and Timmerman, D. (2010). Acoustic streaming cannot discriminate reliably between endometriomas and other types of adnexal lesion: a multicenter study of 633 adnexal masses. *Ultrasound Obst. Gyn.*, 35(3):349–353.

Vanyo, J. P. (1991). A geodynamo powered by luni-solar precession. *Geophys. Astrophys. Fluid Dyn.*, 59:209–234.

Vargaftik, N. B., Volkov, B. N., and Voljak, L. D. (1983). International tables of the surface tension of water. *J. Phys. Chem. Ref. Data*, 12(3):817–820.

Veldman, A., Gerrits, J., Luppes, R., Helder, J., and Vreeburg, J. (2007). The numerical simulation of liquid sloshing on board spacecraft. *J. Comput. Phys.*, 224(1):82–99.

Vergniolle, S. and Brandeis, G. (1996). Strombolian explosions 1. a large bubble breaking at the surface of a lava column as a source of sound. *J. Geophys. Res.*, 101 (B9):20433–20447.

Vergniolle, S. and Caplan-Auerbach, J. (2004). Acoustic measurements of the 1999 basaltic eruption of shishaldin volcano, alaska. 2. precursor to the subplinian phase. *J. Volcanol. Geoth. Res.*, 137:135–151.

Vette, J. I. (1973). Summary of space observations. In Coffey, H. E., editor, *Collected Data Reports on August 1972 Solar-Terrestrial Events*, volume 28, Part II. U. S. Dept. Commerce, Nat. Oceanic & Atmos. Admin.

Vogelsberger, M., Marinacci, F., Torrey, P., and Puchwein, E. (2020). Cosmological simulations of galaxy formation. *Nat. Rev. Physics*, 2(1):42–66.

Waleffe, F. (1990). On the three-dimensional instability of strained vortices. *Phys. Fluids A: Fluid Dyn.*, 2(1):76–80.

Wallace, D. W. R. and Wirick, C. D. (1992). Large air-sea gas fluxes associated with breaking waves. *Nature*, 356(6371):694–696.

Wenz, G. M. (1962). Acoustic ambient noise in the ocean: spectra and sources. *J. Acous. Soc. Am.*, 34(12):1936–1956.

Whitham, G. B. (1999). *Linear and Nonlinear Waves.* Wiley.

Wood, W. W. (1966). An oscillatory disturbance of rigidly rotating fluid. *Proc. Roy. Soc. Lond. A*, 293(1433):181–212.

Worthington, A. M. (1883). On impact with a liquid surface. *Proc. Roy. Soc. Lond.*, 34(220-223):217–230.

Wu, J. (2018). Acoustic streaming and its applications. *Fluids*, 3(4):108.

Wu, J., Ross, J. P., and Chiu, J.-F. (2002). Reparable sonoporation generated by microstreaming. *J. Acous. Soc. Am.*, 111(3):1460–1464.

Xu, Z., Chen, Y., and Xu, Z. (2019). Optimal guidance and collision avoidance for docking with the rotating target spacecraft. *Adv. Space Res.*, 63(10):3223–3234.

Yang, M., Blomquist, B. W., Fairall, C. W., Archer, S. D., and Huebert, B. J. (2011). Air-sea exchange of dimethylsulfide in the Southern Ocean: Measurements from SO GasEx compared to temperate and tropical regions. *J. Geophys. Res. - Oceans*, 116.

Zelinka, M. D., Klein, S. A., Taylor, K. E., Andrews, T., Webb, M. J., Gregory, J. M., and Forster, P. M. (2013). Contributions of different cloud types to feedbacks and rapid adjustments in cmip5. *J. Climate*, 26(4):5007–5027.

Index

9780367271640